ZEMCH International Research 2020

ZEMCH International Research 2020

Editors

Jun-Tae Kim
Masa Noguchi
Haşim Altan

MDPI • Basel • Beijing • Wuhan • Barcelona • Belgrade • Manchester • Tokyo • Cluj • Tianjin

Editors
Jun-Tae Kim
Department of Architectural Engineering, Kongju National University
Cheonan, Korea

Masa Noguchi
Faculty of Architecture, Building and Planning, The University of Melbourne
Melbourne, Australia

Haşim Altan
Department of Architecture, Faculty of Design, Arkin University of Creative Arts and Design
Cyprus

Editorial Office
MDPI
St. Alban-Anlage 66
4052 Basel, Switzerland

This is a reprint of articles from the Special Issue published online in the open access journal *Sustainability* (ISSN 2071-1050) (available at: https://www.mdpi.com/journal/sustainability/special_issues/ZEMCH_international_research_2020).

For citation purposes, cite each article independently as indicated on the article page online and as indicated below:

LastName, A.A.; LastName, B.B.; LastName, C.C. Article Title. *Journal Name* **Year**, *Volume Number*, Page Range.

ISBN 978-3-0365-2850-2 (Hbk)
ISBN 978-3-0365-2851-9 (PDF)

Contents

About the Editors

Jun-Tae Kim is a professor in built environment and renewable energy at the Department of Architectural Engineering and Head of the Graduate School of Energy Systems Engineering, Kongju National University, Cheonan, Korea. He is leader of the Green Energy Technology Research Center and the sole Korean expert for the ongoing International Energy Agency (IEA) Photovoltaic Power Systems (PVPS) Programme Task 15. He participated in previous IEA Solar Heating and Cooling (SHC) Programme Task 41 and IEA SHC/EBC Task40/Annex 52 Net-Zero-Energy Solar Buildings (NZEB) as well. With over 24 years of industrial and academic experience, Prof. Dr. Jun-Tae Kim has developed a cross-disciplinary network nationally and internationally, across both academia and industry, and has built a reputation for innovative and adaptable solutions towards sustainable buildings integrated with solar-harnessing renewable energy technologies.

Masa Noguchi is an associate professor in environmental design at the Faculty of Architecture, Building and Planning, University of Melbourne, Australia. He is a chartered engineer, environmentalist, and technological product designer, registered, respectively, with the Engineering Council, Society for the Environment, and the Institution of Engineering Designers in the UK. In 2002 he also became a member of the Royal Architectural Institute of Canada, and today he serves as a certified passive house designer, registered with the Passive House Institute in Germany. Dr. Noguchi is the founding coordinator of the ZEMCH Network (www.zemch.org), which consists of over 850 partners from nearly 40 countries and has initiated a number of industry–academia knowledge transfer events. At the Melbourne School of Design, he leads Zero-Energy Mass Custom Home (ZEMCH) design courses, i.e., "Travelling Studio" and "Design Thesis Studio" within the graduate program. Dr Noguchi leads ZEMCH engineering design research for the delivery of socially, economically, environmentally, and humanly sustainable built environments in global contexts.

Haşim Altan is a professor of sustainable design and architectural engineering at the Department of Architecture and Dean of the Faculty of Design at Arkin University of Creative Arts and Design (ARUCAD) in Kyrenia, Cyprus. He is a chartered architect (RIBA) and a chartered engineer (CIBSE) with over 20 years of academic and practice experience through architectural engineering and sustainable design for the built environment in the UK, Europe, and the MENA region. As well as having supervised 16 successful PhD students, Prof. Dr. Altan has published over 250-refereed journal and conference papers in addition to edited books and chapters in related fields.

Preface to "ZEMCH International Research 2020"

The built environment continuously accounts for a significant share of energy use and associated carbon emissions in many countries. One contributing factor to these two aspects is the long operational lifespan of buildings. Generally, buildings are used for decades. For instance, a greater percentage of buildings that will still be in use by the year 2050 are the ones already built today. This being the case, the overall building energy consumption profile is substantially evident in the whole energy footprint of the economy. Secondly, the rate at which new buildings replace older buildings is very low. In many countries, this rate is less than an average of 2% per year. Thirdly, although necessary, strict building energy codes are either obsolete with regard to existing buildings or become obsolete in a short time span with regard to new buildings. Additionally, in a very conservative industry such as building construction, the monitoring and implementation of building energy codes is a notable challenge, not to mention the intrinsic limitations of building energy codes. In addition, factors such as occupant behavior, user trends, and climate change resulting in severe cold and hot seasons in certain regions are difficult, if not impossible, to control. Therefore, in an effort to move towards energy efficiency in the built environment, topics covering user choice and behavior, on-site energy generation and utilization in buildings, energy-efficient systems and solutions, responsive and dynamic building systems, innovative artificial intelligent solutions towards conservation and management of resources, reduction of carbon emissions, sustainable systems and environments, and mass customization, among others, have become key global issues being tackled by numerous research institutions.

In response to market needs and demands for socially, economically, environmentally, and humanly sustainable built environments in developed and developing countries to accommodate people with different socio-economic backgrounds of all ages and abilities, the Zero-Energy Mass Custom Home (ZEMCH) was instigated. The ZEMCH Network was officially established in 2010 after a number of international industry–academia collaborative study tours were organized in order to observe the state-of-the-art production and sales facilities of leading low-to-zero energy or carbon dioxide emissions sustainable housing manufacturers in Japan that also practiced inclusive design. Currently the ZEMCH Network consists of 858 global partners from academia, industry, and government, based in over 45 countries. The Network organizes international conferences across the globe to create a networking platform as well as to disseminate information on current developments related to sustainable built environment.

This book compiles recent contributions from 14 author groups that have been published under ZEMCH International Research 2020. A full range of ZEMCH topics, including building envelope evaluations, occupant choice and experience, indoor environmental quality, automated control systems, mass customization, and integration of renewable energy, on both building and urban scales are covered. The aim is to address current questions as well as present challenges and opportunities for the continuous development of built environments for all users with diverse socio-economic backgrounds and cultural differences in developed and developing countries. Housing is a complex system of energy and environment. To deliver a marketable and reliable near-zero-energy/emission-conscious mass custom home, various key design, technological, production and marketing, and delivery and operational parameters need to be optimized harmoniously.

Jun-Tae Kim, Masa Noguchi, Haşim Altan
Editors

Article

Effect of Triangular Baffle Arrangement on Heat Transfer Enhancement of Air-Type PVT Collector

Ji-Suk Yu [1], Jin-Hee Kim [2] and Jun-Tae Kim [3,*]

1 Zero Energy Building Laboratory, Graduate School of Energy Systems Engineering, Kongju National University, Cheonan 31080, Korea; sook1991@smail.kongju.ac.kr
2 Green Energy Technology Research Center, Kongju National University, Cheonan 31080, Korea; jiny@kongju.ac.kr
3 Department of Architectural Engineering, Kongju National University, Cheonan 31080, Korea
* Correspondence: jtkim@kongju.ac.kr; Tel.: +82-41-521-9333

Received: 4 August 2020; Accepted: 7 September 2020; Published: 10 September 2020

Abstract: A Photovoltaic Thermal (PVT) Collector is a device that produces electricity and simultaneously uses a heat source transmitted to back side of the Photovoltaic (PV). The PVT collector is categorized into liquid-type and air-type according to the heating medium. As an advantage, air-type PVT system is easy to manage and can be directly used for heating purposes. The performance of air-type PVT collector is determined by various factors, such as the height of air gap and air flow path (by baffles) in the collector. Baffles are installed in the PVT collector to improve the thermal performance of the collector by generating turbulence. However, the air flow that affects the performance of the PVT collector can vary depending on the number and placement of the baffles. Thus, the flow design using baffles in the collector is important. In this study, the performance of an air-type PVT collector due to the arrangement of triangular baffles and air gap height at the back of the PV module is analyzed through a simulation program. For this purpose, Computational Fluid Dynamics (CFD) analysis was performed with an NX program to compare and analyze the optimum conditions to improve the performance of the collector.

Keywords: air-type PVT collector; CFD (computational fluid dynamic); thermal performance; triangular baffles

1. Introduction

Globally, the use of sustainable energy is increasing, and among renewable energy systems, solar energy and wind power systems are widely used. Especially, solar energy system is classified into a PV system, a solar thermal system, and a photovoltaic/thermal (PVT) system, and it operates by converting solar energy into electricity and thermal energy. However, a system that uses solar energy cannot be completely dependent on power generation and heat collection as it is not uniform. Therefore, it is applied as a method of stably supplying electricity and thermal energy to a storage device by charging it to a consumer [1,2], and a method of increasing system efficiency, such as harvesting more energy by hybridizing with a fossil fuel system such as natural gas [3].

PVT collectors are device that use heat generated at the back of a PV at the same time as the electricity produced at the front of the PV. PVT collectors are classified into air and liquid types according to the fluid used as the heat transfer medium. Air-type PVT collectors have the advantage of being easy to manage. Previous studies focused on improving the collector's own thermal and electrical efficiency through design, simulation, modeling and experimentation of air-type PVT collectors.

For instance, one review paper of previous studies considered results of air flow and single/double flow paths of various air-type PVT systems and various absorbers (i.e., fin, V-groove, round tube, etc.) [4,5]. Conclusions showed that the electrical efficiency was 10%–25% and the thermal efficiency was 40%–70%. Exergy efficiency was also in the range of 5%–25%. Riffat et al. [6] studied parameters affecting the electrical and thermal performance of various types of PVT systems. Parameters examined included the optimal flow rate inside the PVT collector, the presence or absence of a thermal absorber plate, and the thickness of the air layer. In particular, the thermal absorber plate was found to have the most influence on the thermal efficiency of the PVT system. In other studies, it was concluded that as the number of baffles increased, the thermal efficiency increased, but this increase in thermal efficiency decreased above a certain number of baffles [7,8]. So, it is necessary to set a proper number of baffles in consideration of the pressure drop.

Based on the CFD program (i.e., ANSYS Fluent), Chaube et al. [9] analyzed shape, viz. rectangular, square, chamfered, triangular and semicircle baffles after simulation in a Reynolds number range of 2900–19,500. Abuska et al. [10] examined the energy, exergy, economics and environmental performance of air-type collectors with V-groove-shaped protrusions. The thermal and exergy efficiencies ranged from 43% to 60% and 6% to 12%, respectively, and the payback period averaged 4.3–4.6 years. Experiments showed that the collector's thermal efficiency was about 6% higher than that of the flat collector. Furthermore, Fudholi et al. [11] studied the exergy and sustainability index of air-type PVT collectors with V-groove shaped protrusions. The exergy efficiency was 13.36% in theory and 12.89% according to experimental results.

In the studies by Promthaisong et al. [12], triangular ribs were applied to the thermal absorber plate and analyzed in the Reynolds number range of 3000–20,000. The influence of blockage ratio and the pitch ratio of the rib were analyzed as variables. Compared to general collectors, the coefficient of friction and Nusselt number increased by 1.01–4.93 times and 1.02–3.86 times, respectively, by promoting heat transfer. Other studies concluded that when there was a gradient baffle, the maximum enhancement in thermal and effective efficiency were 22.4% and 18.1% in the mass flow rate of 0.045 kg/s compared to general Solar Air Heater (SAH) [13]. In the studies by Bhagoria et al. [14], an air heater with a wedge (triangle) type baffle was designed, and the heat transfer and friction coefficients were analyzed through experiments. The parameters of the baffle were set as the baffle height, angle and baffle spacing; Nusselt number increased up to 2.4 times compared to that of the collector without baffles. In addition, the friction coefficient increased as the angle of the baffles increased, and the heat transfer performance was highest when the baffle angle was 10°. Yadav et al. [15] investigated the heat transfer inside the collector by installing a triangle baffle on the absorber plate using CFD (ANSYS Fluent). Nusselt number increased with increasing Reynolds number, and was 1.4–2.7 times higher than without a baffle collector. To optimize the triangular baffle, thermal performance was compared and analyzed through CFD simulation using the baffle height, pitch, height ratio and distance ratio between the baffles as variables. The Nusselt number decreased as the distance ratio increased, and the friction coefficient increased by 3.356 times compared to the reference. For an equilateral triangle baffle with Reynolds number of 15,000, distance ratio between baffles of 7.14 and height ratio of 0.042, the Nusselt number was 3.073 times that of reference [16]. Bensaci et al. [17] performed numerical/experimental studies according to the baffle arrangement in the collector. The thermal efficiency improved with increasing number of baffles, and the coefficient of friction decreased as the Reynolds number increased. Choi et al. [18] performed CFD analysis under the same conditions by installing several resistors inside an air-type PVT collector. In terms of heat transfer performance, the intersection of triangular baffles led to an improvement of up to 1.86 times.

As described above, the application of a triangular baffle in an air-type PVT collector improves the heat transfer performance. This effect is mainly caused by the generation of a vortex in the air layer by increase of the Nusselt and Reynolds numbers. However, in order to prevent dead space and pressure drop, it is necessary to design the baffle so that the flow inside the air-type PVT collector is smooth. Consequently, baffle must be designed in consideration of the placement conditions of the baffle, so that the flow inside the collector can be smooth and the heat transfer performance can be improved.

In this study, triangular baffles were installed on a thermal absorber plate in the air gap inside an air-type PVT collector; the thermal characteristics of the baffle arrangement condition were analyzed. For this, the heat transfer, pressure drop, and thermal efficiency of the collector were scrutinized through a simulation program (i.e., NX CFD). The primary purpose was to investigate the effects according to the triangular baffle arrangement applied in the air-type PVT collector.

2. Air-Type PVT Collector Model for Simulation

2.1. Model Design

The air-type PVT collector designed for this study is shown in Figure 1. The front of the collector is covered with a general PV module; the dimensions are 1011 mm × 1520 mm. The PV module has a cell covering the front surface, as in a conventional module (electrical efficiency is 17%), and consists of about 54 mono crystal cells between two layers of glass (G/G module). The air layer (height: 34 mm) in the collector has a triangular baffle (width: 82 mm, length: 75 mm, height: 16 mm, angle: 15°) that acts as an air flow obstruction, and has a certain spacing and arrangement inside the collector. Thus, the obstacle baffle generates turbulence and is used as an element to increase heat transfer performance.

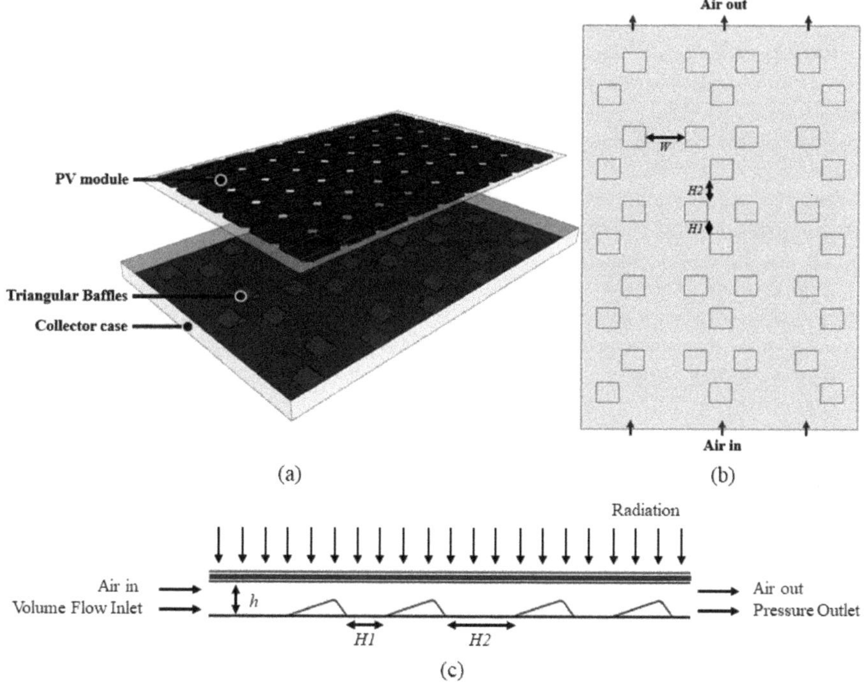

Figure 1. Concept of Air-type PVT collector with triangular baffles (**a**) 3D blown-up picture, (**b**) layout showing baffle positioning, and (**c**) cross sectional profile showing air flow direction.

The input values of lateral spacing (W) and longitudinal spacing (H1, H2) of the baffle, which affect the baffle thermal performance, are presented in Table 1. To adjust the placement of baffles in the collector, three parameters were set: W (62, 82.5, 144 mm), H1 (0, 47 mm) and H2 (83, 130, 176.3 mm). In addition, a total of 10 cases were simulated and compared, including a reference to a PVT collector with a baffle-free air layer.

Table 1. General variables for collector parametric simulation.

Description	Collector Size	W [mm]	H1 [mm]	H2 [mm]	Number of Baffles [B.No.]
Case 1	1011 × 1520 × 34 mm	Reference			
Case 2		62	47	83	35
Case 3		82.5			
Case 4		144			
Case 5		62	0	130	35
Case 6		82.5			
Case 7		144			
Case 8		62	47	176.3	28
Case 9		82.5			
Case 10		144			

2.2. Energy Balance Equations

The heat transfer process along the cross section of the air-type PVT collector is depicted in Figure 2, in which it can be seen that heat is transferred in the order of the top glass, EVA, PV cell, EVA, bottom glass, air layer, baffle and insulation. However, in this simulation, the heat transfer effect of EVA was insignificant, so only the top and bottom glass layers and PV cells were considered. In Equations (1)–(9), $h_{cv,T}$, $h_{cv,B}$, $h_{rd,T}$ and $h_{rd,B}$ are the top and bottom surface convection and radiative heat transfer coefficients of the collector, respectively, T_i is the temperature at point i, T_{sur} is the surface temperature around the collector, T_a is the ambient temperature, $R_{i\text{-}j}$ is the thermal resistance between points i and j, $h_{rd,i\text{-}j}$ is the radiative heat transfer coefficient between points i and j, S_i is the amount of solar radiation absorbed at point i, P_{PV} is the amount of power produced by the PV cells, $h_{cv,fT}$ and $h_{cv,fB}$ are the top and bottom surface heat transfer coefficients of the air layer, respectively, and $q''_{u,T}$ and $q''_{u,B}$ represent the thermal energy recovered from the top and bottom of the collector air layer. Thermal resistance between ft and fb was not calculated. The flow in the air-type PVT collector enters at the inlet, sweeps the baffles in the cavity, and passes through it. Therefore, the thermal energy recovered (q) is a factor that affects the thermal performance of the collector more than up and down thermal conduction.

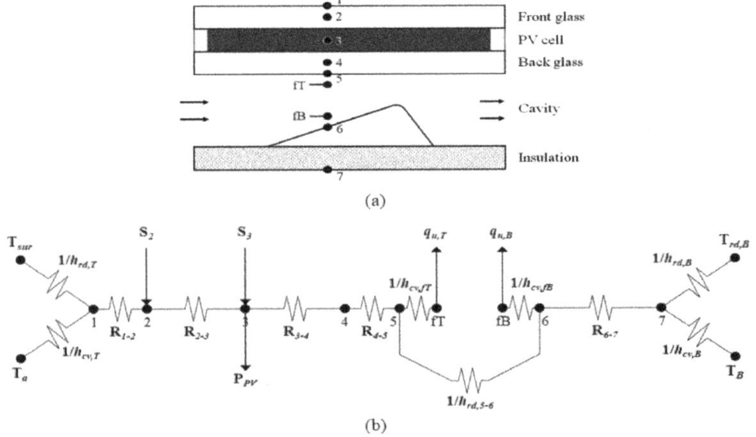

Figure 2. Heat transfer mechanism: (**a**) cross sectional view of collector and (**b**) schematic diagram showing interrelated heat transfer components.

By performing energy balances in the different nodes of Figure 2, Equations (1)–(12) are obtained [19–23].

The heat transfer equation for the front glass top surface (point 1) is given by Equation (1).

$$h_{cv,T}(T_1 - T_a) + h_{rd,T}(T_1 - T_{sur}) + \frac{T_1 - T_2}{R_{1-2}} = 0 \quad (1)$$

The heat transfer equation for the front glass middle surface (point 2) is given by Equation (2).

$$\frac{T_1 - T_2}{R_{1-2}} + S_2 = \frac{T_2 - T_3}{R_{2-3}} \quad (2)$$

The heat transfer equation for the PV layer (point 3) is given by Equation (3).

$$\frac{T_2 - T_3}{R_{2-3}} + S_3 - P_{PV} = \frac{T_3 - T_4}{R_{3-4}} \quad (3)$$

The heat transfer equation for the back glass middle surface (point 4) is given by Equation (4).

$$\frac{T_3 - T_4}{R_{3-4}} = \frac{T_4 - T_5}{R_{4-5}} \quad (4)$$

The heat transfer equation for the back glass bottom surface (point 5) is given by Equation (5).

$$\frac{T_4 - T_5}{R_{4-5}} = h_{cv,fT}\left(T_5 - T_{fT}\right) + h_{rd,5-7}(T_5 - T_7) \quad (5)$$

The heat transfer equation for the fluid in the top portion of the air cavity (point *fT*) is given by Equation (6).

$$q''_{u,T} = h_{cv,fT}\left(T_5 - T_{fT}\right) \quad (6)$$

The heat transfer equation for the fluid in the bottom portion of the air cavity (point *fB*) is given by Equation (7).

$$q''_{u,B} = h_{cv,fB}\left(T_7 - T_{fB}\right) \quad (7)$$

The heat transfer equation for the triangle baffle top surface (point 6) is given by Equation (8).

$$h_{rd,5-6}(T_5 - T_6) = h_{cv,fB}\left(T_6 - T_{fB}\right) + \frac{T_6 - T_7}{R_{6-7}} \quad (8)$$

The heat transfer equation for the insulation outer surface (point 7) is given by Equation (9).

$$\frac{T_6 - T_7}{R_{6-7}} = h_{cv,B}(T_7 - T_B) + h_{rd,B}\left(T_7 - T_{rd,B}\right) \quad (9)$$

The thermal resistances for between the top surface and middle of the front glass layer, R_{1-2}, and the middle and bottom surface of the back glass layer, R_{4-5} are given by Equation (10). th_g, th_{PV}, $th_{tri\text{-}baffle}$ and $th_{insulation}$ are the thickness of glazing, PV cell, triangular baffles and insulation layers, respectively. k_g, k_{PV}, $k_{tri\text{-}baffle}$ and $k_{insulation}$ are the thermal conductivity of glazing, PV cell, triangular baffles and insulation layers, respectively.

$$R_{1-2} = R_{4-5} = \frac{1}{2}\frac{th_g}{k_g} \quad (10)$$

The thermal resistances for between the middle of the front glass layer and PV cell layer, R_{2-3}, and the PV cell layer and middle of the bottom glass layer, R_{3-4}, are given by Equation (11).

$$R_{2\text{-}3} = R_{3\text{-}4} = \frac{1}{2}\frac{th_g}{k_g} + \frac{1}{2}\frac{th_{PV}}{k_{PV}} \quad (11)$$

The thermal resistances for between the triangular baffles and insulation layer, $R_{6\text{-}7}$ are given by Equation (12).

$$R_{6\text{-}7} = \frac{th_{tri-baffle}}{k_{tri-baffle}} + \frac{th_{insulation}}{k_{insulation}} \quad (12)$$

2.3. Initial Modeling and Validation

The air-type PVT collector was installed on the rooftop of an educational building, located in the Cheonan campus of Kongju National University (36.85 N, 127.15 E). The collector was installed on a 2-axis tracker that can be adjusted horizontally (east-west) and vertically to introduce solar radiation into the normal plane. To evaluate the thermal performance, the collector's inlet/outlet temperature, inlet/outlet flow rate, solar radiation, outside temperature, and wind direction/wind speed were measured. The size (1011 × 1520 × 37 mm) and the triangular baffle arrangement case (W = 104 mm, H1 = 74.8 mm, H2 = 94.8 mm and B.No. = 35) of the collector used in the experiment are shown in Figure 3a,b respectively.

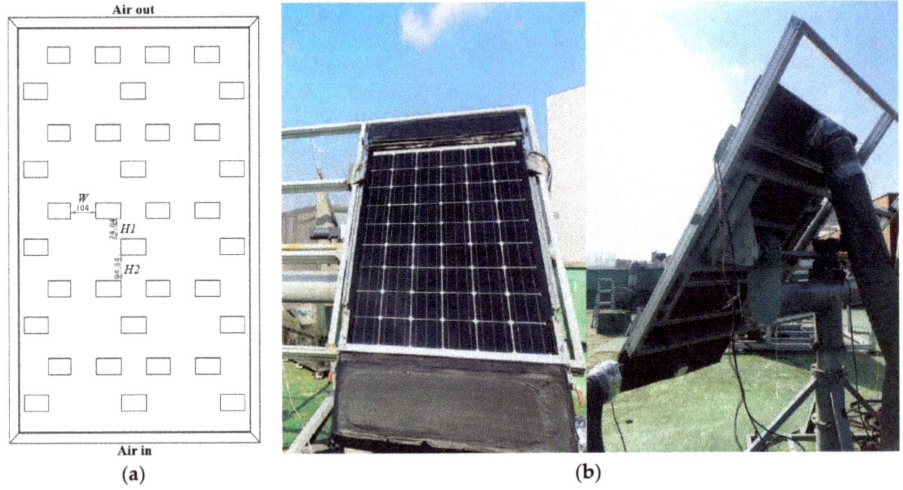

Figure 3. Air-type PVT collector (**a**) and (**b**) Outdoor performance experiment.

Outdoor performance test was performed based on ISO 9806 (Solar energy—solar thermal collectors—test methods). Notably, the solar radiation intensity incident on the collector slope was 700 W/m^2 or higher, and the inlet and outlet flow rates were measured at 100 m^3/h according to the test method. Experimental data were collected and analyzed for 10 min when the inlet/outlet flow rate and temperature, and the outside air temperature were in a steady state. Table 2 illustrates a comparison of experiment and numerical results of outlet temperature of the collector.

Table 2. Comparison of experiment and numerical results.

	G [W/m^2]	T_a [°C]	T_{in} [°C]	T_{out} [°C]	ΔT [°C]
Experimental	837.83	1.5	4.86	12.79	7.93
Numerical (CFD)	837.83	1.5	4.86	13.04	8.18
Error [%]	-	-	-	1.92%	3.05%

Experimental and numerical data showed that the outlet temperature was 12.79 °C and 13.04 °C respectively under the same outdoor conditions, which was almost consistent with an error range of about 1.92%.

G is solar radiation (W/m2), Ta is ambient temperature (°C), Tin is inlet air temperature of PVT (°C), Tout is outlet air temperature of PVT (°C), and ΔT is Inlet and outlet temperature difference (°C)

2.4. Modeling Conditions and Methodology

The NX program used for this study has the capability to simulate fluid flow effects, such as CFD modeling, by quickly creating flow zones for complex geometry and performing computational fluid dynamics. In addition, a combination of thermal analysis functions for conduction, convection, radiation and complex heat transfer is performed [24].

Figure 4 shows the mesh of the computational domain, and the tetrahedral method was selected for the meshing strategy. A tetrahedral mesh is set because it has high solution accuracy and has the advantage of reducing simulation time. In addition, to analyze the airflow characteristics around the triangular baffle, the mesh around the baffles was set densely unlike the mesh of the air layer in the collector.

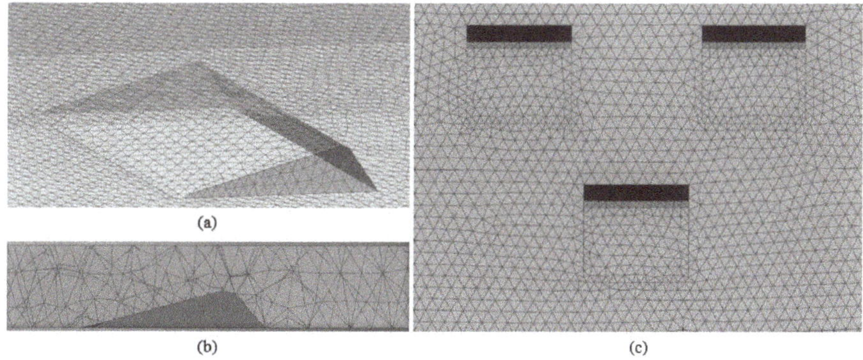

Figure 4. Demonstrating the mesh of the computational domain: (**a**) isometric view, (**b**) cross sectional view, and (**c**) rear view of the collector.

A grid independence test was performed using the outlet temperature of an air-type PVT collector modeled with W = 62 mm, H1 = 47 mm, H2 = 83 mm, B.No. = 35 as the variables. For the test, the mesh elements were set to four cases (62,434, 119,915, 362,677, and 2,359,170), and the outlet temperature values of the collector were 23.53, 23.32, 23.30 and 23.28, respectively (see Table 3). The result values between mesh elements have a percentage difference of less than approximately 1%. Considering the calculation time, a grid of 362,677 elements was selected as an appropriate grid.

Table 3. Grid independence with the variables (W = 62 mm, H1 = 47 mm, H2 = 83 mm and B.No. = 35).

Mesh Elements	Outlet Temperature (T_{out})	Percentage Deviation of T_{out}
62,434	23.53	0.9775
119,915	23.32	0.0858
362,677	23.30	-
2,359,170	23.28	0.08591

To perform the CFD analysis using the NX program, 10 cases were set with different baffle arrangement as the variables (see Table 1), and the heat transfer, pressure drop, and thermal efficiency of the air-type PVT collector were compared and analyzed. To simulate the collector, the thermal energy of the collector was set to be supplied equally as 700 W/m^2 at 1.54 m^2. Furthermore, it was input so that the outside air temperature of 7 °C with mass flow rate of 67 m^3/h was entered into the inlet of collector. The ambient temperature was input to consider the surface heat loss of the collector, and gravity and buoyancy were applied for modeling the actual air flow. The boundary conditions are summarized in Table 4.

Table 4. Boundary conditions of air-type PVT collector.

Boundary Condition Parameter	Value
Turbulent model	*Standard k—ε* model
Heat flux	700 W/m^2 on the PV surface
PV area (L × H)	1.54 m^2 (1011 × 1520 mm)
Fluid	Air
Inlet, Outlet area of collector	0.0344 m^2
Air in (temperature)	67 m^3/h (7 °C)
Air out	Pressure Outlet
Air density	1.225 kg/m^3
Buoyancy	Application
Gravity	9.81 m/s^2
Free convection to environment	Application

The turbulence model used was the standard K-Epsilon model. It is the most common turbulence model in CFD simulation, and k is the turbulent kinetic energy and ε is the dissipation rate of turbulent energy. This model is based on the time-averaged Navier–Stokes equations, which assume that the time-varying velocities of turbulence can be divided into time-averaged velocities and velocity-dependent velocities [25]. Therefore, in the CFD evaluation for this study, air flow analysis was performed using the standard k-ε turbulence model because the analysis is not complicated and its accuracy is good in terms of computational convergence through air flow simulation data accumulated for a long time.

3. Analysis of Simulation Results

3.1. Heat Transfer Performance

Figure 5 shows the flow velocity distribution for cases with and without baffles inside the air-type PVT collector. Based on the results, the flow velocity of the reference collector passed to the outlet without stagnation. The collector with a baffle (W = 144 mm, H1 = 47 mm, H2 = 83 mm, B.No. = 35) tended to have a weak local flow rate at the back of the baffle, but the flow rate was significantly faster in the spaces between baffles. Therefore, compared with the reference collector, the internal flow rate of the collector with the baffle was faster and the outlet flow rate increased.

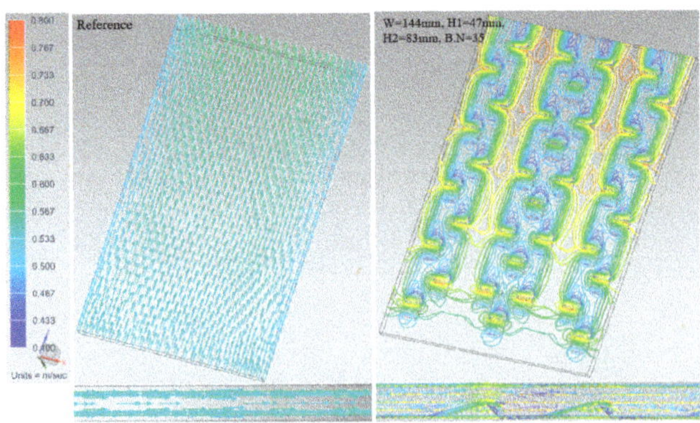

Figure 5. Velocity distribution inside air-type PVT collector: without baffles (**left**) and with baffles (**right**).

Figure 6 shows the temperature distribution inside the air-type PVT collector for cases with and without baffles. In the collector with triangular baffles, the air temperature rose due to locally low flow rate on the back of the baffle, but the space between the top of the baffle and the back of the PV was narrow, causing the air to sweep away quickly. Therefore, it can be deduced that air did not stagnate locally on the back of the baffle and passed quickly to the outlet. As a result, the collector with baffles had a faster outlet flow rate and a higher outlet temperature than the reference collector without baffles, so the former's heat transfer performance was advantageous.

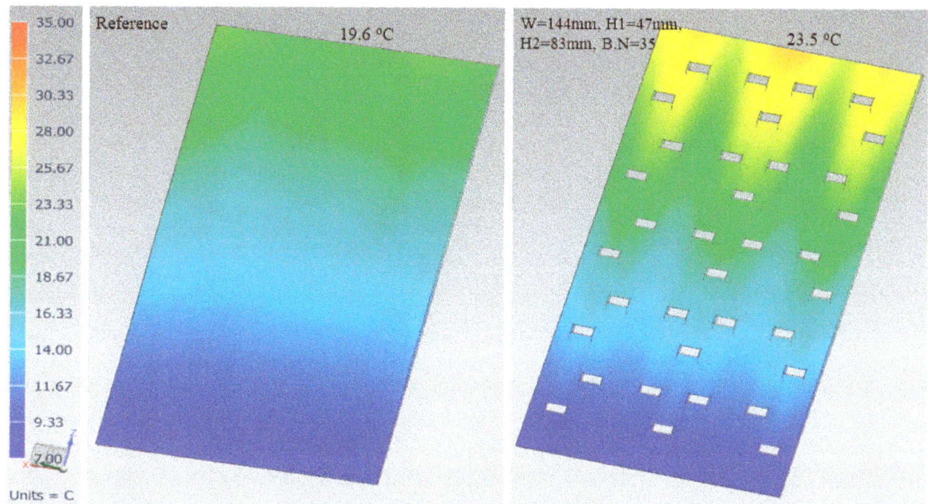

Figure 6. Temperature distribution inside air-type PVT collector: without baffles (**left**) and with baffles (**right**).

Figure 7 graphically illustrates the outlet velocity for each baffle arrangement condition of the air-type PVT collector with triangular baffles. The outlet velocity of the collector with the baffle was higher than that of the reference collector (0.564 m/s). It was confirmed that the smaller the left and right spacing variable (W) of the baffle was, the faster the outlet velocity and the higher the related average Reynolds number were. This is because only the arrangement was changed in a state in which the baffle size was fixed; as the length of the lateral direction (W) decreased, the flow path became narrower. However, the longer H2 was, the farther the front and back between the baffles were and the more space was created in the transverse direction. Consequently, the possibility of vortex generation by the baffle was reduced. Consequently, it was seen that the difference in the average Reynolds number range also decreased. Furthermore, the smaller the number of baffles was, the faster was the outlet velocity, but with a correspondingly lower Reynolds number. This is attributed to the flow in the air layer, which was non-uniform; the air flow rapidly passed through the flow paths in the left and right longitudinal directions in the collector with length of W.

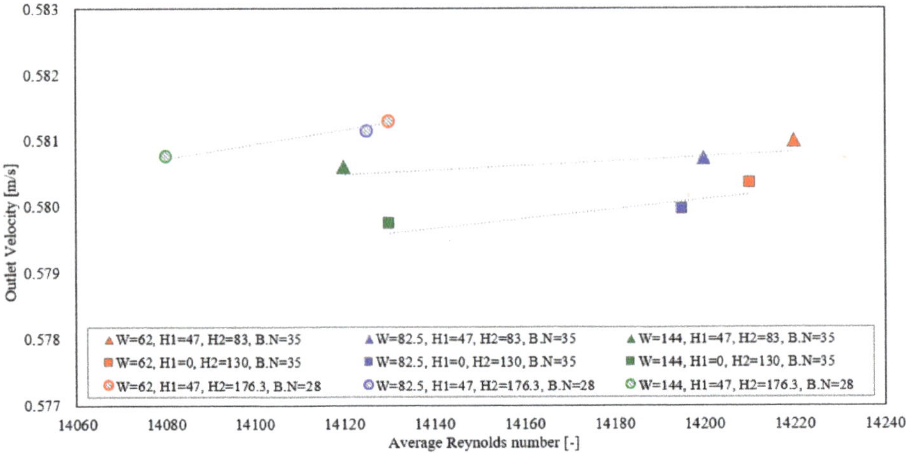

Figure 7. Average Reynolds number and outlet velocity of Air-type PVT collectors.

3.2. Pressure Drop and Thermal Efficiency

Figure 8 shows the pressure drop by baffle placement conditions of the air-type PVT collector with triangular baffles. Since the reference collector had no baffles, turbulence did not occur, resulting in a low value of pressure of 0.2302 Pa. The PVT collector with triangular baffles had a pressure drop caused by the baffles. There was a difference in pressure drop value according to the length of W; it can be seen that the longer the distance was between W, the more the vortex area decreased and the smaller the pressure drop value was. In addition, the value of number of baffles (B.No.) of 35 led to relatively lower pressure drop than did the value of 28. It can be concluded that the number of baffles was large, but stagnation areas of air flow occurred less frequently due to the relatively shorter length of H2. Therefore, it was found that the arrangement of the baffles was more favorable than the number of baffles in terms of pressure drop, and the longitudinal spacing of the baffles (H1, H2) was the main influencing factor.

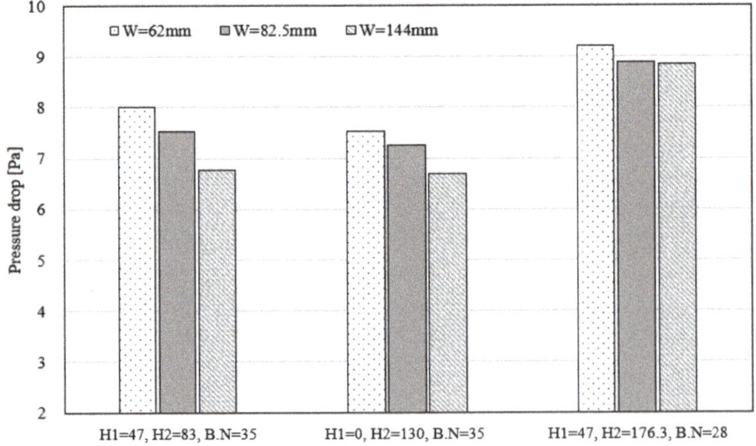

Figure 8. Pressure drop of air-type PVT collector.

Table 5 shows the thermal efficiency and heat gain according to the arrangement parameters of the air-type PVT collector with triangular baffles. The thermal efficiency of the air-type PVT collector can be calculated by Equation (13) [26], which was deduced by applying the collector's flow rate and inlet/outlet temperature. Based on Equation (10), the reference efficiency without baffles was 26.7% and the heat gain was 287Wth. The thermal efficiency of the air-type PVT collector with triangular baffles was in the range of 34.3–35%, and the heat gain was analyzed to be 369–377 Wth. The collector's thermal efficiency was up to 13.5% higher than that of the reference collector, and the heat gain was 89.4 Wth. According to the baffle arrangement parameters, the thermal efficiency and heat gain were highest when the variable conditions are W = 144 mm, H1 = 47 mm, H2 = 83 mm and B.No. = 35.

$$\eta_{th} = \frac{Q_2}{Q_1} = \frac{\dot{m}\, C_p\, (T_{outlet} - T_{inlet})}{A_{PVT}\, G} \tag{13}$$

A_{PVT} is surface area of the collector (m^2), G is solar radiation (W/m^2), $\dot{m}$ is mass flow rate (kg/h), C_p is specific heat of air at a constant pressure (J/kg °C), T_{outlet} is outlet air temperature of PVT (°C), T_{inlet} is inlet air temperature of PVT (°C), and η_{th} is thermal efficiency (-).

Table 5. Thermal efficiency and heat gain by case of the air-type PVT collector.

W [mm]	H1 [mm]	H2 [mm]	B.No.	T_{out}	$Eff._{th}$	Heat Gain [W_{th}]
-	-	-	-	19.60	0.267	287.29
62				23.30	0.345	371.50
82.5	47	83	35	23.29	0.345	371.30
144				23.52	0.35	376.70
62				23.24	0.344	370.30
82.5	0	130	35	23.21	0.344	369.57
144				23.47	0.349	375.40
62				23.28	0.345	371.16
82.5	47	176.3	28	23.23	0.344	370.07
144				23.32	0.346	372.07

4. Conclusions

To confirm the impact of the arrangement of triangular shaped baffles fitted in an air-type PVT collector, the airflow characteristics in the collector was analyzed through a validated CFD model in this study. The heat transfer characteristics, pressure drop, and thermal efficiency of the collector were examined, and the results can be summarized as follows.

- Depending on the variables, the outlet temperature increased by 3.6–3.9 °C and the heat gain increased by 1.28–1.31 times compared to the collector without baffles. Therefore, thermal performance of the collector improved up to 31%.
- The thermal performance of the air-type PVT collector improved when the horizontal spacing of the baffles was wider and the vertical spacing was narrower. Furthermore, it was confirmed that greater number of baffles resulted in higher thermal energy yield of the collector.
- The heat transfer performance increased 1.03 times of the maximum outlet velocity and 1.05 times of the average Reynolds number; according to the triangular baffle placement, the pressure drop increased by about 6.78 Pa under these conditions. Furthermore, variable conditions in which the baffles were more evenly placed (i.e., the case of W = 144 mm, H1 = 47 mm, H2 = 83 mm and B.N = 35) resulted in relatively small pressure drop and high thermal efficiency; thus, these conditions are judged to be advantageous for improving the performance of the air-type PVT collector.

Based on the results of this study, it can be deduced that the thermal performance improvement according to the staggered arrangement spacing of triangular baffles in the air-type PVT collector was not significant, but it was advantageous in terms of heat transfer in the collector when the baffle spacing was even.

Author Contributions: Conceptualization, J.-S.Y., J.-T.K., and J.-H.K.; methodology, J.-H.K., and J.-S.Y.; software, J.-S.Y.; validation, J.-T.K., J.-H.K. and J.-S.Y.; formal analysis, J.-S.Y.; investigation, J.-H.K. and J.-S.Y.; resources, J.-T.K.; data curation, J.-H.K. and J.-S.Y.; writing—original draft preparation, J.-S.Y.; writing—review and editing, J.-T.K., J.-H.K. and J.-S.Y.; visualization, J.-T.K., J.-S.Y.; supervision, J.-T.K.; project administration, J.-S.Y.; funding acquisition, J.-T.K. and J.-H.K. All authors have read and agreed to the published version of the manuscript.

Funding: This research was funded by the Korea Institute of Energy Technology Evaluation and Planning (KETEP) and the Ministry of Trade, Industry and Energy (MOTIE) of the Republic of Korea, grant number 20188550000480 and 20173010013420.

Conflicts of Interest: The authors declare no conflict of interest.

References

1. Mahamad, M.; Ramadan, M.; Olabi, A.G.; Pullen, K.; Naher, S. A review of mechanical energy storage systems combined with wind and solar applications. *Energy Convers. Manag.* **2020**, *210*, 1–14.
2. Lai, C.S.; Jia, Y.; Lai, L.L.; Xu, Z.; McCulloch, M.D.; Wong, K.P. A comprehensive review on large-scale photovoltaic system with applications of electrical energy storage. *Renew. Sustain. Energy Rev.* **2017**, *78*, 439–451. [CrossRef]
3. Rashid, K.; Safdarnejad, S.M.; Powell, K.M. Dynamic simulation, control, and performance evaluation of a synergistic solar and natural gas hybrid power plant. *Energy Convers. Manag.* **2019**, *179*, 270–285. [CrossRef]
4. Rukman, N.S.B.; Fudholi, A.; Taslim, I.; Indrianti, M.A.; Manyoe, I.N.; Lestari, U.; Sopian, K. Electrical and thermal efficiency of air-based photovoltaic thermal (PVT) systems: An overview. *Indones. J. Electr. Eng. Comput. Sci.* **2019**, *14*, 1134–1140. [CrossRef]
5. Mustapha, M.; Fudholi, A.; Yen, C.H.; Ruslan, M.H.; Sopian, K. Review on energy and exergy analysis of air and water based photovoltaic thermal (PVT) collector. *Int. J. Power Electron. Drive Syst.* **2018**, *9*, 1367–1373.
6. Riffat, S.B.; Cuce, E. A Review on Hybrid Photovoltaic/thermal Collectors and Systems. *Int. J. Low-Carbon Technol.* **2011**, *6*, 212–241. [CrossRef]
7. Bakari, R. Heat transfer optimization in air flat plate solar collectors integrated with baffles. *J. Power Energy Eng.* **2018**, *60*, 70–84. [CrossRef]
8. Amraoui, M.A.; Aliane, K. Numerical analysis of a three dimensional fluid flow in a flat plate solar collector. *Int. J. Renew. Sustain. Energy* **2014**, *3*, 68–75.
9. Chaube, A.; Sahoo, P.K.; Solanki, S.C. Effect of roughness shape on heat transfer and flow friction characteristics of solar air heater with roughed absorber plate. *Trans. Eng. Sci.* **2006**, *53*, 43–51.
10. Abuska, M.; Sevik, S. Energy, exergy, economic and environmental (4E) analyses of flat-plate and V-groove solar air collectors based on aluminium and copper. *Sol. Energy* **2017**, *158*, 259–277. [CrossRef]
11. Fudholi, A.; Zohri, M.; Rukman, N.S.B.; Nazri, N.S.; Mustapha, M.; Yen, C.H.; Mohammad, M.; Sopian, K. Exergy and sustainability index of photovoltaic thermal (PVT) air collector: A theoretical and experimental study. *Renew. Sustain. Energy Rev.* **2019**, *100*, 44–51. [CrossRef]
12. Promthaisong, P.; Eiamsa-ard, S. Fully developed periodic and thermal performance evaluation of a solar air heater channel with wavy-triangular ribs placed on an absorber plate. *Int. J. Therm. Sci.* **2019**, *140*, 413–428. [CrossRef]
13. Sivakandhan, C.; Arjunan, T.V.; Matheswaran, M.M. Thermohydraulic performance enhancement of a new hybrid duct solar air heater with inclined rib roughness. *Renew. Energy* **2020**, *147*, 2345–2357. [CrossRef]
14. Bhagoria, J.L.; Saini, J.S.; Solanki, S.C. Heat transfer coefficient and friction factor correlations for rectangular solar air heater duct having transverse wedge shaped rib roughness on the absorber plate. *Renew. Energy* **2002**, *25*, 341–369. [CrossRef]
15. Yadav, A.S.; Bhagoria, J.L. A CFD Analysis of a Solar Air Heater Having Triangular Rib Roughness on the Absorber Plate. *Int. J. Chem. Tech. Res.* **2013**, *5*, 964–971.
16. Yadav, A.S.; Bhagoria, J.L. A CFD based thermos-hydraulic performance analysis of an artificially roughened solar air heater having equilateral triangular sectioned rib roughness on the absorber plate. *Int. J. Heat Mass Transfer.* **2014**, *70*, 1016–1039. [CrossRef]

17. Bensaci, C.; Moummi, A.; Flor, F.J.S.; Jara, E.A.R.; Rincon-Casado, A.; Ruiz-Pardo, A. Numerical and experimental study of the heat transfer and hydraulic performance of solar air heaters with different baffle positions. *Renew. Energy* **2020**, *155*, 1231–1244. [CrossRef]
18. Choi, H.U.; Fatkhur, R.; Kim, Y.B.; Yoon, J.I.; Son, C.H.; Choi, K.H. CFD analysis on the heat transfer performance with various obstacles in air channel of air-type PV/Thermal module. *J. Korean Sol. Energy Soc.* **2018**, *38*, 33–43.
19. Othman, M.Y.; Yatim, B.; Sopian, K.; Bakar, M.N.A. Performance studies on a finned double-pass photovoltaic-thermal (PV/T) solar collector. *Desalination* **2007**, *209*, 43–49. [CrossRef]
20. Khani, M.S.; Baneshi, M.; Eslami, M. Bi-objective optimization of photovoltaic-thermal (PV/T) solar collectors according to various weather conditions using genetic algorithm: A numerical modeling. *Energy* **2019**, *189*, 1–16. [CrossRef]
21. Zhang, G.; Ding, X.; Li, T.; Pu, W.; Lou, W.; Hou, J. Dynamic energy balance model of a glass greenhouse: An experimental validation and solar energy analysis. *Energy* **2020**, *198*, 1–18. [CrossRef]
22. Zhao, Y.; Meng, T.; Jing, C.; Hu, J.; Qian, S. Experimental and numerical investigation on thermal performance of PV-driven aluminium honeycomb solar air collector. *Sol. Energy* **2020**, *204*, 294–306. [CrossRef]
23. Nourdanesh, N.; Hossainpour, S.; Adamiak, K. Numerical simulation and optimization of natural convection heat transfer enhancement in solar collectors using electrohydrodynamic conduction pump. *Appl. Therm. Eng.* **2020**, *180*, 1–10. [CrossRef]
24. Siemens PLM Software, Explore NX and Discover Your Solution. Available online: http://www.plm.automation.siemens.com/products/nx/about-nx-software.shtml (accessed on 4 August 2020).
25. Launder, B.E.; Spalding, D.B. The numerical computation of turbulent flows. *Comput. Methods Appl. Mech. Eng.* **1974**, *3*, 269–289. [CrossRef]
26. ISO 9806:2017. International Organization for Standard. In *Solar Energy—Solar Thermal Collectors—Test Methods*; ISO: Geneva, Switzerland, 2017.

Article

Confinement Effect of Reinforced Concrete Columns with Rectangular and Octagon-Shaped Spirals

Hyeong-Gook Kim, Chan-Yu Jeong, Dong-Hwan Kim and Kil-Hee Kim *

Department of Architectural Engineering, Kongju National University, Cheonandaero, Seobuk, Cheonan 1223-24, Korea; anthk1333@kongju.ac.kr (H.-G.K.); ssbachany@kongju.ac.kr (C.-Y.J.); kimdh@kongju.ac.kr (D.-H.K.)

* Correspondence: kimkh@kongju.ac.kr; Tel.: +82-41-521-9335

Received: 29 August 2020; Accepted: 24 September 2020; Published: 26 September 2020

Abstract: Conventional spiral-type transverse reinforcement is effective at increasing the ductility and the maximum strength of reinforced concrete (RC) columns because it confines the inner concrete and the longitudinal reinforcement. However, when arranging crossties in a RC column with spirals, problems such as mutual interference with longitudinal reinforcement, overcrowding of reinforcement, and deterioration of constructability occur. Furthermore, the loosening of 90 and 130-degree standard hooks due to the lateral expansion of concrete causes buckling of the longitudinal reinforcement. This paper describes the ability of a newly developed spiral-type transverse reinforcement with various yield strengths to confine RC columns subjected to cyclic lateral load and constant axial load. The ductility capacity, energy dissipation, and effective stiffness of RC columns confined by the developed spiral-type transverse reinforcement were compared with those of RC columns confined by typical rectangular reinforcement. The experimental results showed that RC column specimens with the developed spiral-type transverse reinforcement have better performances in terms of ductility capacity and energy dissipation, even though the amount of reinforcement used for the specimens decreased by about 27% compared with the specimen with typical rectangular reinforcement.

Keywords: reinforced concrete column; confinement effects; energy dissipation

---

1. Introduction

Reinforced concrete (RC) columns subject to both central axial load and flexural load undergo a rapid deterioration of strength due to lateral expansion of inner concrete after delamination of the concrete cover. At this point, transverse reinforcement of RC columns confines the lateral expansion of core concrete, thus increasing the compressive strength and ductility under lateral load. The lateral confinement performance of RC columns is influenced by the strength ratio of transverse reinforcement and concrete, the amount and shape of transverse reinforcement, and the shear span to depth ratio [1–10]. Many models have been developed based on experiments to predict the strength and behavior of RC columns and concrete cylinders with shear reinforcement [11–18]. Extensive research has also been conducted not only on conventional crossties, but also on interlocking spirals to suppress lateral expansion of core concrete and buckling of longitudinal reinforcement, as well as shear reinforcement of RC columns using Fiber-Reinforced Polymer (FRP) [19–27].

Transverse reinforcement of RC columns can be largely classified into crossties and spiral reinforcement. Compared to crossties, spiral reinforcement can effectively confine the inner concrete of RC columns, and is thus more advantageous in enhancing ductility. However, due to difficulties in bar arrangement during construction, there is a higher demand for columns with rectangular cross-sections than for those with circular cross-sections. As such, most columns have shear reinforcement in the form of crossties, which are also more common than spiral reinforcement even in rarer columns with circular cross-sections. In terms of ductility enhancement, introducing sub-ties is more efficient than decreasing the spacing of transverse reinforcement. Sub-ties are effective at suppressing the lateral expansion of core concrete, and also prevent the buckling of longitudinal reinforcement. The use of conventional sub-ties with 90-degree and 135-degree standard hooks can result in problems such as mutual interference with longitudinal reinforcement, overcrowding of reinforcement, poor filling of concrete, and deterioration of constructability. When subject to repeated lateral forces such as seismic loads, the loosening of 90-degree hooks causes a decrease in effective lateral confinement, which may contribute to buckling and weakening of longitudinal reinforcement.

This study proposed a new type of transverse reinforcement with spirally arranged crossties to improve the constructability of RC columns and to resolve structural issues associated with conventional crossties. Cyclic loading tests were performed on RC columns with the new spiral-type transverse reinforcement and subject to a central axial load. The effects of transverse reinforcement shape and yield strength on crack formation, ductility capacity, energy dissipation capacity, and effective stiffness in relation to drift angle were assessed. In addition, the constructability of RC column members was evaluated by measuring time consumed in arranging the proposed transverse reinforcement.

2. Experimental Program

2.1. Materials

The concrete mixture specifications are given in Table 1. Ready-mixed concrete with a design strength of 24 MPa was used to manufacture the specimens, as described in Table 1. Concrete cylinders with dimensions of ϕ100 mm × 200 mm were manufactured in accordance with ASTM C31/C31M. The compressive strength of the concrete was tested according to ASTM C39/C39M. The mean compressive strength of concrete measured in the cylinder test was 22.4 MPa. This value was used to predict the shear strength of specimens.

Table 1. Proportions of concrete mixture.

f'_c	G_{max}	W/C	S/a	Unit Weight(kg/m^3)					Slump
(MPa)	(mm)	(%)	(%)	W	C	S	G	AD	(mm)
24	25	49.7	48.5	82	214	872	936	69.3	120

f'_c: compressive strength of concrete, G_{max}: maximum size of coarse aggregate, W/B: water binder ratio, S/a: fine aggregate modulus, W: water, C: cement, S: fine aggregate, G: coarse aggregate, and AD: water reducing admixture.

Two types of reinforcing bars were used to manufacture the specimens. D19 (286.7 mm^2) deformed bars with a yield strength of 523 MPa were used for longitudinal reinforcement of all specimens. D10 (71.3 mm^2) deformed bars with yield strengths of 540 MPa, 554 MPa, 788 MPa, and 1328 MPa were used for transverse reinforcement. Table 2 shows the physical properties of the reinforcing bars.

Table 2. Mechanical properties of reinforcing steel.

Specimen	Reinf. Bar	f_y (MPa)	ε_y	E_s (MPa)	Remarks
H-F	D10	540	0.0027	200,000	Transverse reinforcement
KSS-5	Φ10	554	0.0028	197,857	
KSS-7	Φ10	788	0.0039	202,051	
KSS-12	Φ10	1328	0.0066	201,212	
All Specimens	D19	523	0.0028	186,586	Longitudinal reinforcement

f_y: yield strength of reinforcement, ε_y: yield strain of reinforcement, and E_s: modulus of elasticity.

2.2. Specimen Details

To evaluate the lateral confinement effect of RC columns in relation to shape and strength of transverse reinforcement, this study fabricated four specimens as shown in Table 3. H-F refers to RC column specimens with rectangular transverse reinforcement, while KSS-5, KSS-7, and KSS-12 are specimens with the proposed KSS-transverse reinforcement comprised of rectangular and octagonal spirals. The numbers 5, 7, and 12 in the KSS specimen name represent the yield strength grade of the proposed transverse reinforcement, that is, 500 MPa, 700 MPa, and 1200 MPa, respectively.

Table 3. Properties of specimens.

Specimen	f'_c (MPa)	ρ_w (-)	f_{wy} (MPa)	$\rho_w f_{wy}$ (MPa)	B (mm)	D (mm)	d (mm)	s (mm)	v (mm^3)
H-F	22.4	0.0051	540	2.75	450	450	400	125	235,504
KSS-5		0.0037	554	2.04					171,387
KSS-7			788	2.91					
KSS-12			1328	4.90					

f'_c: compressive strength of concrete, ρ_w: volume ratio of transverse reinforcement, f_{wy}: yield strength of transverse reinforcement, s: spacing of transvere reinforcement, and v: volume of transvere reinforcement.

Figure 1 shows details of bar arrangement of H-F specimens with conventional rectangular transverse reinforcement and KSS specimens with the proposed spiral-type shear reinforcement. As shown in Figure 1, rectangular crossties were spaced 125 mm apart in H-F specimens, and sub-ties having 90-degree and 135-degree bending angles in longitudinal and lateral directions as specified in the ACI design code were arranged with the same spacing. As shown in Figure 1b, KSS specimens had rectangular crossties and octagonal sub-ties arranged spirally to facilitate confinement of longitudinal reinforcement at the edges and inner longitudinal reinforcement. The rectangular crossties were given the same spacing as the crossties of the H-F specimens.

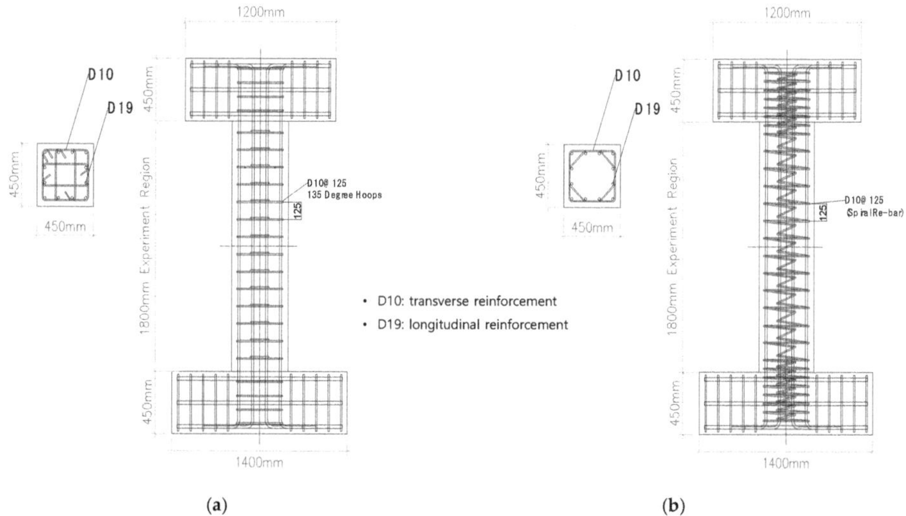

Figure 1. Details of specimens (Unit: mm): (**a**) H-F; (**b**) KSS.

All specimens had square cross-sections with width (B) of 450 mm and column depth (D) of 450 mm; the shear span to depth ratio (a/d) was set to 2.0. The effective depth (d) was set to 400 mm in consideration of the concrete cover and crosstie diameter. All specimens had four longitudinal reinforcing bars (D19) with yield strength of 523 MPa on each side to prevent shear failure and induce flexural failure due to yielding of longitudinal reinforcement before other types of failure. Transverse reinforcements (D10) were arranged at a spacing of 125 mm. In Table 3, the reinforcement ratio ($\rho_{w.H}$) of each specimen was calculated using the following:

$$\rho_{w.H} = \frac{A_{s.H}}{B \cdot s} \tag{1}$$

$$\alpha = \frac{v_{KSS}}{v_H} \quad (2)$$

$$\rho_{w.KSS} = \alpha \cdot \rho_{w.H}. \quad (3)$$

Here, $\rho_{w.H}$ and $\rho_{w.KSS}$ are the transverse reinforcement ratios of specimens with rectangular crossties and specimens with the proposed spiral-type crossties. $A_{s.H}$ is the cross-sectional area of the rectangular crossties, v_H and v_{KSS} are the volume of rectangular crossties and of the proposed spiral-type crossties for transverse reinforcement with spacing s, and α is the ratio of v_{KSS} to v_H.

Since the proposed spiral-type crossties have octagonal sub-ties arranged spirally between rectangular crossties, this study used v_{KSS}/v_H instead of the volume ratio of spiral reinforcement comprised of square or circular steel to calculate the transverse reinforcement ratio $\rho_{w.KSS}$ of KSS specimens, expressing it in terms of the transverse reinforcement ratio of the H-F specimens. Through Equations (1)–(3), KSS specimens were found to have the same ρ_w of 0.0037. These were more advantageous in that the amount of reinforcement was 27% less than that of H-F specimens with the same crosstie spacing s.

2.3. Test Setup and Instrumentation

Using a hydraulic pressure system, the test specimens were subjected to reversed cyclic bending, shear, and axial load in a setup with vertically fixed top and bottom stubs. Lateral force was applied to the loading frame connected to the upper stub. The lateral force actuator, with a loading capacity of 1000 kN, was located so that point of contra flexure is produced at the midspans of the specimens. An axial force corresponding to 15% of the compressive strength of the column was continuously applied using a vertical actuator with a loading capacity of 2000 kN until the end of the test. Figure 2a presents details of the loading and measurement system. Several linear variable displacement transducers (LVDT) were installed to measure the drift angles of specimens. Two LVDTs of 300 mm were installed on the upper and lower stubs of the specimens; average measurements were used to calculate the drift angle. The strain in the transverse and longitudinal reinforcement was measured used strain gauges attached to the reinforcing bar surface. Figure 2b shows the loading protocol used in this testing program. The specimens were loaded monotonically up to the first yield drift angle, δ_y, followed by a series of drift-controlled loading cycles comprising two full cycles with specified drift angles of about $\pm 2\delta_y$, $\pm 3\delta_y \cdots$. The tests were terminated when the lateral force in the post-peak load-deformation curve dropped to approximately 85% of the peak-recorded load.

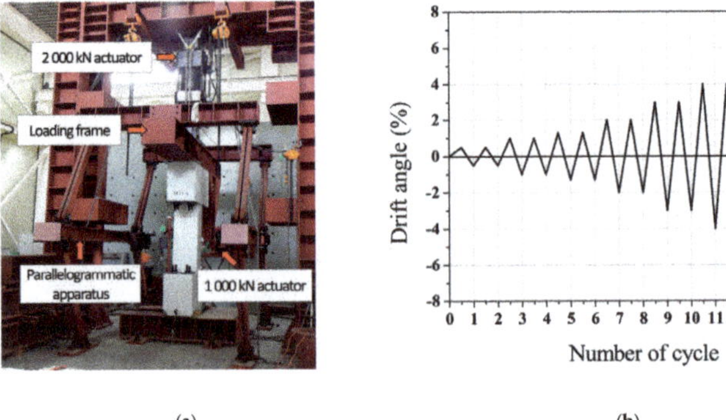

Figure 2. View of test setup and loading history: (**a**) Test setup; (**b**) Loading history.

3. Experimental Results and Discussions

3.1. Load Versus Drift Angle Relations

The lateral load vs. drift of the specimens are presented in Figure 3 Quantitative values of measured yield and maximum load, and drift angles, are given in Table 4. It was observed that the longitudinal reinforcement of specimen H-F yielded at a drift angle of −1.24%, and the load reached the maximum value at −441.4 kN at the drift angle of −1.88% in the negative direction. At the drift angle of −4.01%, where the load dropped below 80% of the maximum load, the test was terminated. On the other hand, longitudinal reinforcements of specimens KSS-5, 7, and 12 yielded at drift angles less than −1.04%, earlier than specimen H-F. The average yield load, P_y, of specimens KSS-5, 7, and 12 was about 3.1% lower than that of specimen H-F, while the average maximum load of the specimens was very similar to that of specimen H-F. The effective stiffness of the specimens with KSS at yield load increased as the yield strength of the transverse reinforcement increased. After peak load, the strength of specimen KSS-7 dramatically decreased, to below 80% of the maximum strength, and thus the test was finished at the drift angle of −3.08%. The observed ductility of specimen KSS-7 is lower than that of the other specimens. It was confirmed that specimen KSS-7 experienced bond failure between longitudinal reinforcement and concrete after maximum load. All specimens showed similar behavior in terms of load vs. drift angle. These experimental results verify that the proposed transverse reinforcement effectively suppressed the lateral expansion of concrete, thereby increasing the maximum strength and ductility of the RC columns.

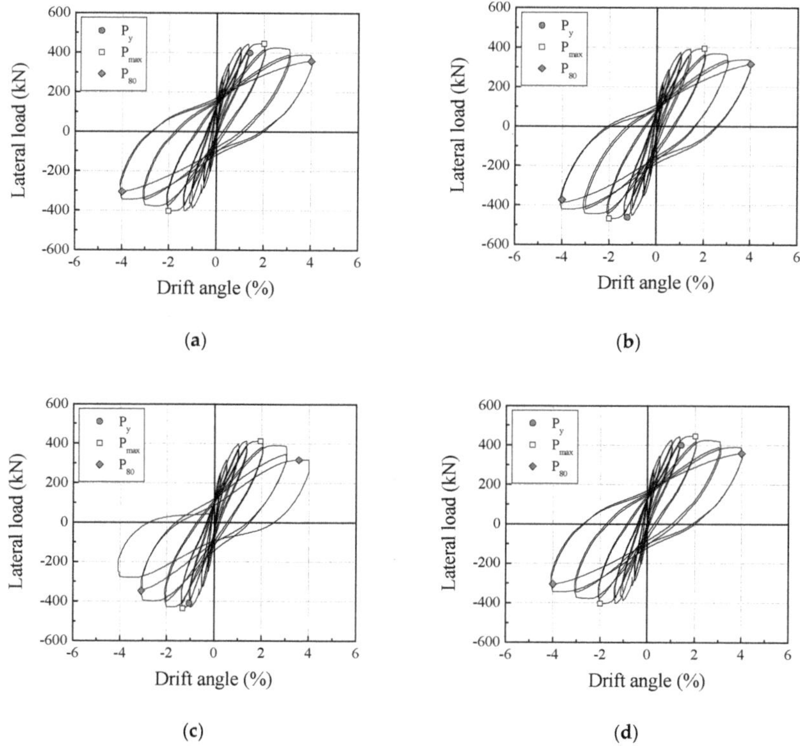

Figure 3. Lateral load versus drift angle relationships: (**a**) H-F; (**b**) KSS-5; (**c**) KSS-7; (**d**) KSS-12.

Table 4. Results of cyclic loading tests.

Specimen	Loading Direction	At Yielding of Reinforcement		At Peak Load		At 0.8 P_{max}		Failure Mode
		P_y (kN)	D_y (%)	P_{max} (kN)	D_{max} (%)	P_u (kN)	D_u (%)	
H-F	Positive	399.3	1.35	407.9	1.87	326.4	4.04	Flexural/Buckling
	Negative	−433.9	−1.24	−441.4	−1.88	−353.1	−4.01	
KSS-5	Positive	380.3	1.49	393.1	1.89	314.5	4.01	Flexural
	Negative	−454.6	−1.16	−467.0	−1.95	−373.6	−4.07	
KSS-7	Positive	409.9	1.64	412.0	1.35	330.1	3.02	Flexural/Bond
	Negative	−409.5	−1.01	−435.8	−1.33	−348.7	−3.08	
KSS-12	Positive	426.8	1.33	439.4	1.78	351.5	4.02	Flexural
	Negative	−418.5	−0.96	−421.6	−1.33	−337.3	−4.10	

Notation-P_y: yield load, P_{max}: maximum load, P_u: ultimate load (0.8 P_{max}), D_y: drift angle at P_y, D_{max}: drift angle at P_{max} and D_u: drift angle at P_u.

3.2. Crack Patterns and Failure Modes

Crack patterns of the specimens at maximum load are shown in Figure 4. In specimens with an axial force ratio of 15%, flexural cracks were first observed at a 0.5% drift angle at both plastic hinge regions. Except for specimen KSS-12, bond cracks appeared along the longitudinal reinforcement, with an increase in the number of flexural cracks when the drift angle Fexceeded 1.0%. In general, bond cracks are observed on RC members when shear span-to-effective depth ratio (a/d) lies within a range of 1.0 to 2.5. In the case of specimen KSS-7, remarkable bond cracks occurred along the longitudinal reinforcement. The bond cracks induced failure of that specimen earlier than for the other specimens after maximum load. Concrete deterioration due to cyclic loading was observed in both plastic hinge regions.

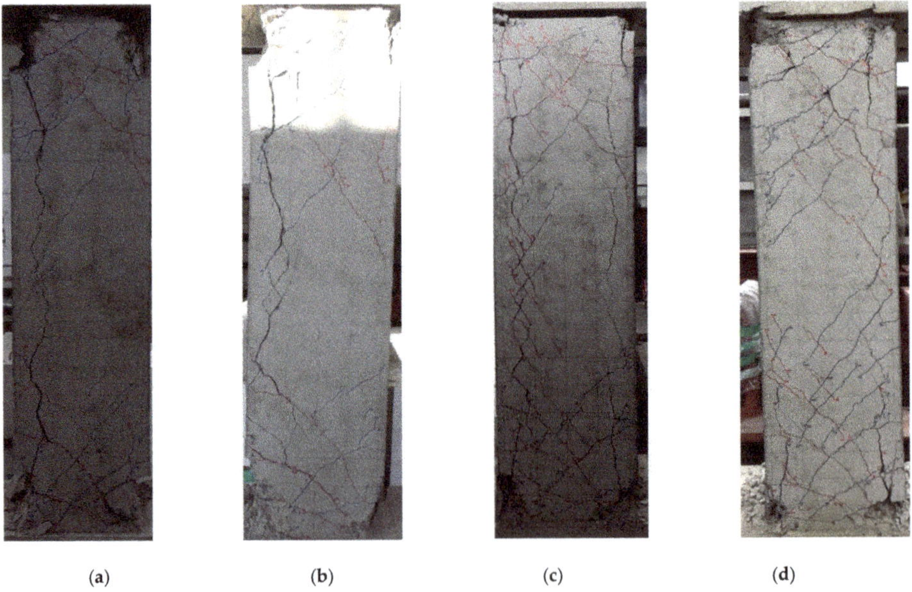

Figure 4. Crack patterns in specimens at failure: (**a**) H-F; (**b**) KSS-5; (**c**) KSS-7; (**d**) KSS-12.

Figure 5 shows the status of reinforcements in the lower plastic hinge region of specimens after the cyclic loading test. It was observed that crossties with 90 and 130-degree standard hooks in specimen H-F became loose due to the lateral expansion of concrete. Furthermore, buckling of the longitudinal reinforcement was observed in the specimen. The buckling of longitudinal reinforcement degrades the load-carrying capacity of RC structures subjected to seismic loads [28]. Compared with specimens KSS, specimen H-F showed remarkable spalling of concrete cover in the plastic hinge region. In the case of specimens KSS, buckling of longitudinal reinforcement was not observed. This means that the rectangular shear reinforcement and the octagon-shaped sub-ties confined the longitudinal reinforcement and inner concrete until failure. Kani et al. [29] found that when a/d of RC members is smaller than 2.5, the shear resistance of RC members increases significantly. It means that the structural performance of RC members with a/d greater than 2.5 is likely to be determined by the bending resistance. It is well known that using spiral reinforcement can greatly improve the bending capacity of RC columns. Thus, it can be understood that the proposed transverse reinforcement (KSS) is effective at improving the strength and lateral load-carrying capacity of RC columns with a shear span-to-effective depth ratio of more than 2.5.

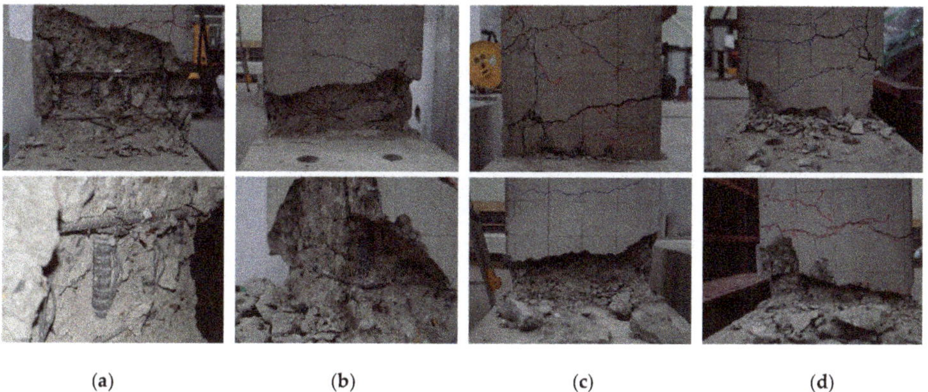

(**a**) (**b**) (**c**) (**d**)

Figure 5. Observation of reinforcements in plastic hinge region: (**a**) H-F; (**b**) KSS-5; (**c**) KSS-7; (**d**) KSS-12.

3.3. *Ductility and Energy Dissipation Capacity*

The ductility and energy dissipation capacity of the specimens were experimentally investigated in this study. The effective stiffnesses, the ductility factor (μ), and the energy dissipation capacity for each specimen are given in Table 5. The ductility factor (μ) was taken as the ratio of ultimate story drift, Δ_u, to story drift corresponding to yield load, Δ_y. In this study, a story drift corresponding to 80% of the maximum load was taken as ultimate story drift Δ_u. The energy dissipation, W, was defined as the sum of the area enclosed by the load-story drift curves.

Table 5. Comparison of ductility factor and energy dissipation.

No. (i)	Specimen	$\rho_w f_{wy}$	Effective Stiffness			Ductility Factor		Energy Dissipation	
			$K_{e.y}$	$K_{e.max}$	$K_{e.u}$	μ	μ_i/μ_1	W	W_i/W_1
			(N/mm)	(N/mm)	(N/mm)	(-)	(-)	(J)	(-)
1	H-F	2.75	19,440.0	13,043.7	4891.9	3.23	1.00	148,702.1	1.00
2	KSS-5	2.04	21,772.0	13,304.8	5099.6	3.51	1.08	150,297.2	1.01
3	KSS-7	2.91	22,524.8	18,203.8	6589.7	3.05	0.94	145,008.7	0.98
4	KSS-12	4.90	24,218.8	17,610.7	4570.5	4.27	1.32	164,628.7	1.11

Notation-i: number of specimens, $K_{e.y}$: effective stiffness at yield load, $K_{e.max}$: effective stiffness at maximum load, $K_{e.u}$: effective stiffness at ultimate load.

Although the amount of transverse reinforcement was reduced by about 27%, the effective stiffness of the specimens with KSS at yield load in the negative direction was greater than that of specimen H-F. $K_{e.y}$ increased as the yield strength of transverse reinforcement increased. Specimen KSS-7, moreover, showed the highest effective stiffness at the maximum load and showed the lowest effective stiffness reduction rate among all specimens. In terms of ductility capacity, while specimen KSS-7 showed a ductility factor similar to that of specimen H-F, specimens KSS-5 and 12 showed a greater ductility factor, which increased in proportion to the yield strength of the reinforcement. The energy dissipation also showed a trend similar to the ductility factor.

Figure 6 uses an index, $\rho_w f_{wf}$, to compare the ductility and energy dissipation capacities of the specimens. Considering that specimen KSS-7 experienced bond failure, it can be understood from Figure 6a that even if the utilized transverse reinforcement ratio, ρ_w, is reduced, the ductility capacity of the RC columns can then be improved by increasing the yield strength of the reinforcement, f_{wf}. Figure 6b presents the ratio of the energy dissipated in specimens with KSS to that dissipated in specimen H-F at each drift angle. As a result of the cyclic loading test, specimen KSS-5, due to its lower effective stiffnesses of up to 3.0% of the drift angle, showed energy dissipations lower than those of specimen H-F; however, both specimens showed similar energy dissipation capacities at the end of the test. Specimen KSS-12 showed the best performance in terms of the ductility and energy dissipation.

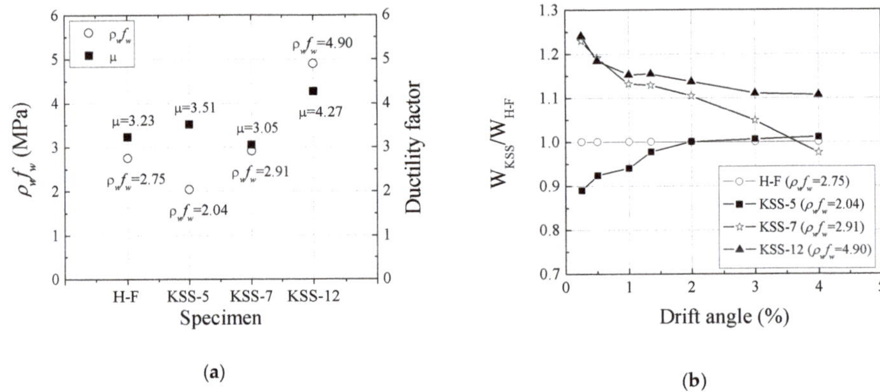

Figure 6. Comparison of ductility factor and energy dissipation: (**a**) Ductility factor; (**b**) Energy dissipation.

It was found that the seismic performance of RC columns with the same cross-sectional property can be enhanced by increasing the yield strength of transverse reinforcement. Furthermore, it is possible to reduce the amount of reinforcing steel used for the construction of reinforced concrete structures if the inner concrete and the longitudinal reinforcement are confined by transverse reinforcement with an appropriate shape.

4. Constructability of Proposed Transverse Reinforcement

This study performed mockup tests to evaluate the constructability of RC columns in relation to the shape of the transverse reinforcement. The arrangement of transverse reinforcement was done by skilled workers; constructability between H-F and KSS specimens was compared based on the assembly time of the transverse reinforcement. The specimens were fabricated considering the cross-sections of columns in actual RC structures. Table 6 presents information on specimen cross-sections and arrangement details, and assembly times of transverse reinforcement measured during the mockup tests. The average assembly time of transverse reinforcement was 56 min 12 s for H-F specimens, and 21 min 12 s for KSS specimens. The assembly time for transverse reinforcement of KSS specimens was 60% faster than that of H-F specimens. This is because KSS specimens pull down transverse reinforcement from the top of longitudinal reinforcement in a spring-like manner to fit the given spacing, whereas H-F specimens introduce 90 and 130-degree standard hooks between transverse reinforcement after completing the arrangement of transverse reinforcement. The evaluation of structural performance and constructability showed that the proposed transverse reinforcement will have advantages over conventional rectangular reinforcement in terms of reduced amount of reinforcement in fabricating column members and improved constructability.

Table 6. Comparison of constructability between H-F and KSS.

Specimens	BxD (mm)	H (mm)	s (mm)	1st	2nd	3rd	Average
H-F	500 × 500	3500	135	43′40″	56′50″	68′06″	56′12″
KSS				20′32″	21′35″	21′30″	21′12″
KSS/H-F				47.0%	38.0%	31.6%	37.7%

5. Conclusions

In this study, an experiment was conducted to evaluate the flexural performance of RC columns confined with a newly developed spiral type transverse reinforcement (KSS). Experimental results were compared with an RC column confined with a conventional hoop reinforcement. The experimental results showed that KSS transverse reinforcement, due to its superior confinement effect for concrete and longitudinal reinforcement, increases the effective stiffness of columns at yield and maximum strength. Moreover, the results showed that the ductility and energy dissipation capacities of RC columns with KSS were improved compared with those of an RC column with conventional hoops, even though the amount of reinforcement used decreased by about 27%. Finally, it was found that KSS is effective at reducing the amounts of steel and time required to arrange transverse reinforcement.

Author Contributions: Investigation, D.-H.K.; Supervision, K.-H.K.; Writing—original draft, H.-G.K.; Writing—review & editing, C.-Y.J. All authors have read and agreed to the published version of the manuscript.

Funding: This work was supported by the Priority Research Centers Program through the National Research Foundation of Korea (NRF) founded by the Ministry of Education (2019R1A6A1A03032988); This research was supported by Basic Science Research Program through the National Research Foundation of Korea (NRF) founded by the Ministry of Education (2018R1A2B3001656); This research was supported by the Basic Science Research Program through the National Research Foundation of Korea (NRF) founded by the Ministry of Education (2019R1I1A3A01058156); This work was also supported by a research grant of the Kongju National University in 2020 (2020-0203-01).

Conflicts of Interest: The authors declare no conflict of interest.

References

1. Marinez, S.; Nilson, A.H.; Slate, F.O. Spirally Reinforced High-Strength Concrete Columns. *ACI Struct. J.* **1984**, *8181*, 431–442.
2. Moehle, J.; Cavanagh, T. Confinement Effectiveness of Crossties in RC. *J. Struct. Div.* **1985**, *111*, 2105–2120. [CrossRef]
3. Mander, J.B.; Priestley, M.J.N.; Park, R. Observed Stress-Strain Behavior of Confined Concrete. *J. Struct. Eng.* **1988**, *114*, 1827–1849. [CrossRef]
4. Yong, Y.K.; Nour, M.G.; Nawy, E.G. Behavior of Laterally Confined High-Strength Concrete Under Axial Loads. *J. Struct. Eng.* **1989**, *114*, 332–351. [CrossRef]
5. Muguruma, H.; Watanabe, F. Ductility Improvement of High-Strength Concrete Columns with Lateral Confinement. *ACI Spec. Publ.* **1990**, *121*, 47–60.
6. Muguruma, H.; Watanabe, F.; Tanaka, H. Ductile Behaviour of High-Strength Columns Confined by High-Strength Transverse Reinforcement. *ACI Spec. Publ.* **1991**, *128*, 877–891.
7. Razvi, S.R. Strength and Ductility of Confined Concrete. *J. Struct. Eng.* **1992**, *118*, 1590–1607.
8. Sheikh, S.A.; Toklucu, M.T. Reinforced Concrete Columns Confined by Circular Spirals and Hoops. *ACI Struct. J.* **1993**, *90*, 542–553.
9. Cusson, D.; Paultre, P. High-Strength Concrete Columns Confined by Rectangular Ties. *J. Struct. Eng.* **1994**, *120*, 783–804. [CrossRef]
10. Razvi, S.R.; Saatcioglu, M. Circular High-Strength Concrete Columns under Concentric Compression. *ACI Struct. J.* **1999**, *96*, 817–825.
11. Mander, J.B.; Priestley, M.J.N.; Park, R. Theoretical Stress-Strain Model for Confined Concrete. *J. Struct. Eng.* **1988**, *114*, 1804–1826. [CrossRef]

12. Cusson, D.; Paultre, P. Stress-Strain Model for Confined High-Strength Concrete. *J. Struct. Eng.* **1995**, *121*, 468–477. [CrossRef]
13. El-Dash, K.M.; Ahmad, S.H. A model for stress-strain relationship of spirally con_ned normal and high-strength concrete columns. *Mag. Concr. Res.* **1995**, *47*, 177–184. [CrossRef]
14. Hoshikuma, J.; Kawashima, K.; Nagaya, K. A Stress-Strain Model for Reinforced Concrete Columns Confined by Lateral Reinforcement. *Doboku Gakkai Ronbunshu* **1995**, *520*, 1–11. (In Japanese) [CrossRef]
15. Razvi, S.; Saatcioglu, M. Confinement Model for High-Strength Concrete. *J. Struct. Eng.* **1999**, *125*, 281–289. [CrossRef]
16. Li, B.; Park, R.; Tanaka, H. Stress-Strain Behavior of High-Strength Concrete Confined by Ultra-High- and Normal-Strength Transverse Reinforcement. *ACI Struct. J.* **2001**, *98*, 395–406.
17. Mendis, P.; Pendyala, R.; Setunge, S. Stress-Strain Model to Predict the Full-range Moment Curvature Behavior of High-Strength Concrete Sections. *Mag. Concr. Res.* **2000**, *52*, 227–234. [CrossRef]
18. Kim, Y.S.; Kim, S.W.; Lee, J.Y.; Lee, J.M.; Kim, H.G.; Kim, K.H. Prediction of Stress-Strain Behavior of Spirally Confined Cocrete Considering Lateral Expasion. *Constr. Build. Mater.* **2016**, *102*, 743–761. [CrossRef]
19. Tanaka, H.; Park, R. Strnegth and Ductility of Reinforced Concrete Columns with Interlocking Spirals. In Proceedings of the Tenth World Conference on Earthquake Engineering, Balkema, Rotterdam, 19–24 July 1992; pp. 4371–4376.
20. Tanaka, H.; Park, R. Seismic Design and Behavior of Reinforced Concrete Columns with Interlocking Spirals. *ACI Struct. J.* **1993**, *90*, 192–203.
21. Khaloo, A.R.; El-Dash, K.M.; Ahmad, S.H. Modeling for Lightweight Concrete Columns Confined by Either Single Hoops or Interlocking Double Spirals. *ACI Struct. J.* **1999**, *96*, 883–891.
22. Kim, J.K.; Park, C.K. The Behavior of Concrete Columns with Interlocking Spirals. *Eng. Struct.* **1999**, *21*, 945–953. [CrossRef]
23. Tan, T.H.; Yip, W.K. Behavior of Axially Loaded Concrete Columns Confined by Elliptical Hoops. *ACI Struct. J.* **1999**, *96*, 967–971.
24. Benzaid, R.; Mesbah, H.; Chikh, N.E. FRP-confined Concrete Cylinders: Axial Compression Experiments and Strength Model. *J. Reinf. Plast. Comp.* **2010**, *29*, 2469–2488. [CrossRef]
25. Yin, S.Y.L.; Wu, T.L.; Liu, T.C.; Sheikh, S.A.; Wang, R. Interlocking Spiral Confinement for Rectangular Columns. *Concr. Int.* **2011**, *33*, 38–45.
26. Chen, Y.; Feng, J.; Yin, S. Compressive Behavior of Reinforced Concrete Columns Confined by Multi-Spiral Hoops. *Comput. Concr.* **2012**, *9*, 341–355. [CrossRef]
27. Guo, Y.C.; Xiao, S.H.; Luo, J.W.; Ye, Y.Y.; Zeng, J.J. Confined Concrete in Fiber-Reinforced Polymer Partially Wrapped Square Columns: Axail Compressive Behavior and Strain Distributions by a Particla Image Velocimetry Sensing Technique. *MDPI Sens.* **2018**, *18*, 4118. [CrossRef]
28. Bechtoula, H.; Kono, S.; Watanabe, F. Experimental and Analytical Investigations of Seismic Performance of Cantilever Reinforced Concrete Columns Under Varying Transverse and Axial Loads. *J. Asian Archit. Build. Eng.* **2005**, *4*, 467–474. [CrossRef]
29. Kani, M.W.; Huggins, M.W.; Wittkopp, R.R. *Kani on Shear in Reinforced Concrete*; Department of Civil Engineering, University of Toronto: Toronto, ON, Canada, 1979.

Article

A Customer Integration Framework for the Development of Mass Customised Housing Projects

Cynthia dos Santos Hentschke *, Carlos Torres Formoso and Marcia Elisa Echeveste

Building Innovation Research Unit (NORIE), Universidade Federal do Rio Grande do Sul, Av. Osvaldo Aranha, 99, 706, Porto Alegre 90035-190, RS, Brazil; formoso@ufrgs.br (C.T.F.); echeveste@producao.ufrgs.br (M.E.E.)

* Correspondence: cynthiahentschke@gmail.com

Received: 23 September 2020; Accepted: 22 October 2020; Published: 27 October 2020

Abstract: Mass customisation is a business strategy that aims to deliver a variety of products that fulfil customer requirements and, at the same time, keep price and delivery time within acceptable limits. It has been adopted in different sectors to increase value generation, including house building. A major challenge in mass customisation is customer integration, i.e., how to improve value generation by understanding and considering requirements from different customers, and defining their involvement in product development. Most studies on this topic tend to be technology-focused, often being limited to methods and digital tools to generate and display product alternatives. The aim of this paper is to propose a framework of decision categories for customer integration and for devising the scope of customisation to support the definition of mass customisation (MC) strategies. Design science research was the methodological approach adopted in this investigation. It was based on a literature review about mass customisation practices and also on an empirical study developed in a residential building company from Brazil. The main contribution of this paper is a framework for customer integration, which contains a set of decision categories related to the definition of the scope of customisation and customer integration, and a list of practices that are applicable to house building. A secondary contribution of this investigation is a set of constructs that have been used to describe the decision categories and their relationships.

Keywords: mass customisation; customer integration; residential; practices

1. Introduction

In the current scenario of the house building industry, there is a fierce market competition in different countries, primarily concerned with costs, demanding strategies to increase productivity [1,2] and, at the same time, to consider customers heterogeneous demands [3]. Understanding customers' needs and preferences is a challenge due to their changing lifestyles and different family structures [4–6]. Therefore, customer requirements must be appropriately understood and communicated to decision-makers, such as investors, developers and designers; otherwise, value generation may be compromised [4]. The progressively increasing diversity of customer requirements has created business opportunities related to product customisation in several different sectors [7,8], including house building [9]. According to Wang et al. [10] this shifting focus from company to customer demand is a driving force in industrial innovation.

Mass customisation (MC) is a strategy that aims to fulfil customer requirements [11–13], and, at the same time, achieve high efficiency and competitive advantage [2,11], through flexible processes and supply chain integration [1,14]. Therefore, companies combine elements of mass and craft production to improve value generation for specific market segments [15–17]. In the house building industry, besides contributing to competitive advantage, the adoption of MC can provide benefits related to environmental and social sustainability, by avoiding waste caused by product changes made after occupancy by users, as well as by increasing their perceived value and sense of ownership [5,18].

Several successful applications of MC in the manufacturing industry have been reported in the literature [7,19,20]. However, its body of knowledge is dispersed and is still growing [7]. According to Piller [21] and Suzic et al. [20], there is a lack of in-depth understanding of the strategies for implementation. Other authors [22,23] argue that the further expansion of the field depends on the development of models and tools to support companies in new product development (NPD). A major challenge in MC is customer integration, i.e., how to improve value generation by understanding and considering requirements from different customers, as well as defining their degree of involvement in NPD [22,24]. Most studies on this topic tend to be technology-focused [19], being often limited to methods and digital tools to generate and display product alternatives, such as configurators and choice menus [22].

In the house building industry, the implementation of MC is still latent [1,25], sparse and more focused on operations [1]. A critical challenge for the adoption of MC in housing is capturing customers' requirements [3,14,25–27], and establishing a balance between offering variety and achieving efficiency and, consequently, housing affordability [1,9,25,27]. Several research opportunities on this topic have been pointed out in the literature, such as the definition of solution spaces, and the support to customers' decision-making during the configuration process [1–3,22,25]. However, Khalili-Araghi and Kolarevic [3] suggest that new methods for customer integration are needed to reduce the trade-offs between customers perceived value and the complexity that results from customisation. Kotha [17] argues that technologies and tools alone are insufficient to achieve MC goals, as the adoption of this strategy requires an organisational context that fosters continuous improvement, learning and knowledge creation.

Some studies have associated the use of MC strategies with prefabricated or industrialised construction methods (e.g., [1,28,29]). However, this strategy has also been explored by companies that adopt traditional construction methods (e.g., [6,9,25,26]). In fact, some of the potential improvements related to MC are not directly related to the type of technology used, such as understanding customer requirements, customer interaction, and visualisation approaches [2,6,25,30]. Rocha [30] suggests that the definition of an MC strategy can be divided into decision categories, and should start by making some core decisions related to the scope of MC, and then move to other areas, including customer integration. Wikner [31] defines decision categories as ways to classify decisions and support the segmentation of complex decision problems into a structured and relatively independent way to facilitate decision-making.

A possible starting point to understand key decision categories is to analyse practices implemented in the industry [20,32,33]. Those practices can be regarded as methods, tools or techniques that have been successfully used in real-life situations for improving performance or solving problems [32]. By understanding the underlying ideas of those practices, they can be adapted to other companies facing similar challenges [33]. This research seeks to further understand practices as an expression of tacit knowledge that can be applied for learning, working, innovating and organising [34].

Therefore, this research study aims to answer the question: How can customer integration in the NPD of mass-customised house building projects be managed? The main outcome of this investigation is a framework of decision categories for customer integration and for devising the scope of customisation to support the definition of MC strategies. It is based on practices identified in the literature and also on an empirical study carried out in a house building company. The framework is meant to be used by companies to support the definition of MC strategies. A secondary contribution of this investigation is a set of constructs that have been used to describe the decision categories and their relationships.

This paper is structured into six sections, including the introduction. In the theoretical background section, MC is discussed, emphasising its core concepts, especially the ones related to customer integration. In the third section, the research method is presented, including the methodological approach and research design. Then, the results of the empirical study are presented in the fourth section. In section five, the framework for customer integration is presented and evaluated. Finally, in section six, the main conclusions and opportunities for future research are presented.

2. Theoretical Background

2.1. Mass Customisation and Related Concepts

According to Silveira et al. [35], the success of MC strategies relies on several internal and external factors, such as customers demand for customisation, market and value chain readiness, technology availability, and knowledge sharing. Other studies [22,28,30] point out that the implementation of MC depends on the coordinated efforts from three different areas of the company: customer integration, product design and operations management. After requirements are captured, the design area must focus on developing product alternatives by translating those requirements into specifications. Finally, operations management is concerned with producing and delivering customised goods, by managing resources and the supply chain to achieve time and cost-effectiveness [22,30].

MC depends strongly on the company's ability to translate customers' demands into new products and services, in which knowledge creation and information sharing play a key role [22,35]. According to Kotha [17,36], knowledge creation in the MC strategy has two primary sources of information: (i) external, from customers, and (ii) internal, related to internal processes and workers' experiences.

Customers inputs into NPD can be communicated in different ways, such as desires and needs, suggestions towards product solutions, and even insights that may lead to radical innovations [37]. According to Piller et al. [24], by translating customer preferences and needs into product requirements, companies are able to transform subjective information into explicit knowledge. This knowledge can be used to understand customer demands and inspire new developments [17,24,36]. Besides, feedback from customers and previous choices can be used by companies to introduce innovations and also provide guidance on whether to limit or expand product variety [17,36]. Furthermore, Wang et al. [10] discuss emerging methods for collection and storage of customers inputs based on "Big Data" and other IT tools to support decision-making. Therefore, different practices can be used to capture such knowledge [37].

The level of customisation is concerned with the range of customisation options to be offered in order to satisfy different customers [13]. However, this decision needs to be based on the analysis of trade-offs between the company's capabilities and customers' demands [7,35,38]. Moreover, customisation can occur at various points in the value chain, from a minor product adaptation to full customisation defined at the design stage [35,39]. Each one of these points may be related to a specific level of customisation, and requires the definitions of how and when customers' needs are translated into product specifications. A number of taxonomies of customisation types have been proposed in the literature based on the level of customisation, such as the MC generic levels proposed by Silveira et al. [35]: design, fabrication, assembly, additional custom work and services, package and distribution, usage and standardisation. Another example is Barlow et al.'s [40] set of strategies for the house building industry (Table 1), which is based on Lampel and Mintzberg [15].

Table 1. House building strategies.

MC Strategies in House Building [40]	Description of the Customisation Level
Pure standardisation	Standardised product. No possibility of changing products
Segmented standardisation	Limited choice focused on aesthetic elements and or based on aggregate knowledge regarding customers' requirements
Customised standardisation	Balance between cost, lead time and choice, associated with postponement and modular practices
Tailored customisation	High variety or availability of choice. The product is fabricated by combining a set of standardised design elements
Pure customisation	Infinite choice, relatively high costs and lead time

Source: adapted from Barlow et al. [40].

The location of the customer order decoupling point (CODP) is essential to define the customisation level [41,42]. It divides the value chain in processes based on forecasts (mostly standardised) and on customer demands (customised according to orders) [24,38,39] (see Figure 1). The CODP also defines which activities are postponed until the customer's specific requirements are captured, and an order is placed [24].

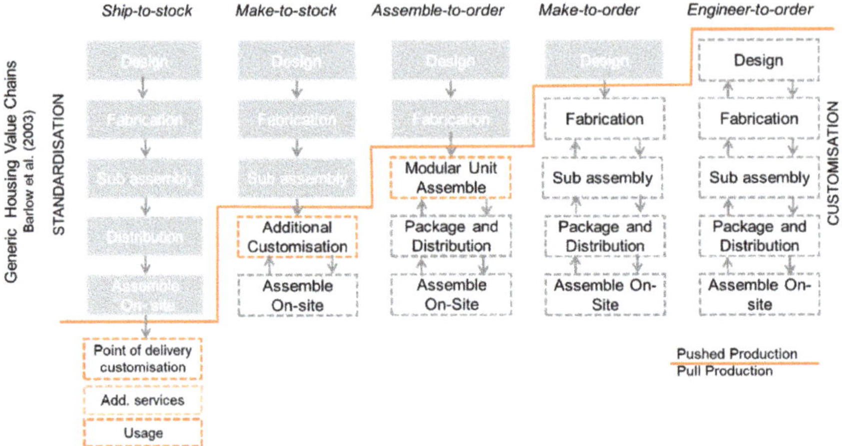

Figure 1. Customer order decoupling point (CODP) in housebuilding. Source: adapted from Barlow et al. [40] and Silveira et al. [35].

Therefore, the extent of customer integration is closely related to the level of customisation [24] and the CODP definition [39]. In fact, the level of customisation usually defines the intensity of customer–company interaction during NPD [24,38]. Moreover, a high customisation level should rely on collaboration with customers from early design stages, while a low one requires less intense participation of customers [38].

When defining the level of customisation, companies should bear in mind that offering too many options not only can make operations inefficient, but also cause customers frustration and confusion, the so-called burden of choice [43,44]. Thus, the definition of a limited solution space plays a key role in MC. The solution space consists of a combination of different customisation units (i.e., customisable attributes and their available options) and rules to combine them, limiting the set of possible product alternatives [30,44]. However, even if there is a limited number of flexible processes, a large number of features and product alternatives may be generated [7,19,21].

Previous studies [14,23,28,30] have pointed out that devising a solution space must be based on the identification of customers' needs and preferences for product customisation, and decide whether and how those will be meet [2,28,44]. It must also be highlighted the importance of post-occupancy evaluations (POE) to capture requirements and provide feedback for the NPD of future house building projects [14,26].

Rocha [30] proposed three core decision categories to define the scope of an MC strategy in house building: (i) the solution space; (ii) customisation units; and (iii) classes of items, which are specific properties of options offered in the customisation units [30]. Additionally, Amorim [45] proposed a decision category named communication of customisation information that defines how the information is made available, when and for whom. This is strongly supported by previous studies [9,25–27,45–47], that highlight the need to improve the effectiveness of information flows between different sectors of the company, in order to facilitate collaboration and improve value generation.

Rocha [30] suggests that the level of customisation should be considered as an operations management related decision category, as it is related to the definition of when and how customisation units are defined. However, Schoenwitz et al. [28] suggest that customers' preferences play a key role in the definition of the customisation level, indicating that there is an interaction between customer integration and operation management decisions. The same authors also pointed out that the definition of a single CODP neglects the possibility of choices to be made separately for different components and attributes, which are made feasible by prescribing multiple decoupling points.

2.2. Customer Integration

According to Franke, Keinz and Schreier [48], the value delivered by mass customised products is driven by the fit, style and functionality, or utility perceived by customers, and the uniqueness of a product. Customers are often willing to pay extra to obtain customised goods [1,21,38]. Furthermore, Piller [21] argues that the willingness-to-pay (WTP) reflects the value perceived in the increment of utility that they gain from a product that better fits their needs rather than the best standard product available. Therefore, customer integration should start from capturing needs and preferences, and estimating the WTP for a customised good [22].

Kumar et al. [7] argue that customer integration embraces not only co-design but also other types of interactions between companies and customers, which can be enabled by modular design, configurators, and elicitation of needs. It means that customers can have an active role in product definition, configuration or modification within a given solution space [19,21]. Thus, premium prices are charged to cover additional costs resulting from customisation, such as higher costs of sales [17] and operations [24]. Moreover, customer integration can also bring some cost-saving results from collecting consistent market information and establishing a close customer–company relationship [24].

In this context, new relationships must be established between customers and companies [3]. Thus, companies can benefit by expanding the use of traditional customer relationship management (CRM) tools [49] to relational marketing ones [50]. These are means to build long-lasting relationships with customers, by improving value generation through interactions, creating trust and increasing loyalty [49–51]. According to Tommaso [50], relational marketing is based on a logic of exchange and learning. It can potentially improve customer experience, which refers to the combination of a number of personal impressions (considering cognitive, affective, behavioural, physical and social aspects of the response), resulting from interactions between a customer and a product or service [50].

According to Silveira et al. [35], the customer–company interface must be tailored to each unique context. Fetterman et al. [25] proposed a set of steps to outline a customer–company interface for the house building industry, which is built on a proposition by Silveira et al. [35]: (i) defining a solution space to be offered to customers; (ii) collecting and storing information on customers choices; (iii) transferring data from retail to production; (iv) translating customers choice into product design features and manufacturing instructions; and (v) delivering customised products and offering post-occupation customisation. In step two, effective ways to present the solution space for customers are needed [30,35], enabling them to deal with the variety of alternatives, avoiding the burden of choice [43].

Rocha [30] suggested two decision categories for customer integration, namely, configuration sequence and visualisation approaches. These are concerned with how the customisation units are presented to customers and how they engage in creating the product. The first one involves defining a sequence of decisions to be made by customers when configuring their product alternatives [30]. The visualisation approaches decision category defines how the customisation units will be displayed and to whom (i.e., customer, company or both), being divided into three types: collaborative, transparent and do-it-yourself [30], similar to the approaches proposed by Gilmore and Pine [16]. For example, in the collaborative approach, both customers and companies are aware of the customisation process and can be applied through choice menus and or a dialogue between the company and customers [30]. However, Rocha [30] only proposed a broad definition of those three approaches, without discussing how to implement or combine them for effectively presenting the solution space to customers.

3. Research Method

Design science research (DSR) was the methodological approach adopted in this investigation. This approach typically involves the development of innovative solution concepts, named artefacts, to solve classes of practical problems, and at the same time contribute to the development of mid-range theories, i.e., theoretical models that apply to a limited range of situations [52,53]. The main reason for choosing DSR is the prescriptive, rather than descriptive character of this investigation. The practical problem addressed by this research work is how house building companies can use customer integration concepts to support the definition of MC strategies and improve value generation for customers.

There are different types of outcomes in DSR, such as models, methods, constructs, instantiations [54] and technological rules [55]. The artefact proposed in this research is a conceptual framework which prescribes a set of core and customer integration decision categories that can be used to support the definition of MC strategies in house building companies. This research work also proposes new constructs and adapts existing ones, which are useful for describing those decisions categories.

Figure 2 provides an overview of the research design, in which the activities are organised similarly to the DSR steps proposed by Lukka [53]: (i) identify a practical problem and understand it from a theoretical perspective; (ii) devise the solution; (iii) test and refine the solution in an empirical study; (iv) analyse the utility of the solution and discuss the theoretical contributions of the investigation.

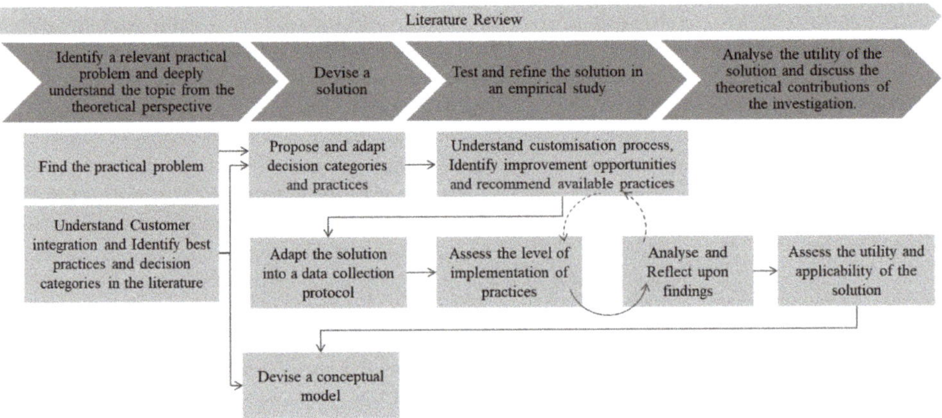

Figure 2. Research design.

A literature review on customer integration and MC practices was carried out in order to obtain a deep understanding of the topic, in the first step of the research (Figure 2). The aim was also to find descriptions of practices that were successfully used for customer integration, by using the snowballing technique, complemented by an advanced search in the Scopus repository. The search was undertaken in journal and conference papers, from 1998 to 2018 and its results were limited to areas relevant for house building such as engineering, management, and environmental science, from which 24 papers were selected. As a result, two sets of practices were identified, one related to the MC core decision categories and the other to customer integration. Information about those practices was stored and further categorised in a database, according to authors, and country of adoption.

In the second step of the research, the selected practices were associated with decision categories (Figure 2). Some of the decision categories considered were identified in the literature review (see Section 2), such as solution space, visualisation approaches, and configuration sequence. Furthermore, the processes of classifying practices into decision categories available in the literature brought to light some gaps, which resulted in the proposition of some additional decision categories.

The third step of the research consisted of the development of an empirical study in a house building company, named Company P, in which the implementation of MC practices and decision categories was assessed (Figure 2). The aim of this study was to understand further the underlying ideas of practices and to test the utility of the proposed decision categories. It was part of a broader research project, in which the MC strategy of the company was assessed, and some improvements were implemented by the company, which took approximately two years.

Company P was founded in the 1970s as a family company, being currently one of the largest construction companies of the South Brasil, with 252.312 m^2 built so far. They have over 20 years of experience in delivering customised residential building projects for upper-middle and middle-class customers. Their products are made from a combination of traditional methods of construction with industrialised components, such as internal drywall partitions and precast façades. This company was chosen because its business strategy was strongly based on the customisation of products to obtain market differentiation. Moreover, the company was willing to take part in this project and had a department entirely dedicated to customising residential projects. The customisation team (CT) had six architects, including a coordinator.

The focus of the empirical study was on a relatively new market segment explored by the company in which a limited solution space was offered to customers. Within this context, the productivity–flexibility trade-off had to be managed carefully in order to increase the perceived value for customers without substantially increasing costs and lead time.

The empirical study started by assessing and analysing the customisation process adopted by Company P, based on multiple sources of evidence (see Table 2). Several semi-structured interviews were carried out with representatives of different departments of the company. These interviews were divided into three sections: (i) company's general information (e.g., business model, customers, competitors, history); (ii) description of NPD and customisation practices; (iii) description of products and customisation options. Additionally, one open-ended interview was carried out with the customers and customisation manager about the role of the customisation department and the MC strategy. Based on the interviews and documents analysis, a customisation process map was devised by researchers and discussed with the CT. Simultaneously, the existing customer integration practices were compared to a preliminary list of practices extracted from the literature, and a gap analysis was then carried out, resulting in the identification of some improvement opportunities. Those improvements were discussed with Company P's representatives in two meetings. Then, the company decided to implement some of the suggested improvements.

Table 2. Sources of evidence used to understand the customisation process and identify improvement opportunities.

Source of Evidence	Details and Participants	Duration
Open-ended interview	Customers and Customisation Manager (civil engineer)	1 h 6 min
Semi-structured interview	Customisation Coordinator (architect), and Customisation Architect	1 h 6 min
	Customisation Architect	58 min
	Project Coordinator (architect), Project Analyst (civil engineer)	1 h 2 min
	Product Development Analyst (architect)	34 min
	Production Manager (civil engineer)	50 min
	Product Intelligence Manager (civil engineer)	40 min
	Marketing Manager (administrator)	53 min
	CRM department coordinator (marketing)	40 min
	Customisation Architect in charge of "point of delivery customisation"	53 min

Table 2. *Cont.*

Source of Evidence	Details and Participants	Duration
Document analysis	Proposed solution space	
	Product catalogues	
	Presentation of customisable attributes for customers	
	Company web site	
	Contracts	
	Post-occupation evaluation questionnaire	
	Project customisation management spreadsheet (containing dates for decision making, residential units customised by customers, customisation units chosen, etc.)	
	Customisation status on-site communication	
Observations	Participant observations of the interaction between the CT and customers during the construction site open day promoted by the CRM department	1 h 30 min
Meetings	One meeting with the CT to discuss their processes, identified practices and improvement opportunities	1 h 22 min
	One meeting with the CT to discuss research findings	1 h
	One meeting with the CT, manager and professionals from other departments of Company P to discuss research findings and the utility of the artefact	1 h 30 min

Approximately one year later, after the implementation of some improvements by the company, a data collection protocol was used to assess Company P's MC strategy regarding core and customer integration categories. This data collection protocol was based on the final set of decision categories and on the full list of practices, being used as a reference to discuss the adoption of practices with the CT (Table 3). This assessment was based on a 5 point scale. Besides, data about the perspective of customers were captured qualitatively during three open days in construction sites, bringing another perspective to the discussions.

Table 3. Sources of evidence used on the assessment of the level of implementation of practices.

Source of Evidence	Details and Participants	Duration
Document analysis	Customised units database	
	Simplified choice menu	
Observations	Two participant observations of the interaction between the CT and customers during the construction site open day promoted by the CRM department	4 h
		5 h
Semi-structured interviews	Ten interviews with customers during events promoted by the CRM department regarding the customisation service, interaction, visualisation tools and customisation units	Approx. 15 min each
	Four interviews with architects from the CT regarding core and customer integration practices and decision categories	During the discussions
Meetings	One meeting with the CT to discuss core decision categories and related practices, and their utility	1 h 56 min
	One meeting with the CT to discuss core decision categories and related practices and their utility	1 h 45 min
	One meeting with the CT to discuss customer integration decision categories and related practices and their utility	1 h 39 min

Analysis and reflection of the research findings were carried out in the fourth step of the research study. The utility of the research outcomes, i.e., decision categories and MC practices, was assessed based on the following criteria: (i) provide underpinnings to the assessment and monitoring of core and customer integration decision categories; (ii) provide support to understand MC related concepts and its underlying ideas; (iii) support decision-making for defining the MC strategy, particularly in terms of integrating customers in customisation processes. The assessment of utility was carried out in six meetings with representatives of the customisation department, as shown in Tables 2 and 3. Finally, the conceptual framework of decision categories for customer integration was devised.

4. Results

4.1. Identification of Practices from the Literature

Table 4 presents the 44 MC practices that were identified in the literature review concerned with core and customer integration areas, organised according to decision categories. It is noteworthy that 35 of those practices were discussed in up to three different papers out of the twenty four reviewed. The maximum number was seven papers per practice. Therefore, this investigation provides a much broader view of customer integration practices than previous studies. Furthermore, these practices do not overlap with each other, so they can be combined to formulate strategies. Some of the practices provide support to decision making regarding the definition of strategies, while some other practices support the operationalisation of the strategic decisions undertaken.

Table 4. List of Practices.

n°	Description of the Practice	Authors
	Decision Category: Knowledge Management	
1	Present effectively customisation options	[30,45]
2	Establish a protocol to register and manage customer order changes	[25,26,45]
3	Carry out routine construction site visits to check customers' orders compliance by the design team.	[26,45]
4	Use product prototyping to test and communicate technical and design solutions to stakeholders	[45]
5	Create a database of customers orders for housing units customisation shared within departments	[30]
6	Standardise project documentation and communication between customer and developers from the company	[6,30]
7	Use specialised information systems for managing production management of customised products	[9,11,27,45,56]
8	Carry out post-occupation evaluation to understand customers' needs, capture new requirements and feedback the new product development	[14,57–59]
9	Establish a complaint management system and definition of continuous improvement procedures	[14]
10	Adopt methods for identifying the demand for customisation and consumers preferences to define solution spaces	[2,25,28,30]
11	Manage information about customisation orders to create knowledge for the company	[2,30,47,58,59]
12	Carry out product and service research to understand which factors contribute to customers satisfaction regarding the housing unit and customisation process	[14,46]
13	Use choice menus as a learning tool, to understand customers' needs and preferences and provide feedback to new product development	[2,3,27]

Table 4. *Cont.*

n°	Description of the Practice	Authors
14	Map the customisation process to find improvement opportunities and potential areas of economy	[28,30]
15	Create metrics that can be used to analyse the trade-offs between flexibility–productivity	[18,25,46]
16	Share information about the customisation sales and profitability performance within different departments	[46]
	Decision Category: Level of Customisation	
17	Define different levels of customisation according to customers' preferences, distinct market segments, and projects	[25,26,28,46,59]
18	Offer of different customisation units and level of customisation according to the project stage, i.e., a multiple customer order decoupling point approach.	[28,59]
19	Use modular components that allow product variations according to customers' requirements	[6,40,46,60]
	Decision Category: Solution Space	
20	Assess the alignment between the solution space and customer demands to improve the cost-effectiveness of the mass customisation strategy	[27,28,60]
21	Define customisation units based on the region and local needs for the projects and their target customers	[46,58]
22	Define a limited solution space to achieve economies of scale	[6,14,30,40,60]
23	Offer of additional services related to the built environment	[14]
24	Offer extra customisation units at the product delivery	[11,14,40,46]
25	Offer innovative customisation units, such as related to sustainability and automation	[58]
26	Promote multidisciplinary discussions, among different stakeholders, for defining the solution space and level of customisation	[46,60]
27	Refine the solution space according to previous experience in other projects	[25]
28	Define the customisation units based on the balance between the potential value-adding to customers and its feasibility and operations costs	[25,28,57,59,60]
29	Adopt information technology tools or a choice menu to support customers' choice and product configuration, which is well integrated into the new product development	[3,6,11,25,27,40,56,61]
30	Offer additional customisation units post-occupancy or substitution of previously chosen components according to customers emerging needs	[25,58]
	Decision Category: Customer Interaction and Relationship	
31	Advertise the possibility of customisation to customers as a competitive differentiation in the market	[26,59]
32	Co-design	[6,40,46,58]
33	Define interactions with customers and display them in a customer journey representation	[14,25,61]
34	Identify potential customers for new projects to establish effective communication with the target audience	[47,59]
35	Have meetings with clients for product configuration and cost estimation	[30]

Table 4. *Cont.*

n°	Description of the Practice	Authors
36	Offer customers precise product specifications and information regarding the customisation status	[59]
37	Establish a dialogue between customers and the company's representatives for configuring the product according to their needs	[6,27,40,61]
38	Adopt methods and tools to collect customer orders in a standardised and systematic way	[11,25–27,45,46]
39	Promote customer interaction with product prototypes to learn about them, their needs and capture requirements	[6,40]
40	Use product catalogues for advertising and informing customers about the product and customisation process	[6,25,40]
	Decision Category: Visualisation Approaches	
41	Use tools, lists, databases of that communicate additional costs for customisation to support customer decision-making during configuration, enabling negotiation and increasing transparency	[6,11,30,45,61]
42	Build a prototype or showroom for showing the customisation units available to customers	[26,30,58]
43	Present standard product specifications through images and information	[30]
44	Use virtual prototyping, e.g., building information management (BIM) models, to show product alternatives to customers	[2,3,11,56]

The descriptions of the decision categories proposed in this investigation are presented in Table 5. Some of them were subdivided into sub-categories or decision domains that characterise sets of processes that depend on similar preconditions [31].

Table 5. Decision categories, source and research contributions.

	Categories	Source	New Definition or Adaptation
Scope of Customisation (Core)	Solution Space	Adapted from Rocha [30]	The solution space decision category was adapted to consider both customisation units and classes of items due to the interdependency among those decisions. They can be regarded as decision domains.
	Level of Customisation	Adapted from Rocha [30]	This decision category is concerned with the definition of the levels of customisation to be adopted by the company. It is closely related to the customer order decoupling point, customer integration and product variants definitions.
	Knowledge Management	Proposed in this investigation	This new decision category addresses how to manage knowledge created by the company, considering customers, processes and workers information, as suggested by Kotha [17], including the communication of information and knowledge created. It allows companies to continuously update competencies, apply practices and routines, promoting organisational learning and continuous improvement [17,36].

Table 5. *Cont.*

	Categories	Source	New Definition or Adaptation
			Customer based knowledge decision domain: it aims to define approaches to assess customers demand for customisation to establish a solution space and to evaluate the delivered products for understanding emerging and evolving requirements. Additionally, it is necessary to establish how is this information will be used to feedback the new product development. It is strongly related to customer integration and value generation.
			Organisational knowledge decision domain: it is concerned with how to make explicit tacit knowledge from workers and processes, and translate it into practices to be adopted. It also encourages the reflection upon practices for disseminating them and refining the MC strategy.
			Communication of customisation information decision domain: it embraces practices that promote transparency and continuous improvement by making relevant customisation information available to stakeholders during new product development. In this research, it is considered as a way to disseminate information and knowledge created, and not limited to the interface between product design and operations, explored by Amorim [45].
Customer Integration	Visualisation Approaches	Adapted from Rocha [30]	Further than just defining who is aware of what is happening in the customisation process, visualisation approaches decision category regards the definition of how the solution space and the customisation units will be presented to customers. Therefore, the "visualisation tools" decision domain was proposed with that aim, specifically for defining tools that portray the solution space.
	Configuration Sequence	Proposed by Rocha [30]	
	Customer Interaction and Relationship	Proposed in this investigation	It regards the definition of approaches to interact with clients during the new product development and develop a close relationship with them throughout their entire journey, for achieving loyalty [49–51]. This decision category is closely related to planning the customer experience [50].

Four core decision categories for MC in house building were defined in this investigation (Table 5). In relation to the previous literature, a new core decision category related to knowledge management was proposed, which is concerned with how to establish a knowledge-creating system to support MC. This decision category was based on contributions from several authors [17,22,24,35,36]. Three decision domains were proposed within the knowledge management category: customer-based knowledge, organisational knowledge, and communication of customisation information.

Three customer integration decision categories were defined, including "visualisation approaches" and "configuration sequence", based on Rocha [30]. The "customer interaction and relationship category" was proposed to address decisions regarding how companies interact with customers, when and for which purpose, and establish a trustworthy relationship, during NPD. By contrast, the decision categories proposed by Rocha [30] were focused on defining the customer–company interface, by broadly specifying who visualises what during the customisation process, and the sequence of decisions to be made by customers when configuring a product. The adapted version of visualisation approaches decision category includes the decision on whether to use visualisation tools for displaying the solution space. Additionally, there seems to be a gap in the literature regarding configuration sequences, since no practices for the house building industry have been found.

4.2. Empirical Study in Company P

4.2.1. Understanding the Customisation Process and Identifying Improvement Opportunities

Company P offers six different product types; each one of them focused on a different market segment with different customisation levels (Table 6). Most of the company's previous experience on customisation is related to A and B product types, which can be classified as tailored customisation. In those market segments, customers may hire their own architects to develop the interior design of their units. However, the focus of this investigation is on the D, E and F product types, in which customers can customise only a limited set of elements, mostly related to the finishings and fixtures of the residential unit. Product types D and E could be classified as a "segmented standardisation" level of customisation and F as a "point of delivery customisation".

Table 6. Company P product types and levels of customisation.

Product Types	F	C, D and E	A and B
Development stage of Customisation	Delivery	Construction	Fabrication
Level of customisation	Point of delivery customisation	Segmented standardisation	Tailored customisation
Available customisation units	Floor finishings, fixed furniture, air-conditioning, kitchen counter and bathroom sink stones	Drywall partitioning, floor finishings, double glazing, kitchen counter and bathroom sink stones, and the laundry tub	Internal layout, ceiling finishings, water and electricity services, air conditioning and internal finishings

The customisation department is in charge of defining the solution space for each project within the boundaries established for each product type by the NPD department. During the conceptual stage, representatives of both departments discuss which customisation units regarding layout and finishings will be offered to customers. At the end of that stage, two customer decision-making deadlines for the layout and finishings are established at the project launch meeting, which involves several departments of the company. These deadlines are included in a brief that is delivered to the project designers. After the project launch into the market, the CT defines different alternatives to be offered as finishings.

The customisation offered involves four main touchpoints with customers, in which different customisation units are available and portrayed by different visualisation tools (Figure 3). At each of these points, the customisation department is in charge of: (i) establishing a dialogue with customers; (ii) collecting and processing customer orders; (iii) making design changes; and (iv) delivering that information to the construction site. The CRM department promotes open days for visits to construction sites by the clients. In those open days, the CT is available at the housing unit prototype to offer customisation services. The CT guides customers through the solution space by using different visualisation tools, such as illustrated blueprints and finishing material catalogues, and informs prices of product alternatives by using simulations based on a simplified choice menu. The visualisation tools highlighted in yellow were, in Figure 3, improvements carried out during the empirical study.

The display of product prototypes in the construction site open days was identified as a key element for the customisation strategy of Company P. These enabled the CT to guide customers to make decisions within the solution space offered, and provided an opportunity for creating a relationship with clients. The CT may also arrange individual meetings in case open days cannot be undertaken or if customers show an interest in product customisation after those events. If the customer opts for a customised unit, an additional contract is signed. During construction, the CT carries out routine visits to the site to check whether customers' orders have been fulfilled in the construction site.

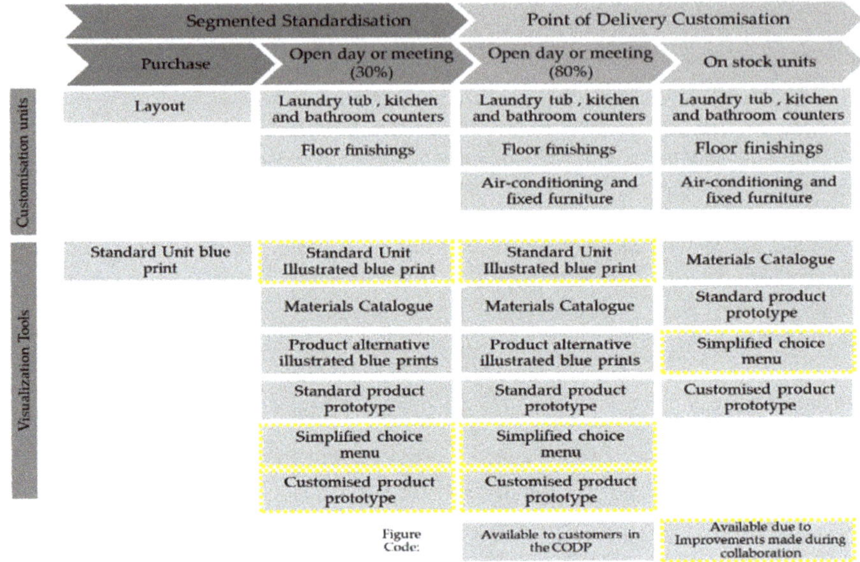

Figure 3. Touchpoints and customisation units.

In Company P, customers initiate their journey with the company when they purchase a housing unit, being registered at the CRM department. That department has three communication channels with customers: (i) an area in the company's web site, (ii) an APP, and (iii) a call-centre that connects customers to different departments. Besides being in charge of promoting construction site visits and events with customers, the CRM department is responsible for carrying out customer satisfaction surveys in different moments: (i) in construction site open days; (ii) post-occupancy evaluation undertaken one-year after project delivery; (iii) after the response of the company to complaints after project delivery; and (iv) when completing five years, considering the possibility of providing references of the company to friends or family.

Table 7 summarises the identified improvement opportunities as well as the improvements implemented by Company P during this research study. Those opportunities were classified according to decision categories and practices. For instance, regarding the "knowledge management", the company carries out a POE, yet, it is mostly concerned with the overall customer satisfaction with the product, but no questions are asked about customised items. Another example is facilitating, standardising and digitalising customer order collection, which was carried out by the CT, who used to handwrite customers' requests during open days, before processing these back at the office and e-mailing them to be confirmed by customers. This opportunity, for instance, inspired the development of a simplified choice menu, which enabled the use of a digital tool for registering customers' orders and simulating the product alternative costs in real-time.

A critical barrier for improvements, identified in interviews and participant observations, was the lack of communication between departments, which occasionally confused customers. For instance, the sales department offered the "point of delivery customisation" of residential units that have not been sold yet, while the customisation department offered other options at different touchpoints. Moreover, different customisation units are offered in each touchpoint, so by making the early announcement of the "point of delivery customisation", the sales department has confused customers regarding the available customisation units, the timing and to whom report their decision.

Table 7. Improvement opportunities identified during the understanding of customisation process.

Decision Category Related	Improvement Opportunities	Practices	Improvements Implemented
Customer interaction and relationship	Ability to better inform customers regarding the customisation offer and configuration process	31, 34, 40	
	Facilitate, standardise and digitalise customer order collection, reducing the processing time of the information and rework Tool for simulation of product alternatives costs to negotiate with customers during open days	29, 37, 38, 41	Development of a simplified choice menu
Solution Space	Solution space is defined based solely on CT expertise; there is an opportunity to enhance its definition by considering customers' preferences and other departments views	20, 21, 26, 28	
Knowledge management	Incentive communication and collaboration between departments	5, 7	Customised units database shared in the company intranet
	Develop graphs and presentation regarding the customisation department performance	15, 16	Starting to report to the manager
	Better understand the customisation process of residential projects and identify improvement opportunities	14	Customisation process map and service blue print in development
	Improve the market research to understand the demand for customisation of housing, providing insights to NPD and definition of a solution space	8	
	Improve POE assessment methods by including customised items and aiming to understand more deeply customers perception regarding product and service	10, 12	
	Process available information regarding customisation of projects and housing units to transform into knowledge	11, 13	

Lastly, the use of traditional construction methods and the outsourcing of product design created barriers for Company P in the adoption of modularity-related practices. As discussed by Fettermann et al. [25], the customisation of buildings that use traditional construction methods usually has little support from modularity, limiting the advantages of scale.

4.2.2. Assessing the Level of Implementation of Practices

The level of implementation of practices was assessed by the CT considering a five-point scale: not applicable (1), not applied with intended adoption (2), partially applied (3), partially applied with intended improvement (4), applied (5). This assessment is presented in Figure 4. It is noteworthy that the adoption of practices depends on the context of each organisation. Therefore, practices that are "not applicable" are the ones that were not considered to be useful to Company P, while the practices that are "not adopted with intended adoption" are the ones that the company recognises the need to implement shortly. Some practices were assessed as "partially applied with intended improvement", meaning the company has adopted it, but there is still room and motivation to improve.

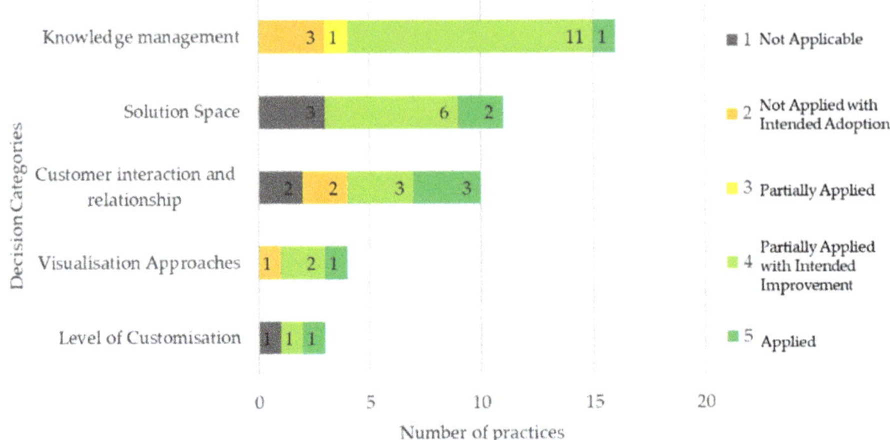

Figure 4. Practices by decision categories according to CT's assessment.

The number of fully applied practices is noticeably low. However, there was evidence that the company is motivated to continue improving, considering that many of the recommended practices changed to partially applied within the time frame of this research project. Further details on each decision category assessment are discussed in subsequent sections.

During the assessment of the level of implementation of MC practices, other improvement opportunities were identified (Table 8). Although many improvement opportunities remained from the previous research stage, the CT seemed motivated to improve. For example, the use of three-dimensional models to display product alternatives as a visualisation tool for meetings with customers was suggested during the discussions and shortly adopted two weeks later.

Table 8. Improvement opportunities identified during the assessment of the level of implementation of MC practices.

Decision Category Related	Improvement Opportunities	Practices	Improvements Implemented
Customer interaction and relationship	Continuous improvement of the customers decision support tools and techniques, facilitating the configuration process and increasing its transparency	29, 37, 38	Use of a simplified choice menu
Solution space	Improve the delimitation of the solution space and establish borders to the flexibility offered	22, 27, 28	Started processing data regarding some projects
Knowledge management	Create cost indexes, based on past projects percentages, to manage loss risk	15	
Visualisation approaches	Need for complementary visualisation tools to aid the solution space and standardised product explanation during meetings with customers	44	Three-dimensional model of the housing unit and customisation units
	Customers have presented some difficulties to envision and understand how the customised product will be delivered	42	New customised product prototype

The CT mentioned some barriers that they face in the adoption of MC practices such as financial and human resources, and tools to develop and implement new solutions. Moreover, a challenge for the customisation department is to be perceived as an innovation and customer-oriented sector as the development of new product ideas is often assigned to them. Thus, the CT must embrace activities that were not always related to their scope of expertise, such as customising non-residential projects. Moreover, the uncertainty of the new product types and attempts to improve the existing ones can be overwhelming, since their scope is continuously increasing.

4.2.3. Analysis of Decision Categories

Knowledge Management

Knowledge management was one of the decision categories which had the largest number of improvements during the empirical study and the highest number of "partially applied with intended improvement". One of the most significant improvements implemented was to share customisation information among departments, by using a customised units database. Initially, data regarding the customisation of housing units were held on by the CT and operations only. After that change, the CT compiles that information and shares it in the company intranet, making it available to sales and other departments. Additionally, any changes in customers' orders are also registered in that database. These improvements resulted in a high level of implementation of practices related to the "communication of customisation information" decision domain (see Table 4 practices one to seven), yet with room for improvement.

Even though the CT considered that many of the partially adopted practices of the "customer based knowledge" decision domain (see Table 4 practices eight to 13) had to be improved, there was much concern with how to operationalise the proposed practices, due to limited resources, and fear of exhausting customers with too many questions.

When discussing practice eight, "POE to understand customers' needs, capture new requirements and feedback the NDP", the CT stated that it would be beneficial to know customers' desires and preferences by including questions related to the scope of customisation on the existing POE. This improvement would avoid the initial concern to overload CT with an additional task and overwhelm customers with too many questionnaires.

Practices 14, 15 and 16 are related to the "organisational knowledge" decision domain, and for the last two of them, the company has plenty of data. However, the data have not been processed to create knowledge. For instance, practice 16, "share information regarding customisation performance ... ", is at its early adoption stage. Another example is practice 15, related to the creation of metrics: the CT argued that they have a large amount of data, but have not been able to establish any metrics yet. The reflections regarding strengths and shortcomings of the company strategy also inspired the proposition of a new practice, named "use methods and discussions to learn from practices adopted in other departments and levels of customisation", fostering the creation of a knowledge creation system and continuous improvement.

Level of Customisation

Practices 17 to 19 (see Table 4) are related to the definition of the level of customisation. The CT reported that they offer options for the attributes defined by the company, according to market segments and CODPs, yet the variation of the solution space offered in different projects is small. Nevertheless, the CT argued that they intend to offer more variety (e.g., painting services), as this would probably contribute to increasing customer satisfaction. However, the decision about the solution space should be carefully defined, as this would also affect operations. Furthermore, the decisions regarding the level of customisation are more strategic, once it might affect different departments, being out of the scope of the CT to be undertaken.

Solution Space

The solution space was identified as a critical area for improvement in the gap analysis. The CT's partially apply six practices that could still be improved. Regarding the assessment of practice 28, "define the customisation units based on the balance between the potential value-adding to customers and its feasibility and operations costs", the CT defines the solution space based on their previous experience with customers, considering general definitions made by the company for the segment and the return of investment. However, Company P has no systematic way to assess the value-adding potential of customisation units, neither discuss its feasibility and operations costs with all stakeholders. This criticism corroborates the findings of Fettermann et al. [25].

Practice 29, "IT tools and choice menu to enable customers to choose, configure and be integrated into the NPD", was assessed as partially applied with improvements to be done. Its application has evolved significantly during this research study, by the development of a simplified choice menu. However, some additional improvements opportunities were identified, regarding the visualisation of the product alternatives.

Some of the identified practices provided insights on how to overcome improvement opportunities. For instance, "Promote multidisciplinary discussions, among different departments and stakeholders" (practice 26), should be used to overcome the poor communication among stakeholders regarding customisation issues. The CT suggested some inter-department seminars to increase awareness about their work. As discussed by Kotha (1995), the information exchange between coworkers and cross-training can support the conversion of tacit into explicit knowledge and foster the adoption of practices and organisational learning. Another practice that was poorly adopted by Company P was "refine solution space according to previous experience in other projects" (practice27), meaning that lessons from previous projects were only learned informally.

Customer Interaction and Relationship

The practices related to customer interaction and relationship have significantly evolved over the empirical study. In fact, this decision category was concerned with an important role played by the customisation department, as the CT had the mission of establishing a good relationship with customers, as well as dealing with some reported problems related to customisation during NPD.

A strength of Company P's customer integration strategy was to "establish a dialogue between customers and the company's representatives for configuring the product ... " (practice 37), which was mentioned by customers in the interviews and by the CT during the meetings. Customers mentioned that having a dialogue with the CT and engineers was an important source of information, which made it easier to choose customisation units and created trust. Additionally, several customers seemed to like the customisation service because of its convenience, reducing the time to move in and the need to deal with further construction works. At the end of the empirical study, this dialogue was aided by the combination of different visualisation approaches, such as the product prototype (practice 42), finishing material samples and the choice menu (practice 41).

The CT pointed out that practice 35, "have meetings with customers for product configuration", was implemented for product types D and E during the collaboration period. During those meetings, the architects explained the solution space and established a dialogue for configuring the unit, but without having the chance to show the prototype for customers.

Three improvement opportunities related to three practices that were considered as "not applied but intended": "advertise the possibility of customisation ... " (practices 31), "use product catalogues for advertising and informing customers about the product and customisation process" (practice 40), and "clearly define interactions with customers and display them in a customer journey" (practice 33). In fact, the possibility of customisation was timidly mentioned in project information at the company website, and it was not always announced in the open day invitations. Interviews and observations in open days confirmed this fact, as several customers had only been informed of the possibility of customising their housing unit during that day, being surprised and confused.

Therefore, the company could improve communication regarding the possibility of customising residential units, to avoid confusion and increase transparency and trust in the relationship with customers. These shortcomings are also related to the lack of clarity about customers' involvement in the customisation process.

Visualisation Approaches

Several improvements have been made to embrace practices related to the visualisation approaches decision category. Currently, the CT offers more precise information regarding the customised units through the customised product prototype (practice 42), and the use of the simplified choice menu to simulate product alternative additional costs (practice 41). During the discussions, the CT architects revealed that they were trying to adapt the choice menu to product type B, in which the range of options is broader than in other product types, highlighting the opportunity of tailoring practices for different market segments.

An intended adoption by the CT can be seen in regard to practice 44 "Virtual prototyping, e.g., building information management (BIM) models, to show product alternatives to customers and ease choice". The initial step was developing three-dimensional models to illustrate product alternatives to be used in meetings with customers.

According to customers, some additional visualisation tools supported decision making during the open days such as standard housing unit prototype, finishing materials catalogue, and, in the third open day, the comparison between the standard and customised housing unit prototypes.

4.2.4. Analysis and Reflection

A low level of implementation of MC practices was identified in Company P, similarly to the results carried out by Fettermann et al. [25] on the MC practices of three Brazilian house building companies. The main improvement opportunities identified in this investigation were also similar to that study, being concerned with the solution space and visualisation decision categories, and customer-based knowledge decision domain. Jensen et al. [2] argue that by understanding customer's needs and preferences and making product recommendations based on the available solution space, companies can save much time in the configuration process, and also increase quality and reduce rework. Additionally, the implementation of MC practices enabled Company P to provide a better service for customers and to improve efficiency in some internal processes.

The CT has pointed out in the discussion meetings that some practices could be adapted for other product types that had a higher degree of customisation. However, this would require the analysis of a different context, in which the complexity of interactions with stakeholders would be much higher. These considerations reinforce the need for devising context-specific practices and implementation guidelines, as suggested by Suzic et al. [20].

The lack of communication, according to Andújar-Montoya [9], can be attributed to the fact that the NPD in housing is often divided into stages, which are not properly integrated. Beyond that, Schoenwitz et al. [60] suggest that this disconnection reflects different degrees of awareness regarding the customisation strategy, similar to what was observed in Company P. For example, the sales department was willing to extend the list of options with the aim of signing a contract, in opposition to the production management team. This often occurs due to different mindsets, concerns and nature of the job [60]. According to CT members', this reflects the lack of a common understanding in Company P of the role and impacts of customisation in house building projects. Thus, by encouraging better communication between departments, companies should be able to build up relationships based on trust, mutual commitment and understanding of others expectations, which might avoid extra costs and delays [9].

According to Gherardi [34], a shared understanding is needed to apply MC practices, i.e., a minimum agreement is necessary for the practice to be adopted and continue to be used.

Therefore, there must be opportunities for increasing the awareness of different stakeholders regarding MC, as well as for negotiation when deciding to adopt MC practices as a way to promote innovation.

4.3. Assessment of the Utility of the Solution

The utility of the proposed decision categories and the list of practices was tested in two different stages of the empirical study, both in terms of identifying improvement opportunities and assessing the evolution of the MC strategy. The decision categories were also used to increase the CT awareness and understanding regarding key concepts, enabling them to provide numerous examples and opinions during the discussions. In fact, the CT stated that through the discussions they were able to perceive underlying ideas that they overpass in daily routine, and that this can also be useful as arguments when discussing with other departments, which contributes to improve collaboration.

The discussions regarding practices and decision categories were also useful to understand the scope of the MC strategy of Company P, and, more specifically, to identify gaps and limits for implementation. For instance, some of the solution space practices were immediately rejected by the CT, due to limitations of MC scope that were defined by existing capabilities, and focus on specific market segments. Moreover, the customer and customisation manager highlighted that the practices identified in this investigation could be useful to support decision making, such as, for refining the solution space based on the choice of users from previously delivered residential projects. Furthermore, the participants pointed out the need to improve the identification of customers' needs and to provide feedback to NPD as two major gaps in the MC strategy of the company, highlighting the importance of the customer based knowledge decision domain.

Several improvement opportunities provided further evidences of the utility of the customer integration and core decision categories. After the first presentation of research findings, many improvements were undertaken, regarding the communications with other departments, customer interaction and relationship, and visualisation approaches. Furthermore, the refinement of the strategy was also influenced by lessons learned from other segments, projects and experiences. An example is the simplified choice menu that was adopted for some market segments, in which the team had more experience. That successful solution inspired the customisation department to adapt it to A and B product types. This example reinforces the need for creating a knowledge system that enables continuous improvement and organisational learning.

The discussions with the CT also brought to light many relevant customer integration aspects. The CT coordinator highlighted the utility of customer integration decision categories in terms of making explicit what the company offers, and how the customer is involved, which makes the decision-making process as straightforward as possible. In fact, some practices related to product visualisation approaches that were implemented by the company along the study, such as the choice menu, the customised product prototype, and 3D models had a positive impact in terms of explaining the solution space to customers.

5. Discussion

The aim of this research was to devise a framework to support the definition of MC strategies by house building companies regarding customer integration. The framework was initially based on a set of practices obtained from the literature review and on some existing MC conceptual frameworks (e.g., [24,28,30,36]). Furthermore, new MC decision categories and some adaptations on the existing ones have been proposed, for the context of house building projects. This research work has two main contributions in terms of new decision categories, namely "knowledge management" and "customer interaction and relationship". The first one sheds light on the relevance of creating knowledge and disseminating it within the company as a core element of an MC strategy. The second decision category expands the vision of previous research, concerned with defining an interface, to establish a long-lasting relationship with customers, by planning interactions and building trust. Tommaso [50] states that comprehensive knowledge about customers is essential to create relationships and manage customers experience, by anticipating behaviour and needs. This statement brings up the inherent connection between those two decision categories.

The resulting set of decision categories and practices, as well as their relationships, are the building blocks for the proposed framework on customer integration. Figure 5 provides an overview of the framework. It is noteworthy that the framework also includes a set of core decision categories at a higher abstraction level, as customer integration and core decision categories are connected by decision making refinement cycles.

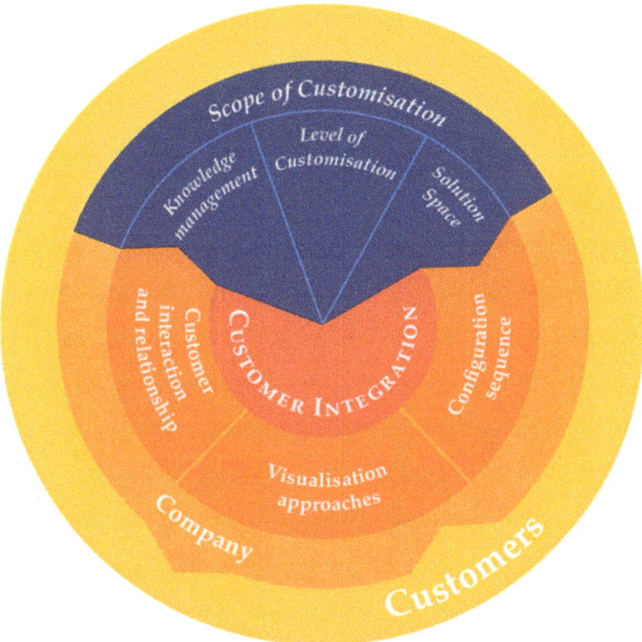

Figure 5. Customer integration framework for customised housing projects.

Firstly, decisions regarding customer based knowledge must be undertaken (Figure 5). Moreover, the definition of the level of customisation and of the solution space must be made, based on understanding the demand for customisation [21,28]. In this research, the level of customisation was assumed to be a strategic decision, being part of a broad definition of product types.

In the construction industry, there are often multiple CODPs, and the level of customisation and the customisation units must be defined for each of them. Therefore, the definition of the solution space follows the level of customisation by specifying the customisation units to be offered in each CODP. The solution space is outstandingly a core element of the MC strategy, as it influences the decisions regarding customer integration. Moreover, both visualisation approaches and configuration sequence decision categories are related to operationalising the solution space offer and supporting customers decision-making regarding the customisation units and product configuration. The customisation level and solution space definition provide directions on how should customer and company interact and establish a relationship.

The development of the framework can also be regarded as a contribution in terms of understanding of MC concepts, decision categories and domains, and their relationships in more detail, as shown in Table 9.

Table 9. Main research contributions—decision categories, source and relationship.

Group	Categories	Main Authors	Key Relationships with Other Decision Categories
Core categories	Solution space	[7,19,21,30,44]	The **solution space** decision category includes the **customisation units** and **classes of items decision domains**. A different set of **customisation units** are offered at each **CODP** defining precisely a **level or different levels of customisation**.
	Level of customisation	[7,21,28,30,35,38,40]	The **level of customisation** has a great influence on the level of **customer integration** and on **operations**. The level of customisation can vary by adopting different **CODPs**. Each **customisation level** defines boundaries for the **solution space** and defines **a CODP**.
	Knowledge management	[9,17,24–27,36,37,45–47]	The definition of a **customer integration** strategy relies on understanding customers demand for customisation, a concern of the **customer based knowledge** decision domain. Furthermore, the same domain influences the definition of the **customisation level and solution space**, by providing systematic information regarding customers needs, preferences and perception of the product in use. Additionally, discussions between departments regarding internal competences and performance of customisation can produce **organisational knowledge** related to the company's capabilities, in order to limit the **solution space**. The **communication of customisation information decision domain** depends on the amount of information produced by the MC strategy, which is closely related to the **level of customisation**. The higher the level of customisation, the higher is the need for sharing information and more intensive collaboration.
Customer Integration	Visualisation approaches	[6,30]	The **visualisation approaches** used to present the **solution space** for customers are defined accordingly to the **customisation level** and **customer interaction and relationship**.
	Configuration sequence	[30]	The **customisation units** can be organised and presented in different **configuration sequences** to facilitate customers' choice.
	Customer interaction and relationship	[14,19,21,22,24,49–51]	The **level of customisation** establishes limits and underpins the **customer interaction and relationship** decision category. **Visualisation approaches and tools** provide support for the implementation of this category.

Finally, as pointed out by some previous studies, customer integration needs to be aligned with operations and product design areas [22,28,30]. Although these areas are not represented in the framework, it is recognised that interactions between decision categories from different areas must be considered when defining strategies. This connection between areas becomes explicit when considering practices, such as, for example, the application of practices 28 and 36 requires information and actions from both customer integration and operations management teams. It means that communication and close collaboration between areas are essential for the successful implementation of practices.

6. Conclusions

The main outcome of this investigation is a framework of decision categories related to customer integration and the definition of the scope of customisation, considering the context of house building projects. These decision categories emerged from a list of MC practices that were identified in the literature and refined in an in-depth empirical study carried out in a Brazilian company that adopted some MC ideas as part of its business strategy. Some of the decision categories have been proposed in previous studies, and refined in this investigation, while two of them, knowledge management, and customer interaction and relationship, have been originally proposed in this research study.

The main theoretical contributions are concerned with exploring the underlying ideas of those practices, which have been used to explain the decision categories and their relationships. Additionally, the list of practices can be used to assess the degree of implementation of core and customer integration practices in house building companies in order to identify gaps in the existing strategy.

This exploration portrays the fruitful context of MC in construction. There are plenty of opportunities to improve value generation, not only for companies that use industrialised construction methods but also in the case of traditional ones. Customer integration seems to be a key area of improvement in house building companies, demanding efforts from different areas, which are not limited to the development of configurators or digital tools.

A major limitation of this investigation is that it was based on a single empirical study. More insights about customer integration could be obtained if other in-depth empirical studies were carried out in companies from other market segments or countries, providing opportunities for refining the framework and the assessment method.

Some opportunities for further research emerged from the discussions of the framework, such as the need to explore the interfaces between functional areas (customer integration, operations management and product design) and also between decision categories. Those interfaces need to be considered as it is expected that effective MC should have a holistic character. Other opportunities include the development of specific frameworks for product design and operations management for mass customised housing.

Another theme to be explored is how the customisation level contributes to different challenges and issues on the adoption of different sets of practices for customer integration. The higher the level of customisation and the degree of integration, the higher the complexity that needs to be dealt with due to the increasing number of stakeholders and product customisable items. Therefore, different types of MC strategies should be explored by considering the need for using different sets of practices or adapting some of them to specific contexts.

Finally, some other future research opportunities were identified regarding specific decision categories. For instance, not much has been explored regarding the configuration sequence decision category and the interdependences between customisation units that need to be considered in the design of choice menus. Another opportunity is the relationship between the solution space and the level of customisation, which has been superficially explored in the literature. Regarding the customer based knowledge decision domain, there are still many opportunities to explore approaches based on information-driven decision making and recommendation systems. Finally, the customer interaction and relationship decision category represents a fertile ground for the further exploration of experience design in mass customised housing.

Author Contributions: Conceptualisation, C.d.S.H.; methodology, C.d.S.H. and C.T.F.; analysis, C.d.S.H.; investigation, C.d.S.H.; writing—original draft preparation, C.d.S.H., C.T.F. and M.E.E.; writing—review and editing, C.d.S.H., C.T.F. and M.E.E.; supervision, C.T.F. and M.E.E. All authors have read and agreed to the published version of the manuscript.

Funding: This research was partially funded by CAPES (Coordination for the Improvement of Higher Education Personnel) an agency under the Brazilian Ministry of Education.

Acknowledgments: We would like to thank Company P for being a partner in this research project, and Luciana Gheller Amorim and Manoela Conte for the cooperation and support during the empirical study.

Conflicts of Interest: The authors declare no conflict of interest.

References

1. Larsen, M.S.S.; Lindhard, S.M.; Brunoe, T.D.; Nielsen, K.; Larsen, J.K. Mass Customization in the House Building Industry: Literature Review and Research Directions. *Front. Built Environ.* **2019**, *5*, 115. [CrossRef]
2. Jensen, K.N.; Nielsen, K.; Brunoe, T.D. Mass Customisation as a Productivity Enabler in the Contruction Industry. *IFIP Adv. Inf. Commun. Technol.* **2018**, *535*, 159–166.
3. Khalili-Araghi, S.; Kolarevic, B. Development of a framework for dimensional customisation system: A novel method for customer participation. *J. Build. Eng.* **2016**, *5*, 231–238. [CrossRef]
4. Formoso, C.; Leite, F.; Miron, L. Client requirements management in social housing: A case study on the residential leasing program in Brazil. *J. Constr. Dev. Ctries.* **2011**, *16*, 47–67.
5. Hentschke, C.S.; Formoso, C.T.; Rocha, C.G.; Echeveste, M.E.S. A method for proposing valued-adding attributes in customised housing. *Sustainability* **2014**, *6*, 9244–9267. [CrossRef]
6. Noguchi, M.; Hernández-Velasco, C.R. A 'mass custom design' approach to upgrading conventional housing development in Mexico. *Habitat Int.* **2005**, *29*, 325–336. [CrossRef]
7. Kumar, A.; Gattoufi, S.; Reisman, A. Mass customisation research: Trends, directions, diffusion intensity, and taxonomic frameworks. *Int. J. Flex. Manuf. Syst.* **2007**, *19*, 637–665. [CrossRef]
8. Hankammer, S.M. Essays on Customised and Collaborative Value Creation from the Perspective of Sustainability. Ph.D. Thesis, RWTH Aachen University, Aachen, Germany, 2018.
9. Andújar-Montoya, M.D.; Gilart-Iglesias, V.; Montoyo, A.; Marcos-Jorquera, D. A construction management framework for mass customisation in traditional construction. *Sustainability* **2015**, *7*, 5182–5210. [CrossRef]
10. Wang, Y.; Ma, H.S.; Yang, J.H.; Wang, K.S. Industry 4.0: A way from mass customisation to mass personalisation production. *Adv. Manuf.* **2017**, *5*, 311–320. [CrossRef]
11. Shin, Y.; An, S.-H.; Cho, H.-H.; Kim, G.-H.; Kang, K.-I. Application of information technology for mass customisation in the housing construction industry in Korea. *Autom. Constr.* **2008**, *17*, 831–838. [CrossRef]
12. Jiao, J.; Ma, Q.; Tseng, M.M. Towards high value-added products and services: Mass customisation and beyond. *Technovation* **2003**, *23*, 809–821. [CrossRef]
13. Fogliatto, F.S.; da Silveira, G.J.C.; Borenstein, D. The mass customisation decade: An updated review of the literature. *Int. J. Prod. Econ.* **2012**, *138*, 14–25. [CrossRef]
14. Barlow, J.; Ozaki, R. Achieving 'customer focus' in private housebuilding: Current practice and lessons from other industries. *Hous. Stud.* **2003**, *18*, 87–101. [CrossRef]
15. Lampel, J.; Mintzberg, H. Customizing Customization. *Sloan Manag. Rev.* **1996**, *38*, 21–30.
16. Gilmore, J.H., II; Pine, B.J. The four faces of mass customisation. *Harv. Bus. Rev.* **1997**, *75*, 91–102.
17. Kotha, S. Mass Customization: Implementing the Emerging Paradigm for Competitive Advantage. *Strateg. Manag. J.* **1995**, *16*, 21–42. [CrossRef]
18. Rocha, C.; Formoso, C.; Tzortzopoulos, P. Adopting Product Modularity in House Building to Support Mass Customisation. *Sustainability* **2015**, *7*, 4919–4937. [CrossRef]
19. Piller, F.T. Observations on the present and future of mass customisation. *Int. J. Flex. Manuf. Syst.* **2007**, *19*, 630–636. [CrossRef]
20. Suzić, N.; Forza, C.; Trentin, A.; Anišić, Z. Implementation guidelines for mass customisation: Current characteristics and suggestions for improvement. *Prod. Plan. Control* **2018**, *29*, 856–871. [CrossRef]
21. Piller, F.T. Mass Customization: Reflections on the State of the Concept. *Int. J. Flex. Manuf. Syst.* **2004**, *16*, 313–334. [CrossRef]
22. Ferguson, S.M.; Olewnik, A.T.; Cormier, P. A review of mass customisation across marketing, engineering and distribution domains toward development of a process framework. *Res. Eng. Des.* **2014**, *25*, 11–30. [CrossRef]
23. Fettermann, D.C.; Echeveste, M.E.S. New product development for mass customisation: A systematic review. *Prod. Manuf. Res.* **2014**, *2*, 266–290.
24. Piller, F.T.; Moeslein, K.; Stotko, C.M. Does mass customisation pay? An economic approach to evaluate customer integration. *Prod. Plan. Control* **2004**, *15*, 435–444. [CrossRef]

25. Fettermann, D.D.C.; Tortorella, G.L.; Taboada, C.M. Mass customisation process in companies from the housing sector in Brazil. In *Managing Innovation in Highly Restrictive Environments Lessons from Latin America and Emerging Markets*; Cortés-Robles, G., García-Alcaraz, J.L., Alor-Hernández, G., Eds.; Springer International Publishing: Cham, Switzerland, 2019; pp. 99–118. ISBN 9783319937151.
26. Tillmann, P.A.; Formoso, C.T. Opportunities to adopt mass customisation—A case study in the brazilian house building sector. In Proceedings of the 16th Annual Conference of the International Group for Lean Construction (IGLC), Manchester, UK, 16–18 July 2008; pp. 447–458.
27. Martinez, E.; Tommelein, I.D.; Alvear, A. Integration of Lean and Information Technology to Enable a Customisation Strategy in Affordable Housing. In Proceedings of the 25th Annual Conference of the International Group for Lean Construction (IGLC), Heraklion, Greece, 9–12 July 2017; pp. 95–102.
28. Schoenwitz, M.; Potter, A.; Gosling, J.; Naim, M. Product, process and customer preference alignment in prefabricated house building. *Int. J. Prod. Econ.* **2017**, *183*, 79–90. [CrossRef]
29. Bock, T.; Linner, T. *Robotic Industrialization: Automation and Robotiv Tecnologies for Customized Component, Module and Building Prefabrication*; Cambridge University Press: New York, NY, USA, 2015; ISBN 9789004310087.
30. Da Rocha, C.G. A Conceptual Framework for Defining Customisation Strategies in the House-Building Sector. Ph.D. Thesis, Universidade Federal of Rio Grande do Sul, Porto Alegre, Brazil, 2011.
31. Wikner, J. On decoupling points and decoupling zones. *Prod. Manuf. Res.* **2014**, *2*, 167–215. [CrossRef]
32. Echeveste, M.E.S.; Rozenfeld, H.; Fettermann, D.C. Customizing practices based on the frequency of problems in new product development process. *Concurr. Eng. Res. Appl.* **2017**, *25*, 245–261. [CrossRef]
33. Kahn, K.B.; Barczak, G.; Moss, R. Establishing a NPD best practices framework. *J. Prod. Innov. Manag.* **2006**, *23*, 106–116. [CrossRef]
34. Gherardi, S. Knowing and learning in practice-based studies: An introduction. *Learn. Organ.* **2009**, *16*, 352–359. [CrossRef]
35. Da Silveira, G.; Borenstein, D.; Fogliatto, H.S. Mass customisation: Literature review and research directions. *Int. J. Prod. Econ.* **2001**, *72*, 1–13. [CrossRef]
36. Kotha, S. Mass-customization: A strategy for knowledge creation and organisational learning. *Int. J. Technol. Manag.* **1996**, *11*, 846–858.
37. Zogaj, S.; Bretschneider, U. Customer integration in new product development: A literature review concerning the appropriateness of different customer integration methods to attain customer knowledge. In Proceedings of the European Conference on Information Systems (ECIS) 2012, Barcelona, Spain, 11–13 June 2012.
38. Jost, P.-J.J.; Süsser, T. Company-customer interaction in mass customisation. *Int. J. Prod. Econ.* **2019**, *220*, 107454. [CrossRef]
39. Rudberg, M.; Wikner, J. Mass customisation in terms of the customer order decoupling point. *Prod. Plan. Control* **2004**, *15*, 445–458. [CrossRef]
40. Barlow, J.; Childerhouse, P.; Gann, D.; Hong-Minh, S.; Naim, M.; Ozaki, R. Choice and delivery in housebuilding: Lessons from Japan for UK housebuilders. *Build. Res. Inf.* **2003**, *31*, 134–145. [CrossRef]
41. Naim, M.; Barlow, J. An innovative supply chain strategy for customised housing. *Constr. Manag. Econ.* **2002**, *21*, 593–602. [CrossRef]
42. Fogliatto, F.S.; da Silveira, G.J.C. Mass customisation: A method for market segmentation and choice menu design. *Int. J. Prod. Econ.* **2008**, *111*, 606–622. [CrossRef]
43. Huffman, C.; Kahn, B.E. Variety for Sale: Mass Customization or Mass Confusion? *J. Retail.* **1998**, *74*, 491–513. [CrossRef]
44. Salvador, F.; De Holan, P.M.; Piller, F. Cracking the Code of Mass Customization. *MIT Sloan Manag. Rev.* **2009**, *50*, 70–78.
45. Amorim, L.G. Análise de Práticas Relacionadas à Gestão da Produção para Apoiar a Customização em Massa em Empreendimentos Habitacionais. Master's Thesis, Universidade Federal do Rio Grande do Sul, Porto Alegre, Brazil, 2019.
46. Machado, A.G.C.; Moraes, W.F.A. De Um framework para a customização. *Rev. Alcance Eletrônica* **2010**, *17*, 295–311.
47. Barlow, J. From craft production to mass customisation? Customer-focused approaches to housebuilding. In Proceedings of the 6th Annual Conference of the International Group for Lean Construction (IGLC), Guaruja, Brazil, 13–15 August 1998; pp. 1–18.

48. Franke, N.; Keinz, P.; Schreier, M. Complementing Mass Customization Toolkits with User Communities: How Peer Input Improves Customer Self-Design. *J. Prod. Innov. Manag.* **2008**, *25*, 546–559. [CrossRef]
49. Theilmann, C.; Hukauf, M. Customer Integration in mass customisation: A key to corporate success. *Int. J. Innov. Manag.* **2014**, *18*, 1440002. [CrossRef]
50. Toaldo, T. Product Customisation and Development through IoT Technologies: An Empirical Study. Master's Thesis, Universita degli Studi di Padova, Padova, Italy, 2019.
51. Kaur Sahi, G.; Sehgal, S.; Sharma, R. Predicting Customers Recommendation from Co-creation of Value, Customisation and Relational Value. *Vikalpa* **2017**, *42*, 19–35. [CrossRef]
52. Kasanen, E.; Lukka, K.; Siitonen, A. The Constructive Approach in Management Accounting Research. *J. Manag. Account. Res.* **1993**, *5*, 243–264.
53. Lukka, K. The constructive research aproach. In *Proceedings of the Case Study Research in Logistics*; Ojala, L., Himola, O.-P., Eds.; Turku School of Economics and Business Administration: Turku, Finland, 2003; pp. 83–101.
54. March, S.T.; Smith, G.F. Design and natural science research on information technology. *Decis. Support Syst.* **1995**, *15*, 251–266. [CrossRef]
55. Van Aken, J.E. Management research based on the paradigm of the design sciences: The quest for field-tested and grounded technological Rules. *J. Manag. Stud.* **2004**, *41*, 219–246. [CrossRef]
56. Wikberg, F.; Olofsson, T.; Ekholm, A. Design configuration with architectural objects: Linking customer requirements with system capabilities in industrialised house-building platforms. *Constr. Manag. Econ.* **2014**, *32*, 196–207. [CrossRef]
57. Hentschke, C.d.S. Método para Identificar Atributos Customizáveis na Habitação Baseado no Modelo Conceitual Cadeia Meios-Fim. Master's Thesis, Universidade Federal do Rio Grande do Sul, Porto Alegre, Brazil, 2014.
58. Linner, T.; Bock, T. Evolution of large-scale industrialisation and service innovation in Japanese prefabrication industry. *Constr. Innov.* **2012**, *12*, 156–178. [CrossRef]
59. Tillmann, P.A. Diretrizes para a Adoção da Customização em Massa na Construção Habitacional para Baixa Renda. Master's Thesis, Universidade Federal do Rio Grande do Sul, Porto Alegre, Brazil, 2008.
60. Schoenwitz, M.; Naim, M.; Potter, A. The nature of choice in mass customised house building. *Constr. Manag. Econ.* **2012**, *30*, 203–219. [CrossRef]
61. Conte, M.; Hentschke, C.d.S.; Formoso, C.T.; Echeveste, M. Developing a choice menu: An investigation on the definition and offer of customisation units. In Proceedings of the Zero Energy Mass Custom Home (ZEMCH) International Conference, Seoul, Korea, 26–28 November 2019; pp. 1–7.

Publisher's Note: MDPI stays neutral with regard to jurisdictional claims in published maps and institutional affiliations.

Article

Mass Customization for Social Housing in Evolving Neighborhoods in Brazil

Luisa Felix Dalla Vecchia [1,*] and Branko Kolarevic [2]

[1] School of Architecture Planning and Landscape, University of Calgary, Calgary, AB T2N 1N4, Canada
[2] Hillier College of Architecture and Design, New Jersey Institute of Technology, University Heights, Newark, NJ 07102, USA; branko.r.kolarevic@njit.edu
* Correspondence: luisa.felixdallavecc@ucalgary.ca

Received: 22 September 2020; Accepted: 28 October 2020; Published: 30 October 2020

Abstract: Mass customization is being adopted in many housing contexts worldwide to provide families with dwellings that suit their individual needs at costs similar to mass-produced items. However, in many social housing contexts, there are barriers that can hinder the adoption of mass customization, despite the benefits it could bring to residents. This is the case in the Brazilian social housing context considering house units for families of the lowest income range. This paper explores the possibilities and limitations of applying mass customization in this context to improve the living conditions in these neighborhoods as they evolve over time. This study analyzes the ecology of the system of provision of social housing for the lowest income range, pre-occupancy, and post-occupancy in the neighborhood's development over time. This study argues that it would be more feasible and bring more and longer-lasting benefits to the stakeholders involved if mass customization were applied post-occupancy.

Keywords: mass customization; social housing; post-occupancy

1. Introduction

This study investigates how mass customization could be applied in Brazil to social housing developments to promote better environments for the families and the neighborhood over time. Mass customization can be defined as "a system that uses information technology, flexible processes, and organizational structures to deliver a wide range of products and services that meet specific needs of individual customers (often defined by a series of options), at a cost near that of mass-produced items" [1]. In housing, individual customization is traditionally seen as hiring an architect to creatively design a unique home, ideal for the family. Mass housing, on the other hand, is when large numbers of identical homes are built and then sold for much less than the uniquely designed homes. Mass customization promises the best of both approaches: uniquely designed products that better fit the user's needs with mass production efficiency and costs [2,3]. It is being adopted in many housing contexts worldwide to provide families with dwellings that suit their individual needs at costs similar to the mass-produced ones.

Mass customization of housing has been linked to environmental and social sustainability [4]. By providing dwellings that better suit the individual needs of the family, there is less need for renovations and the waste it creates. Furthermore, construction for mass customization often adopts practices, such as prefabrication, which also reduce waste and water and energy consumption. The custom home also increases the users' sense of identity and ownership towards it. However, most cases of mass customization in housing do not consider the spatial needs of the user; the user is limited to choosing their preferences in elements such as surface materials, colors, and finishes [5]. Several examples from industry and research that have addressed mass customization in housing have also considered needs in terms of space, including the kind of space, the relationships between them, and how much space the users need or want [4,6–8]. This approach that includes the spatial aspects shows the greatest potential to avoid unnecessary demolitions and renovations and even to allow families to stay in the same home for longer. This is especially relevant when moving is difficult or not an option, such as for lower-income families for whom mass housing is often the norm.

Several studies have shown how the concept of mass customization and its tools could be used to satisfy each family's needs in mass housing [8–11]. However, in many social housing contexts, there are added barriers that can hinder the adoption of mass customization, despite the benefits it could bring for families. This is the case in the Brazilian context with house units for the lowest income range of the population. Some construction companies in Brazil have started mass customizing their housing products for higher-income housing developments, allowing the customers to choose from standard interior and very few exterior elements. This approach makes a significant difference in terms of the customer expressing their individuality and territoriality and increasing their sense of ownership. However, in most cases, the different spatial needs are not addressed since the definition of how much space and how it is organized stays the same. Furthermore, despite studies suggesting that there would be benefits in adopting mass customization in the context of the lowest income range of social housing and showing potential ways to overcome the identified barriers [12–15], mass customization has not yet been adopted in any cases in this context. This indicates that for the lowest income range of social housing in Brazil, there are still gaps in the literature about the identification of barriers and the alignment of benefits of such a mass customization strategy with the interests of the stakeholders.

The fact that post-occupancy processes are usually not considered in studies of mass customization in housing is also a gap in the literature, especially for the lowest income range of social housing developments in Brazil, which has produced more than 1,500,000 housing units in the last decade [16]. The post-occupancy processes in this social housing context include major geometric changes to the housing units, such as significantly increasing the dwelling's area. How the units are built affects if and how the families can make such changes. Likewise, the intention to allow and facilitate such changes affects how the initial units should be built. Families consider such changes to the units as essential since they seldom have the option of moving, even though their life circumstances often evolve. Therefore, these post-occupancy processes should also be taken into account when considering mass customization strategies for this context. This would allow groups (developers, families, governments, etc.) to maximize the social, environmental, and economic benefits of a mass customization strategy to last over time.

Therefore, this study focuses on how mass customization could be applied in Brazil to social housing developments for the population with the lowest income range to promote better environments for the families and the neighborhoods over time. This research intends to address the identified gaps by analyzing the whole ecology of the system of provision of social housing for the lowest income range of housing programs in Brazil, including the post-occupancy processes, in concert with the possibilities, approaches, and tools of mass customization to indicate directions of how they could be applied in this context to bring lasting benefits to the stakeholders. This research investigates the operational aspects regarding the deployment of mass customization in this context, considering the capabilities and interests of the stakeholders involved, how to maximize the potential for such a strategy being adopted, and ways to benefit the largest number of stakeholders and broader society over time while requiring the minimum changes in policy and regulation. Although this research acknowledges the customization of surface materials and finishes as relevant, it focuses on the spatial needs of the user, as it is the spatial changes to housing units that are most difficult for the users to make and cause the most significant problems. Thus, this aspect is considered most important to customize in this case. While the research analyzes the ecology of the system to determine what would be possible within this context, the benefits of customizing are brought through the customization of individual house design. Thus, the study considers mainly the scale of the house and the changes in processes that could improve the design outcome of individual houses in this social housing context.

Through the results of this research, for this context, broader considerations can be drawn, including for other contexts as shown in the discussion.

2. Materials and Methods

This research is of a qualitative nature; therefore, it takes on the characteristics often present in qualitative research. In particular, this study focuses on interpretation and meaning, in which the researcher plays "an important role in interpreting and making sense of [the collected] data", adopting "practices that embrace interpretation and meaning in context" [17]. What is also relevant is being holistic when considering multiple perspectives and the many factors involved [17]. It starts from the premise that two main factors contribute to the problems that emerge from the evolving built environments in neighborhoods of the lowest income range of social housing programs in Brazil: (1) the need to change and add to the original housing unit given that it often does not satisfy the needs of the user from the outset; (2) the lack of design involved in the creation of changes to the housing units over time. The concepts of mass customization and housing adaptability are considered as significant contributors to the development of solutions that could be applied in this context. The study was divided into four stages, as shown in Figure 1.

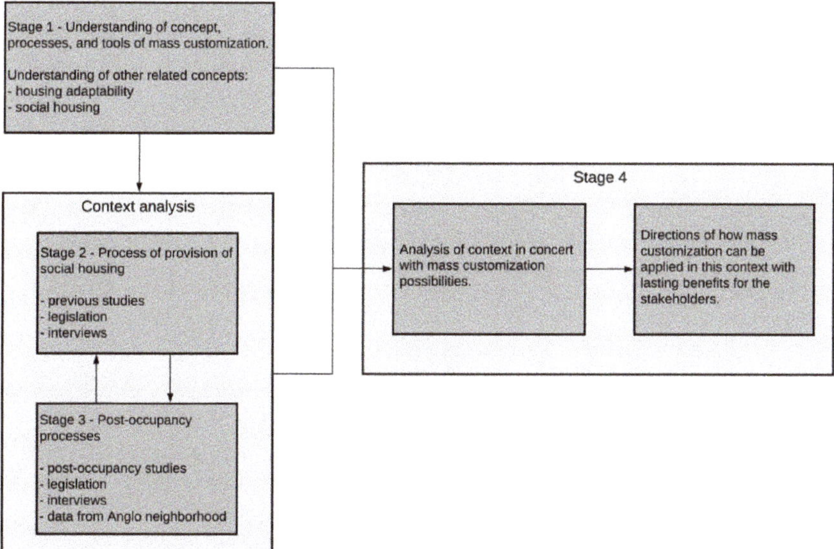

Figure 1. Research stages.

The first stage refers to the understanding of mass customization and its application in housing contexts as well as the understanding of other concepts considered relevant to this research, such as social housing and housing adaptability. The concept of mass customization is used as a guide into processes for the production of individually customized products at prices that "approach, and sometimes beat" those of mass-produced goods [2]. Housing adaptability is used to guide aspects relating to design and technologies, aiming to facilitate a fit between the user's space need and their home, especially being able to change their home after occupancy [18–20]. This stage draws on theoretical references related to the concepts involved in the research as well as previous studies and cases that use and apply those concepts.

The second and third stages refer to the understanding of the context and processes of social housing production in Brazil, focusing on the lowest income range of housing programs. Stage two (shown in Section 3) refers to the process of provision of social housing and stage three (shown in

Section 4) refers to the post-occupancy processes. In these stages, documentation research also was drawn on, including bibliographies on social housing processes, relevant legislation and policies, and post-occupancy studies. Data from a specific social housing neighborhood, the Anglo neighborhood in Pelotas in the south of Brazil, were also analyzed. These data were made available to the authors by the research group Naurb [21] and include information regarding the specific process for the neighborhood's implementation, from application for funding to completion of construction, and data from post-occupancy studies. The latter included demographic data and the changes made by the families to the housing units.

As one of the objectives is to identify the limitations and potentials present in the current social housing processes in concert with the strategies and tools of mass customization, in some cases, the information available in the literature is insufficient to make these connections. Therefore, this study also draws on semi-structured interviews with key stakeholders. A total of 11 interviews were carried out, divided into six categories of stakeholders. The stakeholders were identified based on the literature about the provision of social housing and legislation. Each person interviewed was also asked to identify other stakeholders they believed were relevant to the process and to be interviewed. Whenever possible, interviews were conducted with more than one representative from each category and from more than one city. These interviews included three city architects and engineers, three city social workers, the owner and manager (also an engineer) of a development company, the national manager for approval of innovative technologies for social housing construction, the regional manager from the financial institution Caixa Econômica Federal (CEF), and two experts in social housing who specifically research processes in the lowest income range of social housing.

The fourth stage refers to the analysis of the specific social housing context and current capabilities of the stakeholders in concert with the possibilities and tools of mass customization with a focus on improving the living environments for the families and the city as the neighborhood evolves. From the results, it is possible to indicate directions of how mass customization could be applied in this social housing context, aiming to improve the living conditions within the unit and the neighborhood as a whole, bringing further benefits to the city as these neighborhoods evolve over time. Relevant considerations from this stage are shown in Section 5.

3. Process of Implementation of Social Housing Developments

The process outlined in this section focuses primarily on urban housing in the My House My Life program—translated from programa Minha Casa Minha Vida (MCMV)—since it has been the largest provider of social housing units over the last decade. However, many of the processes and relationships between stakeholders described here apply to other programs too. Furthermore, the interests and motivations of the stakeholders are also maintained across different programs. Most importantly, the solutions in terms of both urban and unit design are also consistent across several different programs, as are the problems they result in and the way the families deal with them in post-occupancy.

Social housing programs aimed at the lowest income range of the population usually produce units that are completely or mostly subsidized by the government. The lowest income range of the MCMV program includes families who earn between zero and up to approximately two times the minimum wage and the subsidies can cover up to 90% of the cost of each housing unit. Figure 2 outlines the main stages for the implementation of a new social housing development and shows the main operational stakeholders involved in this process. Although there may be some variations from city to city, especially regarding how much time each phase takes, the overall process falls within this structure for most cities.

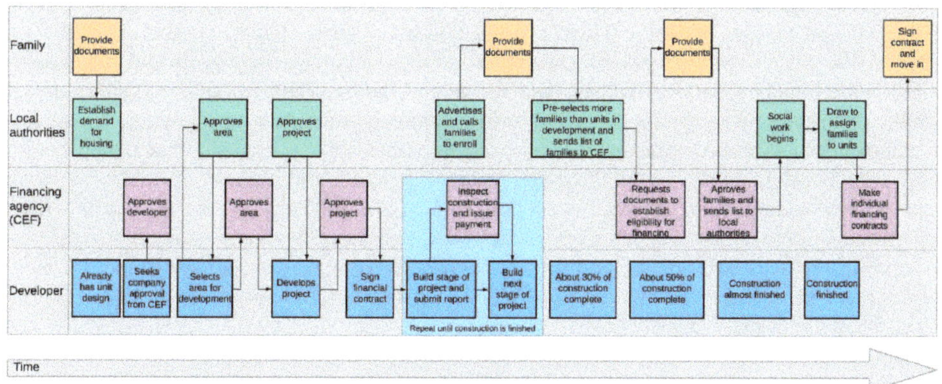

Figure 2. Flowchart of main stages of the process of provision of social housing.

A first step for enabling a new social housing development is registering families who are eligible and interested in receiving a social housing unit. The cities' housing departments usually do this on an ongoing basis. At this point, the families indicate their interest in being considered for a social housing unit and provide the necessary documents to show their eligibility. The local authorities assist the families in registering in the federal government's social programs system (Cadastro Único). The information kept within this system includes identification of each of the family members, their work condition, the family's income, and their living conditions, among others [22]. Information regarding the family's current living conditions is collected during this meeting, both to establish their need for housing and to determine where within the city they currently live.

An initial step taken by the developer is to consult the financing agency Caixa Econômica Federal (CEF) to determine if there is funding available for social housing developments in the chosen municipality. They also seek pre-approval from CEF as a capable company to build within federal programs. This is done to ensure the developer has the financial means to keep construction going since the developer must first build each stage of the construction, foundations for example, with their own capital, after which CEF will verify if that stage is up to their standard to then pay for that part of the construction. Thus, considering the scale of these developments, a considerably large amount of resources is needed to start the process and keep construction going, which smaller companies may not have.

To initiate a new development project, the literature has identified two main distinct possibilities. One possibility is for the city to provide land for the new development and call for interested developers to produce a project. However, the most common scenario is for the city to wait for a developer to show interest in producing housing for this income range. In this case, the developer proposes to the city where they intend to build the new development. At this stage, the process only keeps going if the city authorities approve the proposed land for the new development. This may seem like an opportunity for the cities to have greater control over the locations where developments get built. However, because the cities are dependent on having developers interested in building for this income range, local authorities are often pressured into approving developments in inconvenient areas for fear of losing the interest of the developer [23,24].

The choice of cheaper land on the outskirts of cities—often in areas previously considered rural—and the standardization of design became the norm [23]. When the developer proposes to build a new development for this income range, they usually already have unit designs for varying sizes of lots so no consideration is given to the differences in location, culture, and family composition [13]. This standardization in design leads to several problems, such as poor urban spaces, lack of internal comfort, high energy consumption, and low satisfaction from the users [18,25]. Local authorities, pressured by the building companies, in many cases, have made this process easier by changing urban policy, urban perimeters, and local construction rules. These changes were made to allow the construction of developments in rural areas, with reduced requirements of green and public spaces, reduced unit requirements, and reduced taxes for the construction companies [23,24]. The ministerial ordinance Portaria No 660 [26] is the national regulation that establishes the minimum standards for housing units and urban parameters for projects within the MCMV program and any other program that uses federal funds. It is often the only regulation followed for design purposes since projects cannot be funded if this regulation is not followed; however, it is usually less restrictive than local urban and building codes. Thus, developers often seek special approval within the city for the project to comply only to the program's standards. For house units, the program's regulation establishes a minimum of 36 m^2 and that every housing unit must have at least two bedrooms, a kitchen, a living room, a bathroom, and a laundry area.

Once the developer has the city approval, they submit the project for approval with CEF. Different from the approval from the city, CEF analyzes the proposed budget for the project and whether the percentages of the total funding designated to each stage of the project are within the permitted standards. Once CEF approves the project and the contract between CEF and the developer is signed, then construction can begin. It is important to highlight that the total amount paid for the development is capped based on the number of units and the relative regional importance and size of the city. Therefore, one way to maximize profits is to save as much as possible on aspects of the budget that are not determined by quantity, such as the amount designated to design the development. Furthermore, having the cost of the land as part of the financing process also benefits the developer since they receive this amount at the start and can often negotiate to pay the previous owner in instalments over time, therefore leaving a significant portion of the amount received for the land to be used to start construction.

Although the city approves the final design of the development, it is difficult for the municipality to demand quality in design considering that they cannot deny approval if the project is within the legislation [27]. This was corroborated through interviews with city workers who state that the role of the city regarding the design is limited to checking if it complies with the legislation. This aspect can be a barrier to quality in design considering that the legislation was highly simplified to facilitate the fast approval of projects and that, in the MCMV program, the municipalities are expected to release barriers and facilitate the action of the private sector [27].

Many reasons can lead to delays during construction. An example includes a site inspection determining that a particular stage of the construction is not up to the standard required by CEF and that some aspects need to be redone. This means that the construction company will not receive the amount designated for that stage until it is redone. Depending on the company's resources, this can mean that the rest of the construction stops until they receive that amount. There are cases in which the resources necessary to bring the construction up to the established standard were too high and led the company to bankruptcy. In these cases, construction stops for longer, until another company can be hired to finish construction, usually with an updated cost estimate which increases the final cost for the development. However, construction can also be delayed for other reasons, such as public authorities requiring that construction stop to review specific permits and taking longer than usual to review them. In such cases, if the prices for construction change significantly, the company can apply to update the cost estimate. Thus, it is not uncommon for this kind of development to end up costing much more than what was the original budget and taking much longer than the original schedule.

When construction is about 30% complete, the city's social housing department starts the pre-selection of families [13]. In some cases, part or all of the housing units may be reserved for families that the city is removing from informal settlements in risk areas, for example. At this point in the process, social workers may use different means—such as interviews with families and visits to their current home—to make a social report. With this report, the city ranks the families according to national and local criteria and selects families to continue the process to receive a unit in the specific development. The number of families selected to continue the process is equal to the number of units the development will have plus 30% [28]. The city sends this list of pre-selected families to CEF when construction is about 50% complete [28]. CEF does their own analysis of the documentation to determine which families are eligible for financing. With this updated list of eligible families, the city informs the families that have been selected to receive a unit. However, at this point, the families have not yet been assigned to a specific unit.

Every social housing development, regardless of the program that led to its implementation, must include funds for social work with the families, which is governed by the ministerial ordinance Portaria No 21 [29]. By knowing which families will live in the development, the pre-occupancy social work with the families can begin. This consists of working with the families to prepare them for formal housing and living in the new community. Most often, the families selected for a development come from different areas around the city. The social workers help them create a community association and carry out educational actions regarding things such as how to deal with their garbage, how to save energy, what is acceptable or not in public spaces of the neighborhood, among others. Preparation and support are also given for the families to apply to other social programs when that is the case and things such as finding and enrolling children in a new school. This social work continues for at least one year after the families move to the new development.

When construction is practically finished, and in some cases after it is finished, the city calls the families for the draw of units. This is a draw to establish which family will receive which unit. The updated list of families with their designated unit is then sent to CEF, which makes the individual contracts of the financing process for the families to sign. Although each family signs their contract individually, cities often wait until all the families have signed the contract to give the families the keys to their units all on the same day in a highly publicized event. The beneficiary family must live in the housing unit for ten years of the financing process in order to receive the subsidy. If the family wishes to settle the remainder of the debt beforehand (to sell the unit, for example), then they must pay the total that is still owed without receiving the subsidy (which can be up to 90% of the total cost of the unit). However, illegal sales do still happen.

4. Post-Occupancy Processes

4.1. The Need for Change

Specifically for this social housing context, Brandão [18] indicates that the following aspects frequently appear in post-occupancy studies as motivators for change: aspects related to function, such as the layout and size of the rooms; the size of the housing unit; aspects related to visual and auditive privacy; aspects related to personalization and definition of territory; changes in the family, such as the size of the family; economic and educational level. Thus, most reasons for making spatial changes in this context fall under the categories identified by Friedman [19]—family transformations, fitting new technologies, and affording in stages—and, therefore, are similar to other contexts. However, how these aspects appear and motivate changes to the housing units in this context can have some specific factors not considered elsewhere.

The initial size of the housing units is a crucial factor perceived by the families as needing change. In a study regarding the perception of the users in three different social housing developments, the authors indicate the main reason for wanting to leave the development was the inadequate size of the unit; however, in one of the three developments, which consisted of house units, the possibility of expanding was seen as a positive factor [30]. The aspect of fitting new technologies is often identified in relation to the size of the unit. This does not necessarily mean that the technologies changed since the unit was initially built, but often, the initial unit already did not consider the technologies available at the time it was built in order to minimize the size of the unit. This is a common complaint by families concerning washing machines, for example [31,32]. Although it is not a new technology, most of the social housing developments do not include a space in the unit for this equipment nor the necessary plumbing to facilitate its use. Similar complaints are also seen in relation to space and electrical outlets available in the kitchen.

The aspect of family transformation is also a significant motivator for spatial changes to the unit. This occurs with normal changes to family through time, but, similar to the technologies factor, the family does not necessarily need to change for the original unit to become inadequate. In many cases, as a result of standardization in design, the original unit is already inappropriate for the size or arrangement of the family when they first move in. In a post-occupancy study, Jorge et al. [32] identified that 40% of the units had six or more people living in them. Furthermore, it is also common to have extended family living in the same housing unit. In the same study, this was the case in 30% of the housing units [32]. Another important factor to consider regarding adapting the unit to changes in the family refers to the work conditions of the families. It is common for the family members to have informal work conditions that is, having an unstable source of income. Thus, initiating and running a small business from home is common in these developments and even encouraged given that workshops for this purpose are provided as part of the social work. This kind of activity often requires space beyond the original unit, thus also being a significant driver of spatial changes.

Affording in stages is also an important aspect to consider in this context. While it is not a reason for change, it significantly influences how change happens, as discussed in the next section.

4.2. How Change Happens

In house units, it is expected that families will make changes and expand their unit after moving in, with the possibility of expansion even required by current legislation [26]. Furthermore, many of the interviews done for this research, with architects, engineers, and social workers, had similar statements saying that for house unit neighborhoods, it is unavoidable that the families will expand the units even when it is not allowed. However, because the units are not built with adaptability in mind, it is even more expensive and difficult for the families to make these changes [23]. Most of the changes are carried out illegally by the homeowners themselves. This includes planning what to change and the construction itself.

It is common for the family to start buying construction materials in small quantities from local, often small, building materials stores and building the extension themselves, a small portion at a time. This process continues over time until all the intended expansion is complete. Thus, it is common in these developments to see partially built rooms, such as the outline of a room built to only half the height, or a space with one wall and the roof, which is used as a veranda until the other walls can be built. It is also common to use temporary materials, such as plastic and plywood, as parts of the walls until they can afford a more permanent solution. Another common scenario is for the family to buy all the necessary materials over time, storing it until they have enough to build the entire intended expansion and then building it very quickly. The storage of materials is usually done in the garden or in front of the house on the street, which can implicate some losses due to weather or theft. Often, neighbors, family, and friends also help with the construction. Hiring a resident of the neighborhood who works with construction is also seen in some cases. Therefore, the aspect of affording in stages does not only refer to expanding when they can afford it but also to dividing an intended expansion into stages distributed over time.

Several post-occupancy studies [18,25,31–33] have demonstrated the changes that families make to their housing units in social housing developments. These studies categorize not only the changes made but also why they were made. Previous authors identified four main categories of reasons associated with the changes made by the users: safety, such as building a wall and adding bars to windows; family need, such as building a pantry or new bedroom; size, such as adding or increasing the area of rooms such as the bedroom, bathroom, and kitchen; finishing, such as changing the windows and the wall finishes, e.g., substituting paint for ceramic tiles [33]. It is important to note that, in that study, the reasons of 'needs' of the family and 'size' account for similar changes, referring to spatial changes, and, in many cases, appear together. These authors further indicate that most changes are made due to the dimensional inadequacy of the spaces to their functions and the domestic needs of the occupants [33]. This is consistent with findings from other studies that indicate that for social housing, the most pertinent kind of change is expansions [31]. This author analyzed several social housing developments, indicating the following changes as most frequent:

- Intervention on the façade, including building a wall;
- Adding a garage;
- Increasing the area of the kitchen;
- Creating or increasing the area of the laundry;
- Creating a separate space for business, studies, or hobbies;
- Creating more bathrooms;
- Creating more space for storage;
- Changing the relationship between the kitchen, area for meals, and living room;
- Adding one more room.

These changes are consistent with those shown in other post-occupancy studies [18,25,32–34].

Specifically referring to spaces for business, several post-occupancy studies show that it is a significant driver of renovation; in some cases, having a space for business was considered important by more than 60% of the families [32]. Although this does not necessarily require changes to the units, often it is reflected in the significant amount of cases in which a room for business was added to the units. Likewise, adding a garage is very frequent in cases in which it is possible to expand to the front or side of the unit. In these cases, the garage and business sharing the same space is also common.

An important consideration that is consistent across many post-occupancy studies refers to the order in which changes to the units are made. Changes that are made immediately after the families move in are more related to demonstrating territoriality and differentiating themselves from the neighbors. These include placing significant objects in the front of the house, planting a garden, and changing the color of the façade. These being the first changes made is understandable given the importance of the feeling of ownership towards the unit and the low cost involved in making them. Other changes that are often made before spatial changes in the unit are related to safety, such as building walls around the lot.

Regarding spatial changes, the studies indicate that increasing the area of the kitchen is often a priority and, usually, it involves increasing the area of the unit. Expansions of other spaces or adding other spaces to the unit appear as the second change made or the second most frequent first change. Several studies highlight the significant amount of cases in which walls with pipes (such as the bathroom and kitchen) had to be demolished and rebuilt elsewhere, making the changes more difficult and expensive than necessary and highlighting the need of carefully placing these elements in the original unit design [25,32,33].

It is important to highlight that while the categories of expansions are consistent across many different developments, the solutions for the expansions (for example, where the expansion is built) vary considerably. Not only it is dependent on the original unit design and where there is room to expand, but the solutions also vary within the same development of equal units. Even when an expansion plan was provided, there are many cases of the families not following such a plan. Some of the reasons for this include the plan not meeting the family's needs or for economic reasons [34].

Another important aspect to consider refers to how long after moving in the changes are made. Several studies conducted up to five years after the completion of the development already show significant changes made to most units. In the Anglo neighborhood, completed in two stages, four and two years before the data collection, 80% of the units already had changes made [32]. Even in an older neighborhood, the authors show that most changes (70%) were made within the first three years after moving in [33]. However, it is still relevant to consider that older neighborhoods present more expansions; this shows that there is continual transformation of the houses over time, even if it does slow down.

4.3. Problems with Changes

In most cases, the families perceive the changes as beneficial and as an improvement in their living conditions. However, these changes often result in inadequate situations, such as encroaching onto the public space, having no ventilation and lighting, blocking off windows of the initial unit, inappropriate discharge of rainwater, dangerous stairs, and opening windows directly onto the neighbor's lot, among others [18,25,31]. These inadequate situations can result in negative consequences not only for the families but also for the city, such as increasing health problems and the burden on the city to provide health care and limiting the city authorities' access to public services, such as public lighting, provision of electricity, and water and sewage, for example. In one case reported during the interviews, expansions were built over the water drainage system. This resulted in frequent flooding inside the houses due to a lack of maintenance of the system. In the city of Pelotas, in an older social housing neighborhood, several families built onto the sidewalk enclosing the public electrical posts; currently, access to most of the public lighting and distribution by the local authorities is difficult. Figure 3 shows cases from the Anglo neighborhood that also encroach onto the public space, eliminating the sidewalk. Initial construction for this neighborhood was completed in 2014.

Figure 3. Example of expansions encroaching onto the public space. Images courtesy of NAUrb-UFPel [21].

Referring to self-construction in informal settlements, Estevão and Medvedovski [35] indicate the inadequacies related to humidity as one of the main factors in which the housing environment can be problematic to the health of the inhabitants. Self-construction in formal social housing neighborhoods shows several of the same risk factors. Several authors highlight the lack of natural light and ventilation as a frequent problem resulting from the expansions in social housing neighborhoods [25,32,33]. One of the most problematic situations from such a scenario is that it aggravates health problems resulting from humidity, such as respiratory and skin problems [35]. Furthermore, it often leads to a lack of thermal comfort, which, as well as aggravating existing conditions, such as hypertension [35], can lead to excessive use of energy to mitigate the condition.

Another factor refers to situations of risk to the physical integrity of the inhabitants. The inappropriate proportions of stairs and the use of ceiling slabs as an extension of the house are responsible for an unimaginable number of serious accidents [35]. These situations are also common in self-built expansions in social housing neighborhoods. An example of inappropriate proportions of stairs can be seen in Figure 4.

Figure 4. Stairs in the Anglo neighborhood in Pelotas. Image courtesy of NAUrb-UFPel [21].

In some cases, the expansions reproduce the logic of precarious housing in which the families lived before—for example, building rooms or walls with temporary materials or accommodating a second family on the same lot without a bathroom; this creates a cycle that threatens the condition of a healthy environment and increases the risk of disease contamination [32]. However, from post-occupancy studies, it is possible to indicate that despite the families' low-income situation, many of them invest significantly in expanding their units in a permanent way. Figure 5 shows an example from the Anglo neighborhood, which increased the area on the ground floor for a garage and business and added a second floor. Furthermore, it is important to highlight that the changes made and how they are made is consistent in many post-occupancy studies in such developments from different cities and built through different programs.

Figure 5. Example of housing unit with expansions. Images courtesy of NAUrb-UFPel [21].

As previously explained, many of these changes bring problems that have an effect beyond the family that lives in the unit. However, very rarely are any actions taken to support the families in making better decisions in their expansions or to discourage problematic situations. Many municipalities have in place legislation with mechanisms to allow the municipal authorities to notify, fine, and, in some cases, demolish illegal constructions, such as those that encroach onto the public space, for example. However, these mechanisms are seldom used. Giving a notification when this kind of illegal construction is identified is the action most used; however, these are usually ignored and no further action is taken. When questioned about this scenario, several of the stakeholders interviewed attribute it to political will. They highlight that local politicians fear looking bad for allowing more severe action to be taken 'against' this vulnerable population.

It is important to consider that given the opportunity to have assistance for these expansions, often, families will seek it, as there is significant demand for this kind of assistance from the very few organizations that provide it free of charge for these families. An example of such organizations are the university outreach offices called "Model Offices" (Escritórios Modelo in Portuguese), which are present in some architecture schools. There are several benefits for the families in having their houses regulated within the municipality; one that is immediately perceived by the families is the possibility of being able to apply for financial assistance for further construction. The benefits which the families themselves might not see immediately, such as avoiding health issues, are equally important and in the best interest of the city as a whole.

This need to assist low-income families in design and construction was recognized by the federal government in 2008 when specific legislation was approved, stating that low-income families should have such assistance without being charged for it [36]. However, there are few examples of municipalities and other entities that provide such assistance for improvements to individual housing units. This is understandable since the costs for professionals to work individually with each family are high and the funding is limited. Moreover, in cases where such technical assistance is available, it is usually only for families living in informal settlements, not for families in social housing neighborhoods. This also is understandable since the latter are considered as already adequately housed and it would not make sense to spend limited public funds twice on the same family when there are others in need. However, there are significant funds destined for social assistance, specifically for these social housing neighborhoods. Thus, if costs could be reduced from having a professional work individually with each family, it would be feasible to provide such assistance as part of the initial funding for the development and other social assistance programs that take place within these neighborhoods.

5. Discussion

This section discusses the implications of some of the exposed social housing processes in concert with the possibilities of mass customization.

A vital consideration refers to how public authorities view social housing production. Through legislation and practices, it is clear that the initial housing unit is seen as the final product. All the legislation regarding the production of social housing units refers to the process as finished once the construction of the unit is complete. Post-occupancy social work is the only thing that goes beyond the finalized construction of the initial unit and it does not consider further construction. In contrast, for the families, the initial house unit is seen as a starting point from which they continue to build. As the neighborhood evolves, the self-designed and self-built additions to the units often bring unintended negative consequences along with the intended benefits. The legislation regarding technical assistance in housing design is broad enough that it could include social housing developments. However, it is not seen in practice as applicable in these cases. Pre-occupancy construction, post-occupancy renovation, and technical assistance are organized and seen as completely separate processes. Thus, one happens without any consideration of the other. However, when considering the limits of the initial production of social housing and the needs of the families over time, it becomes clear that for social housing developments, it should all be seen as part of one continuous process, including for funding purposes. Broadening the context considered for the use of funds can lead to a more efficient allocation of those funds [37].

This is particularly relevant when considering mass customization for this context. Previous studies have indicated that small changes in the current process of provision of social housing could create opportunities for mass customization of the initial units [14]. However, by including the post-occupancy processes in the analysis of potential mass customization possibilities and benefits for this context, it is possible to indicate that customizing the initial unit is not the most sustainable option as it may not have lasting results in terms of satisfying the families' needs and avoiding problematic situations over time.

An important consideration is that mass customizer companies usually profit by attracting more customers with customized products, gaining their fidelity, or taking advantage of their willingness to pay a premium for a custom product [2]. However, these options are not available in this social housing context. The amount the company can receive per unit is capped and the families do not have a choice of developer—they are assigned to a unit by the city. Stakeholders from different spheres of government are more interested in a higher number of units built than higher quality in design. This appears in all publicity, and even official program websites state that priority will be given to proposals that reach a higher number of families [38]. Therefore, the companies still would not be motivated to mass customize, even if the added effort and cost were only marginally higher.

Regulations around the numbers and types of rooms that the units must have, combined with the restricted floor area to keep the costs low, mean that there is very little that can be customized in terms of the families' space needs. Although these regulations are necessary to establish a minimum standard, they can also be a barrier to mass customization. Furthermore, most changes made to the housing units post-occupancy add area to the house. This shows that in most cases, a large part of the reason for the unit not satisfying the families' needs refers to it being too small. This aspect would still be present even if the initial units were customized. On a different approach, the initial area of the housing units could be increased, making it more meaningful to customize them. However, this would require significant changes in the current process of provision, especially increasing funding. Therefore, this is not a feasible option. Investing in flexibility could be a way to allow for the families to satisfy their needs after occupancy better, while still building all initial units equal within current regulations. However, the way the families currently change and expand their units also leads to problems for themselves, the neighborhood, and the city. Thus, providing assistance for the families in this renovation process is necessary. It is also important to highlight that the needs of the families change over time. Thus, having a strategy to customize the initial unit and not addressing this as a process over time could still lead to the same problems, especially with the expansions.

This study argues that it would be more feasible and have a higher potential for bringing significant benefits over time—not only for the families but also to the city and other stakeholders—if a mass customization strategy were applied with post-occupancy differentiation. This way, the initial units built could still be small and equal within the development, not requiring significant changes to current social housing programs, their associated policies, and the capabilities of the stakeholders. Post-occupancy, the families would use a mass customization configurator, a co-design system, to interact with and visualize the design of expansions for their homes. Such configurators have the capacity to validate the customer's solutions, including in cases that involve spatial aspects in housing design, as shown by previous authors [7,8]. Only solutions within the legal parameters for the city would be validated. The design validated by the configurator could receive automatic approval by the city. This would significantly benefit the families and the city by avoiding problematic changes to the units. The families would have the opportunity to manipulate and visualize different design solutions and receive feedback before engaging in construction.

Providing this assistance to the families to have validated design solutions could also serve as a justification for cities to take more severe action, such as fines and demolitions, towards problematic construction situations. The interviews done for this research indicate that there is currently a perception that any actions taken in this regard would be hurting an already vulnerable population and, thus, that such actions are not socially acceptable. This perception is intensified by the fact that most cities do not offer an alternative to the current practices of self-design and construction, making it morally and politically challenging to take serious action against such illegal construction. However, if processes such as those proposed in this study are in place and design assistance is available to the families, then it would be a choice of the family to ignore such help and build problematic solutions. This could remove some of the perceived barriers around taking action to correct problematic situations;

for example, demolishing construction that encroaches on public space. In other words, if the city is providing the conditions for families to build legally and these families still choose to build illegal solutions, then it could become socially and politically acceptable for the city to fine or demolish such problematic solutions.

To allow post-occupancy differentiation as easily and as much as possible, increased adaptability of the initial unit is essential. Most cities do not have the resources to critically evaluate and suggest design changes during the approval process of new social housing developments, with city authorities limited to checking for compliance with legislation. One of the results of this limitation is that most projects merely achieve the minimum required by legislation. In terms of adaptability, the regulation has one sentence that states that the housing unit shall be designed to enable its future expansion without loss to the lighting conditions and natural ventilation of the existing rooms [26]. Given that this regulation provides little guidance, it is usually fulfilled by showing that one room can be added. Often this means that the unit was designed to allow only one room, in a specific place, to be added. This is not enough to satisfy the amount and variety of changes the families need.

In contrast, several researchers have developed extensive adaptability guidelines specifically for social housing. However, most of such guidelines depend on subjective design judgment, making it difficult to incorporate in regulation and to check for compliance. Although changing the way projects are approved in cities to include their design interests and feedback would be ideal, it is currently not feasible for most cities; thus, providing such thoughtful feedback is outside their capability. An option that could help include more adaptability in design would be to include guidelines in the regulation of the program in a way that would make compliance to them easy to check. One way this could be done would be to use the adaptability guidelines developed in previous studies, for example, by Brandão [18], but translating them as much as possible into quantifiable acceptable ranges that are easy to demonstrate. For the project to be approved, the developer would then have to demonstrate compliance to at least three of the guidelines in addition to the one already in the regulation. Table 1 shows a selection of guidelines that would be feasible to implement within the regulations of the MCMV program and could be checked easily within the current process of approval. The table also shows possible wording for the regulation, such that it would be easy to demonstrate and check for within the approval process. Figure 6 shows an example floor plan inserted in the existing lots of the Anglo neighborhood that was developed in this study following such guidelines. The walls with pipes and structural columns are shown in gray, and the houses are meant for semi-detached typology. Figures 7 and 8 demonstrate some of the expansions possible with this design, which significantly increase the area of the house, demonstrating the design's compliance to the guidelines in Table 1.

This study proposes that the main agent of the mass customization strategy, in this case, should be the local authorities and not a company as in most mass customization cases. This would allow the city to have more power over the parameters to be validated within the configurator and a building system to be used, also making automatic approval feasible. Furthermore, local authorities are more motivated to keep in mind the best interests of the families and the city and could also incorporate the use of the configurator in post-occupancy social assistance within its current format.

To maximize the potential of such a mass customization approach to bring benefits over time, it is essential to consider the families' current post-occupancy processes. For example, distributing the costs of construction over time and being able to self-build to save even more money are important aspects to be maintained. This has direct implications regarding construction for the mass-customized product. For example, pre-fabricating entire rooms to be combined on-site would not be a feasible solution. Pre-fabricating panels that can be combined, allowing families to build at their own pace, as well as using materials that are familiar to them, would be more feasible. Having the local authority as the main agent would also allow several construction companies to be linked to the strategy to fabricate the necessary components.

Table 1. Adaptability guidelines.

Guideline from Brandão [18]	Proposed for Writing in Regulation	How It Could be Demonstrated
Set the height of the ridge suitable for expansions.	Set the height of the ridge at a height that allows adding rooms of at least 2 m in length, continuing along the slope without needing to change the angle of the roof.	Include, in architectural drawings, a section showing the added room with the expanded roof maintaining the same angle and complying with the minimum heights within the room. Example in Figure 8.
Allow the creation of new roof slopes without affecting the functionality.	Allow the creation of new roof slopes without affecting the functionality.	Include architectural drawings with the new roof slopes.
Separate, if possible, structure from walls.	Separate structure from walls, such that the walls are non-load-bearing.	Highlight, in the architectural drawings, the separation between load-bearing elements and non-load-bearing elements. Include the loads used to calculate the structure. Example in Figures 6–8.
Prepare structure to receive one or more floors.	Prepare structure to receive one or more floors.	Include, in the plans submitted for approval, as well as all the structural elements included to allow more floors, the loads used to calculate the structure and foundations.
Provide permanent walls with pipes.	Place walls with pipes in such a way that the kitchen and other rooms can be expanded and that other rooms can be added without needing to destroy such walls.	Include drawings of the expanded rooms around the walls with pipes and of added rooms showing that those walls do not need to be destroyed. Example in Figure 7.
Provide setback which allows expansion to the front.	Provide setback of at least 2 m.	Shown in drawings.
Adopt broader front for individual sites, if possible.	Adopt front for individual sites of at least 7 m.	Shown in drawings.

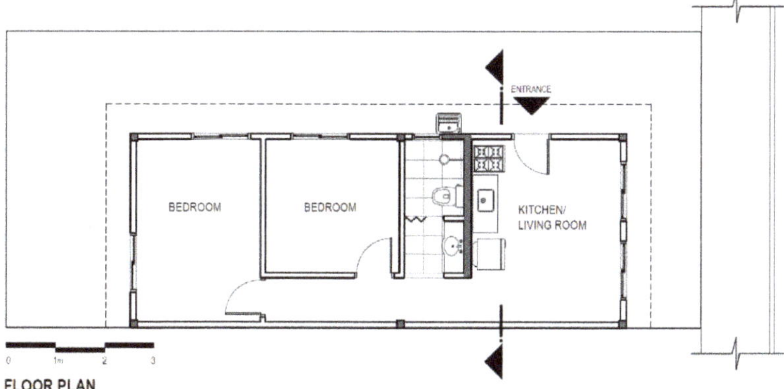

Figure 6. Floor plan of an example unit considering the guidelines for adaptability.

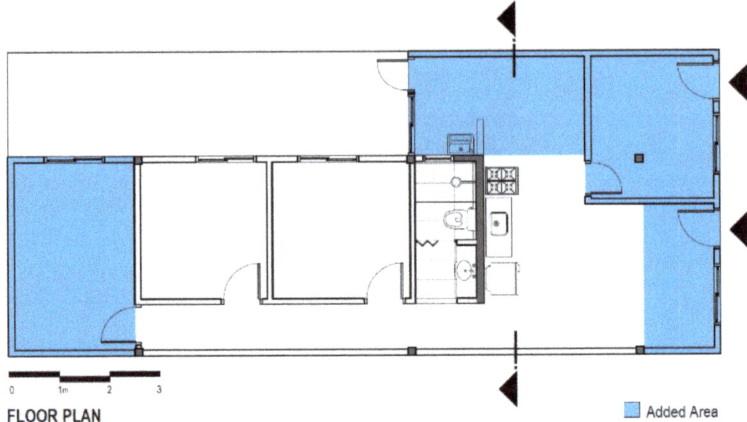

Figure 7. Floor plan of an example unit with expansions, adding several rooms without the need to destroy walls with pipes from the original unit.

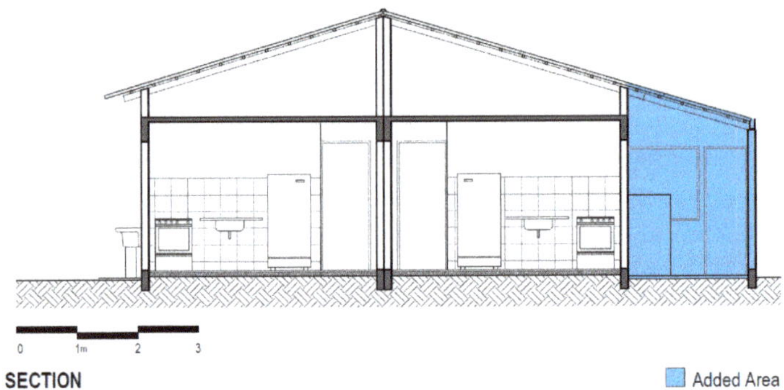

Figure 8. Section of two semi-detached example units. The unit on the right shows an expansion along the slope of the roof maintaining the same angle.

Although this study refers to a specific context, applying mass customization with post-occupancy differentiation could be valuable in many other housing contexts. The aspect of the initial unit not being seen as the final product could be easily extended to other housing contexts. In some cases, it may even be easier to overcome some of the challenges faced in the context considered for this study. In North America, for example, lighter building systems, such as wood-framing, are widely accepted. Furthermore, pre-fabrication within those systems is more widely available. As in some examples shown by Smith [39], a company may seek local pre-fabrication options close to each of the sites where they build to keep the costs low. This could be an opportunity for companies to provide an initial small affordable unit combined with a mass customization system of co-design and prefabricated parts or panels of the same building system that the user continues to engage (buy from the company in this case) over time at their own pace.

Aspects of dimensional or geometric co-design in housing are also relevant to discuss. It has been shown that an aspect that can add value to mass customized products refers to the satisfaction of the customer in perceiving themselves as the creator of the product, the "I designed it myself effect" [40]. However, design complexity and substantial investment in the product have shed doubt on whether this factor can be considered in housing. One of the reasons for dimensional mass customization of housing not being more widely available refers to the customers not having enough confidence or knowledge to take responsibility for the design of their homes; thus, it refers to social and cultural reasons and not technological limitations [5,41]. Many people would prefer to buy the house they perceived as professionally designed, even if they had the option of co-designing. Counteracting this is the need imposed by the context within the lowest income social housing scenario. Many post-occupancy studies [25,32,42] in this context have demonstrated the users' willingness to design their own homes, given that the alternative is not building. Thus, with the use of a co-design system, the resulting designs may still be perceived as more 'professionally designed' than the alternative.

Given the contextual conditions, these social housing neighborhoods could be the ideal place to begin the adoption of such geometric co-design processes, given the reduced emphasis of those social and cultural factors considered as significant challenges to the broader adoption of geometric mass customization in housing. Seeing such processes and their results within this context could be encouraging for developers to adopt geometric co-design in other contexts. Furthermore, it could also be encouraging for the users in overcoming some of their insecurities with what they may perceive as self-design.

6. Conclusions

A significant contribution of this study refers to the evaluation of the possibilities from mass customization in concert with the broader context and interests of the stakeholder in the provision of social housing for the lowest income range of housing programs in Brazil. Some previous studies have considered the benefits that mass customization could bring in the social housing context and indicated some paths for its implementation. However, most of these studies focus on only a few aspects of the provision process or do not refer to post-occupancy as part of the process and, thus, start from the premise that differentiation of the product must be for the initial unit. In contrast, this study argues that it would be more feasible and would bring more and longer-lasting benefits to the stakeholders involved if mass customization were applied with differentiation of the units happening post-occupancy.

Author Contributions: L.F.D.V. collected and analyzed the data and drafted the paper. B.K. provided guidance, editing, and reviewing for the paper. Both authors have read and agreed to the published version of the manuscript.

Funding: This research received no external funding.

Conflicts of Interest: The authors declare no conflict of interest.

References

1. Da Silveira, G.; Borenstein, D.; Fogliatto, F.S. Mass customization: Literature review and research directions. *Int. J. Prod. Econ.* **2001**, *72*, 1–13. [CrossRef]
2. Pine, B.J. *Mass Customization: The New Frontier in Business Competition*; Harvard Business School Press: Boston, MA, USA, 1993; ISBN 0875843727.
3. Tseng, M.M.; Jiao, J. Mass Customization. In *Handbook of Industrial Engineering*; Salvendy, G., Ed.; Wiley: Hoboken, NJ, USA, 2007; pp. 684–709. ISBN 9780470172339.
4. Rocha, C.; Formoso, C.; Tzortzopoulos, P. Adopting Product Modularity in House Building to Support Mass Customisation. *Sustainability* **2015**, *7*, 4919–4937. [CrossRef]
5. Kolarevic, B. Metadesigning Customizable Houses. In *Mass Customization and Design Democratization*; Kolarevic, B., Duarte, J.P., Eds.; Routledge: Abingdon-on-Thames, UK, 2019; ISBN 978-0-8153-6061-2.
6. Barlow, J.; Childerhouse, P.; Gann, D.; Hong-Minh, S.; Naim, M.; Ozaki, R. Choice and delivery in housebuilding: Lessons from Japan for UK housebuilders. *Build. Res. Inf.* **2003**, *31*, 134–145. [CrossRef]
7. Khalili-Araghi, S.; Kolarevic, B. Development of a framework for dimensional customization system: A novel method for customer participation. *J. Build. Eng.* **2016**, *5*, 231–238.
8. Lo, T.T.; Schnabel, M.A.; Gao, Y. ModRule: A User-Centric Mass Housing Design Platform. In *Computer-Aided Architectural Design Futures. The Next City—New Technologies and the Future of the Built Environment*; Celani, G., Sperling, D.M., Franco, J.M.S., Eds.; Springer: Berlin/Heidelberg, Germany, 2015; pp. 236–254.

9. Duarte, J.P. Customizing Mass Housing: A Discursive Grammar for Siza's Malagueira Houses. Ph.D. Thesis, Massachusetts Institute of Technology, Cambridge, MA, USA, 2001.
10. Benros, D.; Duarte, J.P. An integrated system for providing mass customized housing. *Autom. Constr.* **2009**, *18*, 310–320. [CrossRef]
11. Noguchi, M.; Hernandez-Velasco, C.R. A "Mass Custom Design" Approach to Upgrading Conventional Housing Development in Mexico. *Habitat Int.* **2005**, *29*, 325–336. [CrossRef]
12. Tillmann, P.A. *Diretrizes para a Adoção da Customização em Massa na Construção Habitacional para Baixa Renda*; Universidade Federal do Rio Grande do Sul: Porto Alegre, Brazil, 2008.
13. Taube, J. *Reflexões Sobre a Customização em Massa no Processo de Provisão de Habitações de Interesse Social: Estudo de caso na COHAB de Londrina-PR*; Universidade Estadual de Londrina: Londrina, Brazil, 2015.
14. Taube, J.; Hirota, E.H. Customização em massa no processo de provisão de Habitações de Interesse Social: Um estudo de caso. *Ambient. Construído* **2017**, *17*, 253–268. [CrossRef]
15. Azuma, M.H. *Customização em Massa de Projeto de Habitação de Interesse Social por Meio de Modelos Físicos Paramétricos*; Universidade de São Paulo: São Carlos, Brazil, 2016.
16. Ministério do Desenvolvimento Regional. Sistema de Gerenciamento da Habitação. 2019. Available online: http://sishab.cidades.gov.br/ (accessed on 16 February 2020).
17. Groat, L.N.; Wang, D. *Architectural Research Methods*, 2nd ed.; John Wiley & Sons: Hoboken, NJ, USA, 2013.
18. Brandão, D.Q. Disposições técnicas e diretrizes para projeto de habitações sociais evolutivas. *Ambient. Construído* **2011**, *11*, 73–96. [CrossRef]
19. Friedman, A. *The Adaptable House: Designing Homes for Change*; McGraw-Hill: New York, NY, USA, 2002; ISBN 0071377468.
20. Schneider, T.; Till, J. *Flexible Housing*; Architectural Press: Oxford, UK, 2007; ISBN 9780750682022.
21. NAURB-UFPel. Available online: https://wp.ufpel.edu.br/naurb/ (accessed on 11 August 2019).
22. Ministério da Cidadania. Cadastro Único. 2015. Available online: http://mds.gov.br/assuntos/cadastro-unico/o-que-e-e-para-que-serve (accessed on 13 January 2020).
23. Rufino, M.B.C. Um olhar sobre a produção do PMCMV a partir de eixos analíticos. In *Minha Casa e a Cidade? Avaliação do Programa Minha Casa Minha Vida em Seis Estados Brasileiros*; Amore, C.S., Shimbo, L.Z., Rufino, M.B.C., Eds.; Letra Capital: Rio de Janeiro, Brazil, 2015; pp. 51–70. ISBN 9788577853779.
24. Ribeiro, C.; Kruger, N.; Oliveira, T. A Cidade e a Moradia: O caso de Pelotas. *PIXO Rev. Arquitetura Cid. Contemp.* **2017**, *1*. [CrossRef]
25. Palermo, C. Avaliação da qualidade no projeto de HIS: Uma parceria com a Cohab/SC. In *Qualidade Ambiental na Habitação: Avaliação Pós-Ocupação*; Villa, S.B., Ornstein, S., Eds.; Oficina de Textos: São Paulo, Brazil, 2013.
26. *Ministério das Cidades Portaria N° 660*; Ministério das Cidades: Brasilia, Brazil, 2018.
27. Cardoso, A.L.; Mello, I.D.Q.; Jaenisch, S.T. A Implementação do Programa Minha Casa Minha Vida na Região Metropolitana do Rio de Janeiro: Agentes, Processos e Contradições. In *Minha Casa e a Cidade? Avaliação do Programa Minha Casa Minha Vida em Seis Estados Brasileiros*; Amore, C.S., Shimbo, L.Z., Rufino, M.B.C., Eds.; Letra Capital: Rio de Janeiro, Brazil, 2015.
28. *Ministério das Cidades Portaria N° 595*; Ministério das Cidades: Brasilia, Brazil, 2013.
29. *Ministério das Cidades Portaria N° 21*; Ministério das Cidades: Brasilia, Brazil, 2014.
30. Bonatto, F.S.; Miron, L.I.G.; Formoso, C.T. Avaliação de empreendimentos habitacionais de interesse social com base na hierarquia de valor percebido pelo usuário. *Ambient. Construído* **2011**, *11*, 67–83. [CrossRef]
31. Digiacomo, M.C. Estratégias de Projeto para a Habilitação Social Flexível. Master's Thesis, Universidade Federal de Santa Catarina, Florianópolis, Brazil, 2004.
32. de Jorge, L.O.; Medvedovsky, N.S.; Santos, S.; Junges, P.; da Silva, F.N. The spontaneous transformation of Anglo social housing complex in Pelotas/RS: Reflections about the urgency of the concept of Adaptable Social Housing. *Cad. Proarq* **2017**, 122–153. Available online: https://cadernos.proarq.fau.ufrj.br/public/docs/Proarq29%20ART%2007.pdf (accessed on 26 August 2019).
33. Marroquim, F.M.G.; Barbirato, G.M. Flexibilidade Espacial em Projetos de Habitações de Interesse Social. In *Colóquio de Pesquisas em Habitação*; EAUFMG: Belo Horizonte, Brazil, 2007.
34. Larcher, J.V.M. Diretrizes Visando a Melhoria de Projetos e Soluções Construtivas na Expansão de Habitações de Interesse Social. Master's Thesis, Universidade Federal do Paraná, Curitiba, Brazil, 2005.
35. Estevão, M.; Medvedovski, N.S. Entrevista com Mariana Estevão: A Prática da Arquitetura e Urbanismo com a Promoção à Saúde da População Brasileira. *Expressa Extensão* **2017**, *22*, 9–12.

36. *Lei N_0 11.888*; Casa Civil: Brasília, Brazil, 2008.
37. Sdino, L.; Castagnino, P. Housing Affordability Index: Real Estate Market and Housing Situations. In *New Metropolitan Perspetctives: The Integrated Approach of Urban Sustainable Development*; Bevilacqua, C., Calabrò, F., Della Spina, L., Eds.; Trans Tech Publications: Stafa, Switzerland, 2014; pp. 527–535.
38. Caixa Habitação Urbana—Minha Casa Minha Vida. 2019. Available online: http://www.caixa.gov.br/voce/habitacao/minha-casa-minha-vida/urbana/Paginas/default.aspx (accessed on 16 May 2019).
39. Smith, R.E. *Prefab Architecture: A Guide to Modular Design and Construction*; John Wiley & Sons: Hoboken, NJ, USA, 2010; ISBN 9780470275610.
40. Franke, N.; Schreier, M.; Kaiser, U. The "I Designed It Myself"; Effect in Mass Customization. *Manag. Sci.* **2010**, *56*, 125–140.
41. Kolarevic, B. From Mass Customisation to Design "Democratisation". *Archit. Des.* **2015**, *85*, 48–53. [CrossRef]
42. Merisio, B.G. *Modificação da Habitação: Uma Avaliação Pós-Ocupação no Conjunto Habitacional de Interesse Social Ewerton Montenegro Guimarães em Vila Velha/Es*; Universidade Vila Velha: Vila Velha, Brazil, 2016.

Article

Performance Evaluation of Photovoltaic/Thermal (PV/T) System Using Different Design Configurations

M. Imtiaz Hussain [1] and Jun-Tae Kim [2,*]

1 Green Energy Technology Research Center, Kongju National University, Cheonan 122324, Korea; imtiaz@kongju.ac.kr
2 Department of Architectural Engineering, Kongju National University, Cheonan 122324, Korea
* Correspondence: jtkim@kongju.ac.kr; Tel.: +82-41-521-9333

Received: 30 September 2020; Accepted: 14 November 2020; Published: 16 November 2020

Abstract: This study summarizes the performance of a photovoltaic/thermal (PV/T) system integrated with a glass-to-PV backsheet (PVF film-based backsheet) and glass-to-glass photovoltaic (PV) cells protections. A dual-fluid heat exchanger is used to cool the PV cells in which water and air are operated simultaneously. The proposed PV/T design brings about a higher electric output while producing sufficient thermal energy. A detailed numerical study was performed by calculating real-time heat transfer coefficients. Energy balance equations across the dual-fluid PV/T system were solved using an ordinary differential equation (ODE) solver in MATLAB software. The hourly and annual energy and exergy variations for both configurations were evaluated for Cheonan City, Korea. In the case of a PV/T system with a glass-to-glass configuration, a larger heat exchange area causes the extraction of extra solar heat from the PV cells and thus improving the overall efficiency of the energy transfer. Results depict that the annual electrical and total thermal efficiencies with a glass-to-glass configuration were found to be 14.31% and 52.22%, respectively, and with a glass-to-PV backsheet configuration, the aforementioned values reduced to 13.92% and 48.25%, respectively. It is also observed that, with the application of a dual-fluid heat exchanger, the temperature gradient across the PV panel is surprisingly reduced.

Keywords: PV/T system; dual-fluid; glass-to-glass; simulation; model validation

1. Introduction

The increasing demand for energy in day-to-day activities causes the excessive use of fossil fuels, which ultimately results in an increase in greenhouse gas emissions [1]. This is where the international organizations for climate control intervene to compel power generation companies to use sustainable energy sources instead. Due to this reason, the generation of electricity and heat from renewable energy sources has dramatically increased in the last decade [2]. It is anticipated that the use of renewable energy for residential, domestic, and commercial sectors will increase in near future. The photovoltaic (PV) module is a device that is used to convert sun rays directly into electricity, whereas both electricity and heat can be harvested using emerging technology known as photovoltaic/thermal (PV/T) technology [3]. The PV/T system can be non-concentrating (flat plate collectors) or concentrating and usually, flat plate collectors do not need a tracking platform to locate the sun position across the day [4].

A large portion of incident solar radiation is ultraviolet and infrared in nature, which leads to an increase in the operating temperature of PV solar cells [5]. Therefore, in the context of lowering PV module temperature and consequently increasing thermal efficiency, an optimized heat exchanger design and working fluid having superior thermal properties are very important. Water and air are the most common fluids that have been used as coolants for PV modules over the decades [6]. It has been observed that the PV/T system using water as a working fluid showed higher power conversion efficiency than that of air. Liquid fluids have always been the best choice to cool PV cells, rather than air, because of their excellent heat transfer capabilities. PV/T technology can also be categorized as glazed- and unglazed-PV/T systems [7,8]. The unglazed PV/T systems are comparatively inexpensive and considered best especially under high ambient temperature conditions compared to glazed PV/T systems. Furthermore, Vats et al. [9] analyzed the influence of packing factor on the overall energy performance of the semitransparent photovoltaic thermal system. For the comparison purpose, they have considered various types of solar cells with different packing factors, based on the results it was concluded that decreasing the packing factor decreases the PV module temperature and increases the sunlight transmission through the non-packing area.

When the quality of energy is a prime concern, the exergy analysis becomes as important as the energy analysis [10,11], especially for a co-generation system which is producing both electricity and heat simultaneously. The literature review revealed that several studies on the exergy analysis of various solar energy systems have been carried out with the intention of developing new methods and equations [12]. Pathak et al. [13] developed a theoretical exergy model to compare the performance of the PV/T system with conventional systems for a limited roof area. Based on climatic data from three different locations, the exergy performance of the PV/T, PV, and solar thermal systems having similar collector areas were predicted and compared. The outcomes from the comparative analysis show that the PV/T system surpassed the exergy efficiency of both PV and solar thermal systems for all locations.

Over the years, the simultaneous application of two fluids as the coolant in the PV/T systems has been gaining popularity among researchers. Tripanagnostopoulous [14] was the first who introduced the concept of utilizing two working fluids for the same PV/T collector. Using this concept, several studies on the dual-fluid PV/T system regarding performance optimization using different fluids and conduit designs have been published [15]. Jarimi et al. [16] developed a 2-D steady-state model of a bi-fluid PV/T system considering a slight modification in the finned air channel. The simulation-based results were validated using indoor experimental data. The introduction of the dual-fluid concept in PV/T technology for the cooling of solar cells is promising in terms of optimizing solar energy use, where a total equivalent efficiency near 90% is achievable [17]. Additionally, with a smaller area, the dual-fluid PV/T system can generate extra thermal energy.

Based on the literature review, it has been observed that several studies have been performed in the field of PV/T technology considering different aspects e.g., single- and dual-fluid channels for air circulation. In addition, many reported articles had discussed liquid fluid with different designs of tubes such as circular, rectangular, and trapezoidal, etc. To the best of our knowledge, no studies have been reported on the dual-fluid semitransparent PV/T system, in which a glass protection underneath the solar cells has been provided instead of a PV backsheet (PVF film-based backsheet). Due to the provision of dual-fluid coolant and glass-to-glass PV protection, additional solar heat from the PV module surface can be extracted, which will result in the lower temperature of the PV cells compared to the glass-to-PV backsheet based PV/T system. In the dual-fluid semitransparent PV/T system, the percentage of harvested energy per unit area is higher than the conventional glass cover PV/T system. It would therefore be an excellent choice to provide energy for the building and industrial sector.

2. Research Methods

2.1. Mathematical Model

In this study, the transient thermo-electric models for the glass-to-PV backsheet and glass-to-glass PV/T systems are developed and proposed. The transient mathematical model for the sheet-and-tube PV/T system reported by Chow [18] was modified and employed for the proposed design. The energy balance equations for various components were solved using an ODE solver in Matlab software. The following assumptions have been considered during mathematical modeling:

(1) There is no change in the physical dimensions and material properties of the collector components.
(2) For the parallel tube heat exchanger, temperature and flow rate in all tubes were taken as same.

(3) The ohmic losses in the PV cells and edge losses are neglected.
(4) All heat transfer coefficients were calculated in real-time [19].
(5) Only the absorption loss of glass is taken into consideration.
(6) For the glass-to-glass case, the glass cover2 serves as a sheet for the copper tube (carrying water), while for the glass-to-PV backsheet case, the PV backsheet works as a sheet for the copper tube considering famous sheet and tube configuration.

Energy balances for glass-to-glass and glass-to-PV backsheet cases
Glass cover1

$$M_{g1}C_{g1}(dT_{g1}/dt) = G\alpha_1 + h_{pg1}A_{pg1}\left(T_s - T_{g1}\right) - h_{wind}A_{g1\infty}\left(T_{g1} - T_\infty\right) - h_{g1\infty}A_{g1\infty}\left(T_{g1} - T_\infty\right) \quad (1)$$

PV cells (glass-to-PV backsheet)

$$M_sC_s(dT_s/dt) = G\alpha_2 - E - h_{sg1}A_{sg1}\left(T_s - T_{g1}\right) - h_{sp}A_{sp}\left(T_s - T_p\right) \quad (2)$$

PV cells (glass-to-glass)

$$M_sC_s(dT_s/dt) = G\alpha_2 - E - h_{sg1}A_{sg1}\left(T_s - T_{g1}\right) - h_{sg2}A_{sg2}\left(T_s - T_{g2}\right) \quad (3)$$

PV backsheet

$$M_pC_p(dT_p/dt) = G\alpha_3 + h_{sp}A_{sp}\left(T_s - T_p\right) - h_{pt}A_{pt}\left(T_p - T_t\right) - A_{pa}\, h_{pa}\left(T_p - T_a\right) - h_{pb}A_{pb}\left(T_p - T_b\right) \quad (4)$$

Glass cover2

$$M_{g2}C_{g2}(dT_{g2}/dt) = G\alpha_4 + h_{sg2}A_{sg2}\left(T_s - T_{g2}\right) - h_{tg2}A_{tg2}\left(T_{g2} - T_t\right) - A_{ag2}\, h_{ag2}\left(T_{g2} - T_a\right) - h_{bg2}A_{bg2}\left(T_{g2} - T_b\right) \quad (5)$$

where G is the irradiance and E is the electrical power generated by the PV cells. h_{wind} is the convection heat transfer coefficient due to wind [20]. $h_{g\infty}$ is the radiation heat transfer coefficient between glass cover1 and ambient air, and h_{sg1} is the conduction heat transfer coefficient between glass cover1 and the PV cells. h_{pt} is the conduction heat transfer coefficient between PV cells and the tube, h_{pa} is the convection heat transfer coefficient between PV cells and the circulating air, h_{pb} is the radiation heat transfer coefficient between the PV cells and the back panel, and h_{sp} is the conduction heat transfer coefficient between PV cells and the tube. h_{sg2} is the conduction heat transfer coefficient between glass cover2 and PV cells. h_{tg2} is the conduction heat transfer coefficient between PV cells and the tube, and h_{ag2} is the convection heat transfer coefficient between glass cover2 and the circulating air.

$$E = GPF\eta_e \quad (6)$$

$$\eta_e = \eta_r\left[1 - \beta_r\left(T_p - T_r\right)\right] \quad (7)$$

where PF is the packing factor and T_p is the PV plate temperature. β_r is the temperature coefficient and η_e is the efficiency of the solar cells. T_r and η_r are the reference cell temperature and efficiency, respectively.

$$h_{wind} = 3u_a + 2.8 \quad (8)$$

$$h_{g\infty} = \varepsilon_g\sigma\left(T_g + T_\infty\right)\left(T_g^2 + T_\infty^2\right) \quad (9)$$

$$h_{pb} = \left(\sigma\left(T_p + T_b\right)\left(T_p^2 + T_b^2\right)\right)/\left(1/\varepsilon_p + 1/\varepsilon_b - 1\right) \quad (10)$$

where ε_g is the emissivity of the glass and u_a is the velocity caused by wind. ε_p and ε_b are the emissivity of the PV plate and back panel, respectively.

Back panel (glass-to-glass)

$$M_bC_b(dT_b/dt) = G\alpha_5 + h_{tb}A_{tb}(T_t - T_b) + h_{bg2}A_{bg2}\left(T_{g2} - T_b\right) - h_{ab}A_{ab}(T_b - T_a) - h_{b\infty}A_{b\infty}(T_b - T_\infty) \quad (11)$$

Back panel (glass-to-PV backsheet)

$$M_bC_b(dT_b/dt) = h_{tb}A_{tb}(T_t - T_b) + h_{pb}A_{pb}\left(T_p - T_b\right) - h_{ab}A_{ab}(T_b - T_a) - h_{b\infty}A_{b\infty}(T_b - T_\infty) \quad (12)$$

where h_{tb} is the radiation heat transfer coefficient between the tube and back panel, and h_{ab} is the convection heat transfer coefficient between the back panel and the circulating air. $h_{b\infty}$ is the heat loss coefficient between the back panel and ambient air. h_{bg2} is the radiation heat transfer coefficient between the glass cover2 and the back panel.

Considering only absorption losses, solar radiation absorbed across the PV/T system components is defined as follows:

$$\alpha_1 = \alpha_{g1} \quad (13)$$

$$\alpha_2 = \left(1 - \alpha_{g1}\right)A_s\, \alpha_s \quad (14)$$

$$\alpha_3 = \left(1 - \alpha_{g1}\right)(1 - \alpha_s)(1 - A_R)\, \alpha_p \quad (15)$$

$$\alpha_4 = \left(1 - \alpha_{g1}\right)(1 - \alpha_s)(1 - A_R)\, \alpha_{g2} \quad (16)$$

$$\alpha_5 = \left(1 - \alpha_{g1}\right)\left(1 - \alpha_{g2}\right)(1 - \alpha_s)(1 - A_R)\, \alpha_b \quad (17)$$

where A_s is the area covered by solar cells and α_s are the absorptivity of the PV cells. A_R is the ratio of area covered by the collector to the PV cells [21]. α_p and α_b are the absorptivity of the PV backsheet and back panel, respectively. In glass-to-glass case, α_{g1} and α_{g2} are the absorptivity of the glass cover1 and glass cover2, respectively.

Tube

$$M_tC_t(dT_t/dt) = h_{pt}A_{pt}\left(T_p - T_t\right) - A_{tf}\, h_{tf}(T_t - T_f) - A_{ta}\, h_{ta}(T_t - T_a) - h_{tb}A_{tb}(T_t - T_b) \quad (18)$$

$$M_tC_t(dT_t/dt) = h_{tg2}A_{tg2}\left(T_{g2} - T_t\right) - A_{tf}\, h_{tf}(T_t - T_f) - A_{ta}\, h_{ta}(T_t - T_a) - h_{tb}A_{tb}(T_t - T_b) \quad (19)$$

h_{tf} is the convection heat transfer coefficient between the tube and the fluid and h_{ta} is the convection heat transfer coefficient between the tube and the circulating air.

Circulating water

$$M_fC_f(dT_f/dt) = \dot{m}_fC_f\,(T_{f,o} - T_{f,in}) + A_{tf}\, h_{tf}(T_t - T_f) \quad (20)$$

$\dot{m}_f$ and C_f are the mass flow rate and specific heat of the pipe fluid. $T_{f,in}$ and $T_{f,o}$ are the inlet and outlet temperature of the pipe fluid. As reported by, the average convective heat transfer coefficient is important because it considers both convection modes (natural and forced) and entrance effects. It can be shown that

$$\frac{1}{h_{tf}A_{tf}} = \frac{1}{h_f\pi D_i L} + \frac{1}{C_{bo}L} \quad (21)$$

h_f is a convective heat transfer, for a fully developed laminar flow which can be given as:

$$h_f = 4.364\frac{k_f}{D_i} \quad (22)$$

for a fully developed turbulent flow, which can be obtained from the Dittus–Boelter equation [22], k_f is the thermal conductivity of the water and D_i is the internal diameter of the tube.

$$Nu_D = 0.023Re_D^{0.8}Pr^{0.4} \quad (23)$$

Nu, Re, and Pr are Nusselt, Reynolds, and Prantle numbers, respectively.

$$C_{bo} = \frac{k_{bo}\,W_{bo}}{\delta_{bo}} \tag{24}$$

where C_{bo} is the bond conductance and k_{bo} is the thermal conductivity of the bond or adhesive. W_{bo} and δ_{bo} are the width and thickness of the bond.

Circulating air

$$M_aC_a(dT_a/dt) = \dot{m}_aC_a\,(T_{a,o} - T_{a,in}) + A_{pa}\,h_{pa}\left(T_p - T_a\right) + A_{ta}\,h_{ta}(T_t - T_a) + h_{ab}A_{ab}(T_a - T_b) \tag{25}$$

$\dot{m}_a$ and C_a are the mass flow rate and specific heat of the circulating air. $T_{f,in}$ and $T_{f,o}$ are the inlet and outlet temperature of the circulating air. The useful thermal energy and efficiency of the dual-fluid PV/T system are given as follows:

$$Q_u = \dot{m}_fC_f\,(T_{f,o} - T_{f,in}) + \dot{m}_aC_a\,(T_{a,o} - T_{a,in}) \tag{26}$$

$$\eta_{th} = \frac{Q_u}{A_cG} \tag{27}$$

where Q_u and η_{th} are the useful thermal energy and efficiency of the dual-fluid PV/T system. The equivalent thermal efficiency can be calculated as:

$$\eta_{PVT} = \eta_{th} + \eta_e/\eta_{pp} \tag{28}$$

η_{PVT} and η_e are the equivalent thermal and electrical efficiencies, respectively. η_{pp} is the electric generation efficiency of the conventional power plant and its value is taken as 38%.

2.2. Exergy Analysis

Exergy is a thermodynamic concept which defines every transformation process that undergoes the loss of a measure of quality, especially considering low-quality energy such as thermal energy (heat) which involves temperature change. Exergy analysis becomes more important when the extraction of the maximum useful work from the system is concerned. The exergy balance for the single-fluid PV/T system given by Agrawal and Tiwari [23], is modified for the dual-fluid PV/T system for this study. The following equations show the inflow and outflow of exergy from the proposed system [24].

$$\sum Ex_o = \sum E_{th} + \sum Ex_e \tag{29}$$

Ex_o is the overall exergy gain, and E_{th} and Ex_e are the thermal and electrical exergy gains, respectively. For a dual-fluid PV/T system, the thermal exergy gain is the sum of thermal exergy against associated with circulating pipe fluid ($E_{th,f}$) and air ($E_{th,a}$), respectively, can be expressed as follows:

$$\sum E_{th} = \sum E_{th,f} + \sum E_{th,a} \tag{30}$$

$$\sum E_{th,f} = Q_f - \dot{m}_fC_f\,(T_\infty + 273)\log\left(\frac{T_{f,o} + 273}{T_{f,in} + 273}\right) \tag{31}$$

$$\sum E_{th,a} = Q_a - \dot{m}_aC_a\,(T_\infty + 273)\log\left(\frac{T_{a,o} + 273}{T_{a,in} + 273}\right) \tag{32}$$

Q_f and Q_a are the useful thermal gain associated with circulating pipe fluid and air, respectively.

$$\sum Ex_e = \left[\frac{\eta_eGA_c}{1000}\right] \tag{33}$$

$$\mathrm{Ex_{in}} = 0.933 * \mathrm{G} * \mathrm{A_c} \quad (34)$$

$$\eta_{\mathrm{Ex}} = \left[\frac{\mathrm{Ex_o}}{\mathrm{Ex_{in}}}\right] * 100 \quad (35)$$

η_{Ex} is the exergy efficiency, Ex_{in} and Ex_o are exergy input and output to the system, respectively. The exergy inflow is dependent on the available solar radiation and the exergy outflow is associated with the thermal output. Therefore, exergy efficiency is more related to outlet temperature. This means that the higher the power output from the PV/T system lower is the entropy generation rate.

2.3. Description of Proposed PV/T Systems

The schematic and cross-section views of glass-to-PV backsheet and glass-to-glass dual-fluid PV/T systems are shown in Figure 1. In the glass-to-PV backsheet case, the PV cells are sandwiched between the glass cover and PV backsheet, whereas in a glass-to-glass case, PV cells are sandwiched between two glass covers. The solar cells are placed at an equal distance across the collector area, such as the distance between two neighboring solar cells, which was maintained by 20 mm. The PV/T system is comprised of two heat exchangers such as parallel arranged tubes to carry water as coolant and an underneath channel for air circulation. A set of baffles was arranged transverse to airflow on the channel surface with the intention to enhance turbulence and to diminish streamline flow. The tubes carrying water coolant were made of copper and the back panel or air channel was made of chlorinated polyvinyl chloride (CPVC). In order to increase the emissivity and heat transfer rate, both the air channel and copper tube were painted jet black. Both glass-to-PV backsheet and glass-to-glass cases had identical physical dimensions and were analyzed under similar operating conditions. Details of components dimensions and other parameters have been shown in Table 1.

Table 1. Parameters details.

PV cells [17]	Length & width	1.62 m & 0.98 m
	Absorptivity	0.9
	Emissivity	0.88
	Specific heat	900 J/(kg K)
	Temperature coefficient	0.0045/°C
	Reference PV panel temperature	298.15 K
	Thickness of EVA+PV cells	1.2 mm
	Thermal conductivity	148 W/(m K)
Glass cover	Glass solar transmittance	92%
	Thickness of tempered glass	3 mm
	Specific heat	670 (J/kg)
	Density	2200 (kg/m^3)
	Extinction coefficient	26 (/m)
PV backsheet	Thickness of PV backsheet	0.5 mm
	Thermal conductivity	0.2 W/(m K)
	Absorptivity of PV backsheet	0.5
Copper tube	Inner diameter	0.008 m
	Thickness	0.0012 m
	Specific heat	903 J/(kg K)
	Density	2702 kg/m^3
	No. of tubes	9
	Tube spacing	0.11 m
	Material	Copper
Back panel	Density	1520 kg/m^3
	Specific heat	840 J/(kg K)
	Thermal conductivity	0.134 W/(m K)
	Thickness of back panel	4 mm
Fluids used	Water & air	-

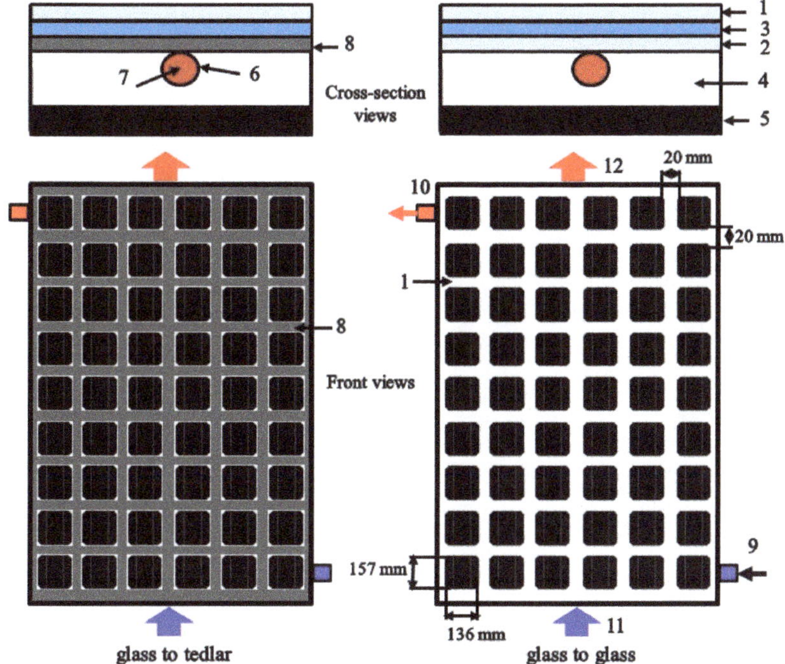

Figure 1. Schematic of dual-fluid photovoltaic/thermal (PV/T) system with glass-to-glass and glass-to-PV backsheet cases.

2.4. Model Validation

The proposed mathematical model of the PV/T system has been validated using solely an air type heat exchanger. The selection of the air type heat exchanger can be explained by the fact that from the previously published studies, the authors found only single-fluid PV/T systems that had used glass-to-glass PV protection. For the purpose of model validation, identical physical dimensions and operating conditions have been used in the mathematical model as presented by Joshi et al. [25]. Figure 2 shows the PV temperatures derived from the proposed mathematical model and measured by Joshi et al. [25]. The depicted measured and predicted PV temperatures varied in accordance with the variable solar radiation reported by Joshi. It is obvious the PV temperature varied directly with the incident solar insolation, but the important point is numerical findings have good agreement with experimental data. In fact, the maximum difference between numerical and measured data is within an acceptable range. It can be deduced from the aforementioned comparison that the proposed model of the PV/T system can be employed for the performance prediction of a physical counterpart.

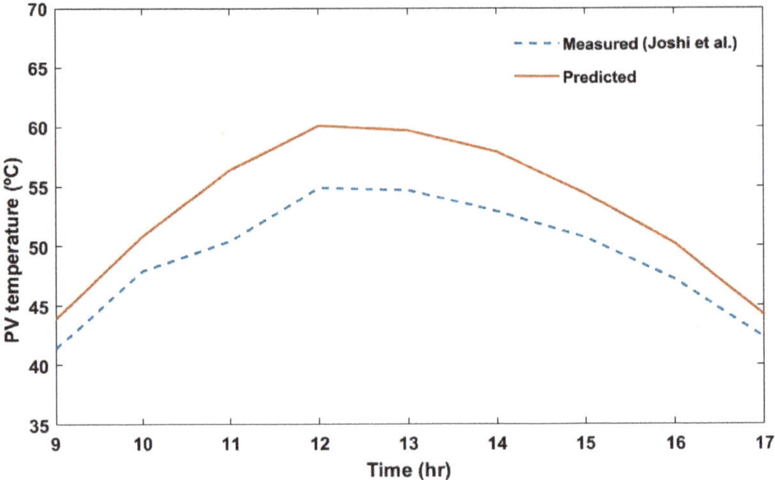

Figure 2. Comparison of numerical and measured PV temperatures.

3. Results and Discussion

It is important to note that during analysis, fixed flow rates for both fluids were used, i.e., 0.024 kg/s and 0.042 kg/s for water and air, respectively. The daily variations of solar radiation and ambient temperature are shown in Figure 3. The interdependence temperature responses of the top glass cover, PV plate, copper tube, and back panel are shown in Figure 4. It can be seen that the variation of temperatures for the PV cells and the copper tube layers are very similar for the glass-to-glass PV/T system, which means there is excellent heat transfer between the aforementioned components. On the contrary, in the glass-to-PV backsheet based PV/T system, the incident solar radiation is trapped in the PV cells which cause a significant increase in its surface temperature. This shows that as the temperature went up, the PV cells lost heat to ambient air at a faster rate than the heat transfer to the copper tube. In addition, compared to the glass-to-PV backsheet case, the higher back panel temperature in the case of the glass-to-glass PV/T system is due to direct solar heating through the non-packing area.

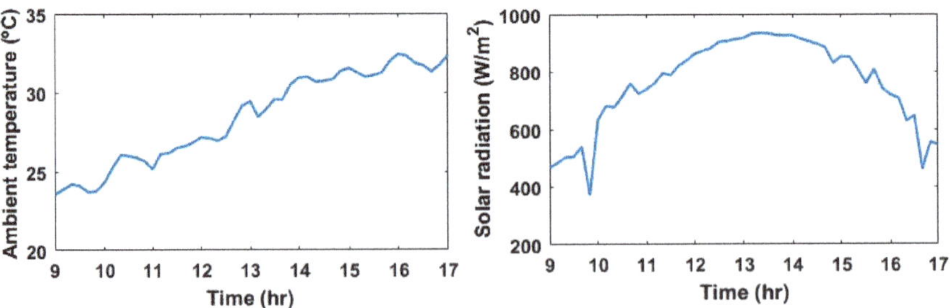

Figure 3. The climatic parameters.

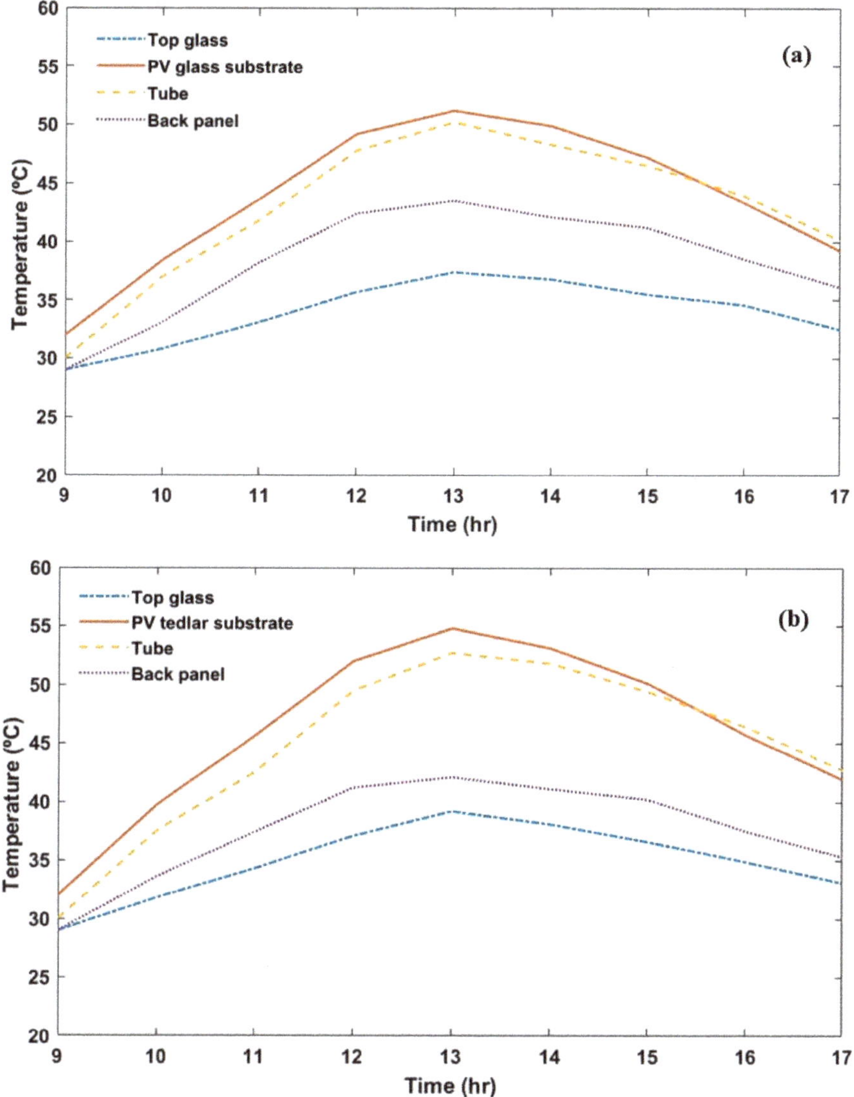

Figure 4. Different layers temperatures across (**a**) glass-to-glass PV/T system (**b**) glass-to-PV backsheet based PV/T system.

The hourly variations of electrical efficiencies from glass-to-glass and glass-to-PV backsheet based PV/T systems are shown in Figure 5. The electrical efficiencies for both systems varied inversely to ambient temperature which is obvious. However, the electrical efficiency for the glass-to-glass PV module is significantly higher than that of the glass-to-PV backsheet case. Installation of the dual-fluid heat exchanger decreases the operating temperature of PV cells and increases the short circuit current and open-circuit voltage, and ultimately the enhancement in electrical efficiency was observed. The value of the average electrical efficiency for the glass-to-glass PV/T system is found to be 15.34%, whereas for the glass-to-PV backsheet case this value reduced to 14.85%. This can be explained by the fact that, due to the opaque nature of the PV backsheet, all of the incident solar radiation is intercepted by PV cells and the PV backsheet surface, which results in the generation of extra heat and hence reduction in electrical performance is observed.

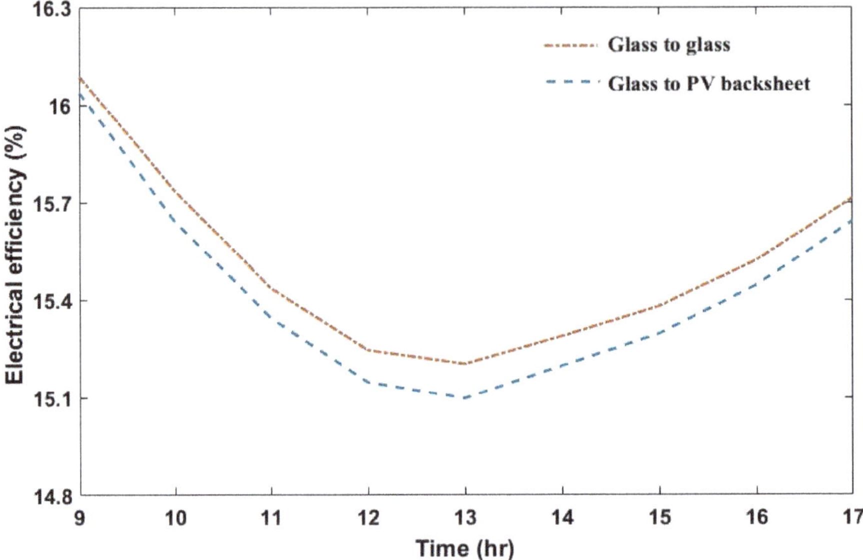

Figure 5. Variations of electrical efficiency against day hours.

Equivalent thermal efficiency terms are used to define the thermal performance of glass-to-glass and glass-to-PV backsheet based PV/T systems. Using dual-fluid as a coolant, the daily variations of equivalent thermal efficiency for both cases are presented in Figure 6. Under similar operating conditions, the daily average equivalent thermal efficiency for glass-to-glass and glass-to-PV backsheet cases are 81.06% and 78.86%, respectively. Whereas, the glass-to-glass PV/T system gives better results compared to the glass-to-PV backsheet case. This may be due to the accumulation or trapping of sun rays at the PV cells and the PV backsheet surfaces. Furthermore, daily useful thermal energy gains for both cases are depicted in Figure 7. The net heat gain depends on ambient temperature; the higher the temperature difference between PV cells and ambient air, the higher the heat losses. The glass-to-glass PV/T system has a maximum useful energy gain of a daily average value of 0.541 kWh, whereas the energy gain for the glass-to-PV backsheet case is 0.422 kWh. In the context of thermal performance, the glass-to-glass-based PV/T system supersedes the glass-to-PV backsheet case due to high heat extraction capacity. Moreover, in the glass-to-glass case, the black painted back panel gets heated directly from sun rays transmitting through the non-packing area of glass and also through conducted heat from the PV cells.

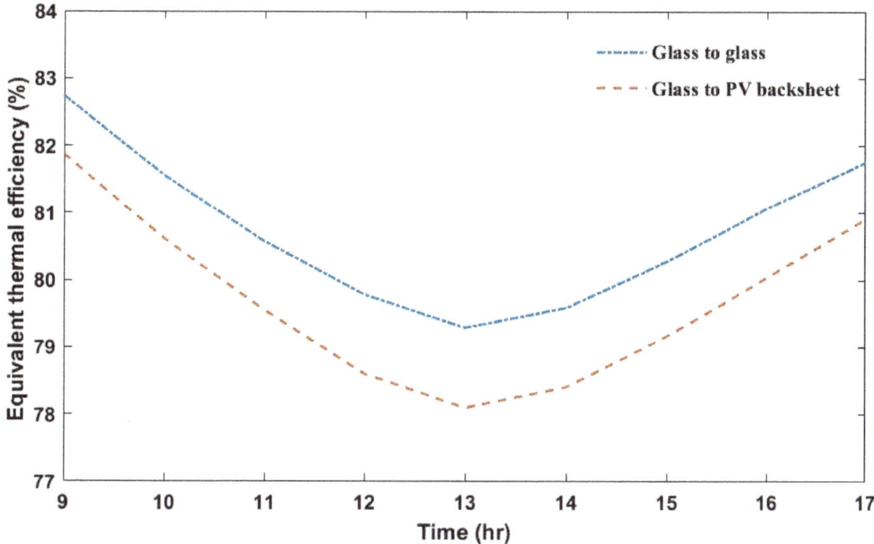

Figure 6. Variations of equivalent thermal efficiency against day hours.

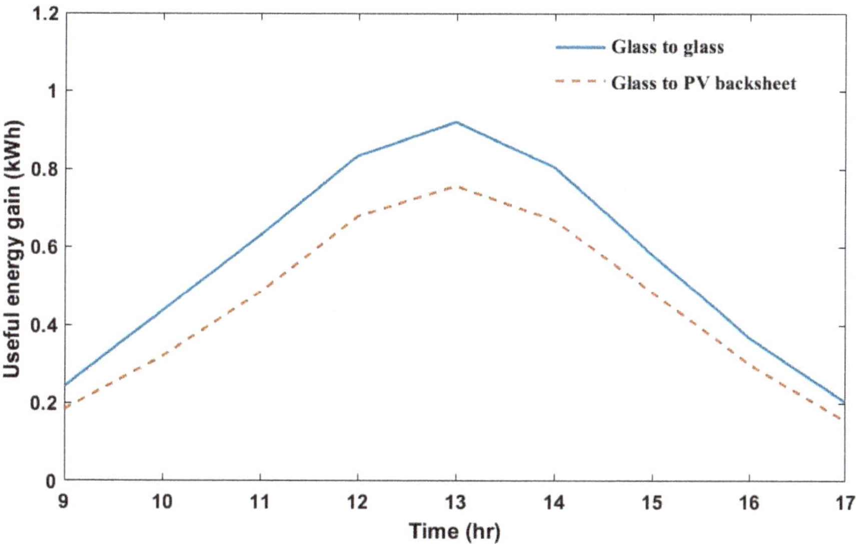

Figure 7. Variations of useful energy gain against day hours.

For the purpose of evaluation and optimization, the exergy analysis is also taken into consideration, which can provide detailed insight into the process for possible improvement in the performance of the dual-fluid PV/T system. In other words, the exergy analysis gives an idea about the maximum possible output that is achievable from the proposed PV/T system. Figure 8 shows the variations of overall exergy efficiency with respect to day time. It can be seen that the exergy efficiency varies linearly with the daily sunlight and depicted the maximum value for both cases during the peak sun intensity hours. The maximum exergy efficiencies for the glass-to-glass case and glass-to-PV backsheet case are 14.25% and 13.87%, respectively. It is observed that the exergy efficiency for the glass-to-glass case is higher than that of the glass-to-PV backsheet case. It can be explained by the fact that the maximum achievable power output or exergy rate from a solar collector varies inversely with the entropy generation rate or irreversibility. As the intensity of solar radiation increases, the PV cell temperature increases. Thereby, the trapped heat in the PV cells accelerates the heat losses to ambient or irreversibility. In glass-to-glass PV protection, the rate of heat extraction by the circulating fluid from the PV cells increased, which ultimately causes a reduction in heat losses.

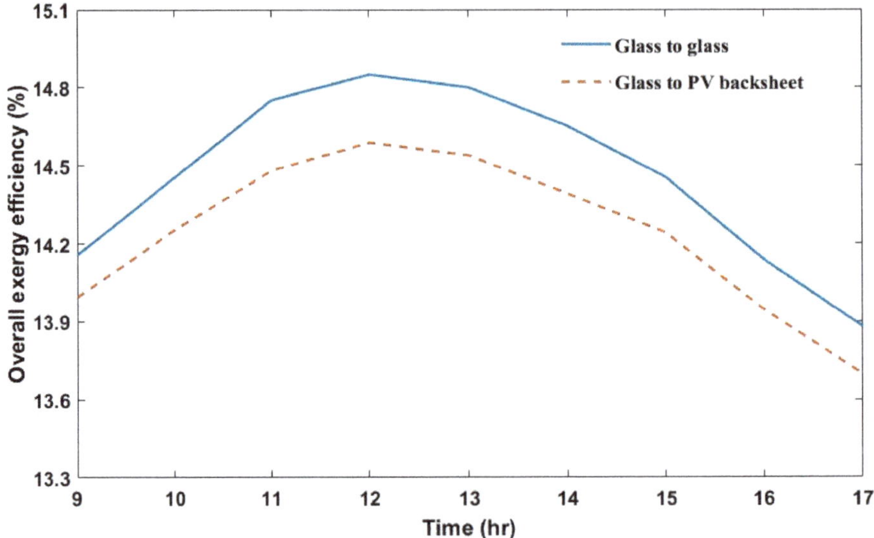

Figure 8. Variations of overall exergy efficiency against day hours.

The long-term performance evaluation of a dual-fluid PV/T system is performed by taking into consideration the monthly average solar radiation and ambient temperature. Figures 9 and 10 show the variation trends of monthly average electrical and thermal efficiencies for both cases across the whole year. The maximum electrical efficiency for glass-to-PV backsheet and glass-to-glass cases are observed in March with values of 13.92% and 14.31%, respectively, whereas in July these values were reduced to a minimum level of 11.87% and 12.18%, respectively. The yearly average total thermal efficiency for glass-to-PV backsheet and glass-to-glass cases are observed to be 48.25% and 52.22%, respectively. Apart from different configurations, both cases produced reasonably good thermal efficiency in comparison with conventional single-fluid exchangers. However, due to direct sun rays transmission in the glass-to-glass PV/T system case, the blackened back panel was heated continuously by the incident solar radiation. Therefore, in the glass-to-glass case, the circulating fluids have a higher temperature and thermal efficiency than that of the glass-to-PV backsheet case. It can be noticed that the maximum overall efficiencies (electrical plus thermal) for both cases were observed in the spring months (March and April). This trend can easily be explained by a higher number of sunshine hours and lower ambient air temperatures.

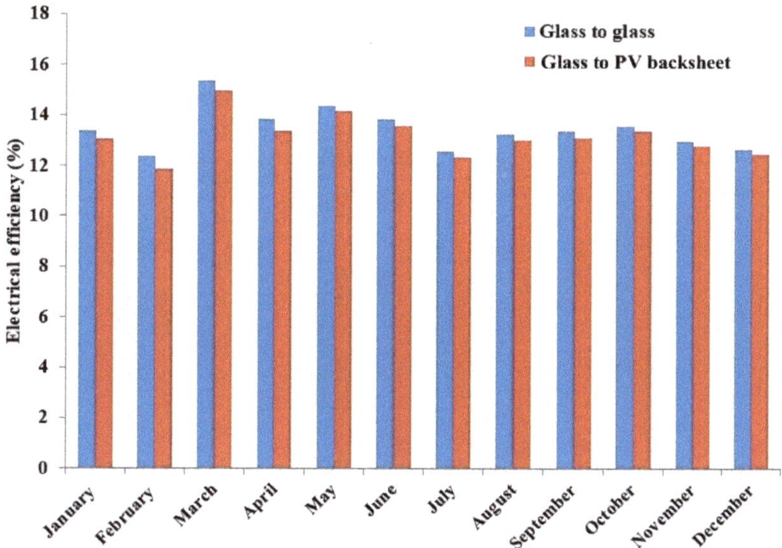

Figure 9. Yearly variations of electrical efficiency of PV/T system with glass-to-glass and glass-to-PV backsheet cases.

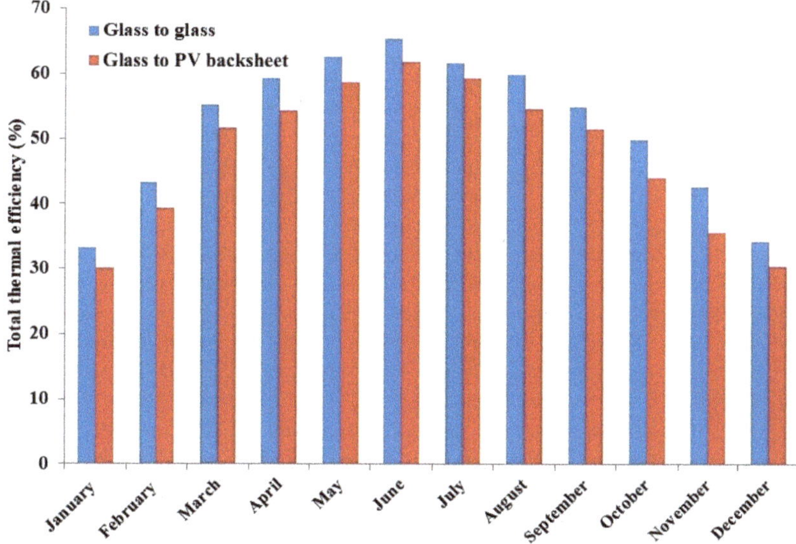

Figure 10. Yearly variations of total thermal efficiency of PV/T system with glass-to-glass and glass-to-PV backsheet cases.

Considering average weather conditions, the yearly (breakdown into months) variations of overall exergy efficiency for both cases are presented in Figure 11. The yearly average exergy efficiency for glass-to-PV backsheet and glass-to-glass cases are 13.23% and 13.85%, respectively. Since the electrical outputs from both PV/T configurations are in the form of exergy energy, therefore, the electrical part is more related to it than the thermal part. Due to this reason, the overall exergy efficiency variation pattern is similar to that of the electrical energy. Furthermore, from the derived results, it can clearly be seen that the glass-to-glass case has higher exergy efficiency than the glass-to-PV backsheet cases.

This is because, due to better heat extraction capabilities, the glass-to-glass case has a lower operating temperature of the PV cells than the latter case. To sum up, the lower the PV cell temperature, the higher the overall exergy efficiency.

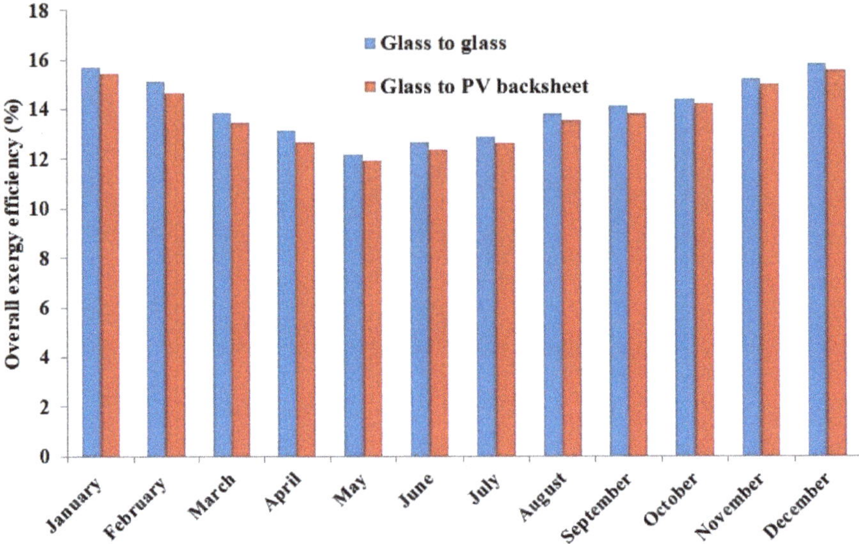

Figure 11. Yearly variations of Overall exergy efficiency of PV/T system with glass-to-glass and glass-to-PV backsheet cases.

4. Conclusions

This study compared two configurations of the PV/T system in the context of evaluating their electrical and thermal performances. It is concluded that a glass-to-glass PV/T system is a better design compared to a glass-to-PV backsheet based PV/T system. The integration of glass-to-glass PV protection with a dual-fluid heat exchanger helps to minimize the PV cells' temperature and consequently, increases the exergy and thermal efficiencies. It is observed that under similar conditions the average electrical efficiency of the glass-to-glass and glass-to-PV backsheet based PV/T systems are 15.34% and 14.85%, respectively. There is an improvement in equivalent thermal efficiency by 2.2% for a glass-to-glass case compared to a glass-to-PV backsheet case. The average useful energy outputs for glass-to-glass and glass-to-PV backsheet based PV/T systems are 0.541 kWh and 0.422 kWh, respectively, whereas yearly average total thermal efficiencies are 52.22% and 48.25%, respectively. The presented transient mathematical model is capable of providing a real-time simulation of the PV/T system similar to what a physical counterpart would. Using glass-to-glass PV protection, the circulated fluid can get direct and indirect solar heat. Additionally, a dual-fluid heat exchanger helps in optimizing the thermal output from the PV/T system, where either fluid can be used according to load requirements. In future studies, the thermal and optical models will be coupled to analyze the performance of a given PV/T system by introducing glazing. The main advantage is, with a smaller area, the suggested system can generate high-temperature heat, and the limitation is the integration of a dual-fluid heat exchanger and additional glass cover in a PV/T unit might cause extra production cost.

Author Contributions: M.I.H.—conception, design, analysis, interpretation of data and the drafting the work; J.-T.K. supported through reviewing the work critically, supervision, and the funding to the research work and its final formatting as an article in its current form. All authors have read and agreed to the published version of the manuscript.

Funding: This work was supported by Korea Research Fellowship Program through the National Research Foundation of Korea (NRF) funded by the Ministry of Science and ICT (2016H1D3A1938222) and The Korea Institute of Energy Technology Evaluation and Planning (KETEP) and the Ministry of Trade, Industry & Energy (MOTIE) of the Republic of Korea (No. 20188550000480).

Conflicts of Interest: The authors declare no conflict of interest. The funders had no role in the design of the study; in the collection, analyses, or interpretation of data; in the writing of the manuscript, or in the decision to publish the results.

Nomenclature

A	surface area (m^2)
C	specific heat (J/kg °C)
E	electrical power (W)
Ex	exergy rate
Ex_o	overall exergy gain
G	solar radiation (W/m^2)
h	heat transfer coefficient (W/m^2 °C)
h_{wind}	wind velocity (W/m^2 °C)
k	thermal conductivity (W/m °C)
M	mass (kg)
$\dot{m}$	mass flow rate (kg/s)
PF	packing factor
Q	energy (W)
Q_u	useful energy gain (W)
T	temperature (°C)
D_i&D_o	tube inner & outer diameters
Nu	Nusselt number
Re	Reynolds number
Pr	Prandtl number
Greek	
α	absorptivity
τ	transmissivity
δ	thickness (m)
σ	stefan-boltzmann constant ($W{\cdot}m^{-2}{\cdot}K^{-4}$)
η	efficiency
ε	emissivity
η_{PVT}	primary energy saving efficiency
Subscripts	
a	circulating air
b	back panel
bo	bond or adhesive
c	collector
e	electrical
f	circulating water
$g1$	glass cover1
$g2$	glass cover2
o & in	outlet & inlet
p	PV backsheet or PVF film-based backsheet
pp	power plant
s	solar cells
t	tube
th	thermal
∞	ambient air

References

1. Jewell, J.; McCollum, D.; Emmerling, J.; Bertram, C.; Gernaat, D.E.; Krey, V.; Paroussos, L.; Berger, L.; Fragkiadakis, K.; Keppo, I. Limited emission reductions from fuel subsidy removal except in energy-exporting regions. *Nature* **2018**, *554*, 229. [CrossRef] [PubMed]
2. Perea-Moreno, M.-A.; Hernandez-Escobedo, Q.; Perea-Moreno, A.-J. Renewable Energy in Urban Areas: Worldwide Research Trends. *Energies* **2018**, *11*, 577. [CrossRef]
3. Chow, T.T. A review on photovoltaic/thermal hybrid solar technology. *Appl. Energy* **2010**, *87*, 365–379. [CrossRef]
4. Charalambous, P.; Maidment, G.; Kalogirou, S.; Yiakoumetti, K. Photovoltaic thermal (PV/T) collectors: A review. *Appl. Therm. Eng.* **2007**, *27*, 275–286. [CrossRef]
5. Taylor, R.A.; Phelan, P.E.; Otanicar, T.P.; Adrian, R.; Prasher, R. Nanofluid optical property characterization: Towards efficient direct absorption solar collectors. *Nanoscale Res. Lett.* **2011**, *6*, 1–11. [CrossRef]
6. Hussain, M.I.; Ménézo, C.; Kim, J.-T. Advances in solar thermal harvesting technology based on surface solar absorption collectors: A review. *Sol. Energy Mater. Sol. Cells* **2018**, *187*, 123–139. [CrossRef]
7. Bhattarai, S.; Oh, J.-H.; Euh, S.-H.; Krishna Kafle, G.; Hyun Kim, D. Simulation and model validation of sheet and tube type photovoltaic thermal solar system and conventional solar collecting system in transient states. *Sol. Energy Mater. Sol. Cells* **2012**, *103*, 184–193. [CrossRef]
8. Rommel, M.; Zenhäusern, D.; Baggenstos, A.; Türk, O.; Brunold, S. Development of glazed and unglazed PVT collectors and first results of their application in different projects. *Energy Procedia* **2015**, *70*, 318–323. [CrossRef]
9. Vats, K.; Tomar, V.; Tiwari, G.N. Effect of packing factor on the performance of a building integrated semitransparent photovoltaic thermal (BISPVT) system with air duct. *Energy Build.* **2012**, *53*, 159–165. [CrossRef]
10. Hosseinzadeh, M.; Sardarabadi, M.; Passandideh-Fard, M. Energy and Exergy Analysis of Nanofluid Based Photovoltaic Thermal System Integrated with Phase Change Material. *Energy* **2018**, *147*, 636–647. [CrossRef]
11. Shahsavar, A.; Ameri, M.; Gholampour, M. Energy and exergy analysis of a photovoltaic-thermal collector with natural air flow. *J. Sol. Energy Eng.* **2012**, *134*, 011014. [CrossRef]
12. Saidur, R.; BoroumandJazi, G.; Mekhlif, S.; Jameel, M. Exergy analysis of solar energy applications. *Renew. Sustain. Energy Rev.* **2012**, *16*, 350–356. [CrossRef]
13. Pathak, M.J.M.; Sanders, P.G.; Pearce, J.M. Optimizing limited solar roof access by exergy analysis of solar thermal, photovoltaic, and hybrid photovoltaic thermal systems. *Appl. Energy* **2014**, *120*, 115–124. [CrossRef]
14. Tripanagnostopoulos, Y. Aspects and improvements of hybrid photovoltaic/thermal solar energy systems. *Sol. Energy* **2007**, *81*, 1117–1131. [CrossRef]
15. Abu Bakar, M.N.; Othman, M.; Hj Din, M.; Manaf, N.A.; Jarimi, H. Design concept and mathematical model of a bi-fluid photovoltaic/thermal (PV/T) solar collector. *Renew. Energy* **2014**, *67*, 153–164. [CrossRef]
16. Jarimi, H.; Bakar, M.N.A.; Othman, M.; Din, M.H. Bi-fluid photovoltaic/thermal (PV/T) solar collector: Experimental validation of a 2-D theoretical model. *Renew. Energy* **2016**, *85*, 1052–1067. [CrossRef]
17. Baljit, S.S.S.; Chan, H.Y.; Audwinto, V.A.; Hamid, S.A.; Fudholi, A.; Zaidi, S.H.; Othman, M.Y.; Sopian, K. Mathematical modelling of a dual-fluid concentrating photovoltaic-thermal (PV-T) solar collector. *Renew. Energy* **2017**, *114*, 1258–1271. [CrossRef]
18. Chow, T. Performance analysis of photovoltaic-thermal collector by explicit dynamic model. *Sol. Energy* **2003**, *75*, 143–152. [CrossRef]
19. Hussain, M.I.; Lee, G.H. Thermal performance comparison of line-and point-focus solar concentrating systems: Experimental and numerical analyses. *Sol. Energy* **2016**, *133*, 44–54. [CrossRef]
20. Hussain, M.I.; Lee, G.H. Numerical and experimental heat transfer analyses of a novel concentric tube absorber under non-uniform solar flux condition. *Renew. Energy* **2017**, *103*, 49–57. [CrossRef]
21. Garg, H.P.; Adhikari, R.S. Transient simulation of conventional hybrid photovoltaic/thermal (PV/T) air heating collectors. *Int. J. Energy Res.* **1998**, *22*, 547–562. [CrossRef]
22. Holman, J.P. *Heat Transfer*; Metric, S.I., Ed.; McGraw-Hill: New York, NY, USA, 1989.
23. Agrawal, S.; Tiwari, G. Energy and exergy analysis of hybrid micro-channel photovoltaic thermal module. *Sol. Energy* **2011**, *85*, 356–370. [CrossRef]

24. Singh, S.; Agrawal, S.; Avasthi, D. Design, modeling and performance analysis of dual channel semitransparent photovoltaic thermal hybrid module in the cold environment. *Energy Convers. Manag.* **2016**, *114*, 241–250. [CrossRef]
25. Joshi, A.S.; Tiwari, A.; Tiwari, G.N.; Dincer, I.; Reddy, B.V. Performance evaluation of a hybrid photovoltaic thermal (PV/T)(glass-to-glass) system. *Int. J. Therm. Sci.* **2009**, *48*, 154–164. [CrossRef]

Publisher's Note: MDPI stays neutral with regard to jurisdictional claims in published maps and institutional affiliations.

Article

Analysis of the Calculation Method for the Thermal Transmittance of Double Windows Considering the Thermal Properties of the Air Cavity

Minjung Bae [1,2], Youngjun Lee [3], Gyeongseok Choi [1], Sunsook Kim [2] and Jaesik Kang [1,*]

1. Department of Living and Built Environment Research, Korea Institute of Civil Engineering and Building Technology, Goyang 10223, Korea; baeminjung@kict.re.kr (M.B.); bear717@kict.re.kr (G.C.)
2. Department of Architecture, College of Engineering, Ajou University, Suwon 16499, Korea; kss@ajou.ac.kr
3. Institute of Environmental Building Facade Engineering, Bel Technology Co. Ltd., Seoul 05548, Korea; leeyj@beltec.co.kr

* Correspondence: jskang@kict.re.kr; Tel.: +82-31-910-0353

Received: 28 October 2020; Accepted: 10 December 2020; Published: 14 December 2020

Abstract: The calculation method for the thermal transmittance (U-value) of double windows as specified by the Korean government (ISO 15099) is often inappropriate. To develop a more suitable calculation method, the thermal properties of the air cavity between the internal and external windows should be considered. Herein, seven cases of double windows were set up. The air cavities were designed in accordance with international standards and computational fluid dynamics (CFD) and used for the calculation of the U-values of the double windows according to ISO 15099 and 10077. All the calculated U-values were compared with experimentally obtained values. In accordance with the ISO 10077-1 method, the thermal resistance of the air cavity calculated using CFD could produce double window U-values that are similar to the experimentally obtained values. In most cases, the difference between the theoretical and experimental U-values was 5% and less than 0.14 $W \cdot m^{-2} \cdot K^{-1}$, implying that the U-values calculated using CFD and the ISO 10077-1 method are approximately equal to the experimentally obtained U-values. Korean regulations do not include ISO 10077-1 for double-window assessment. However, these criteria can provide a solution in improving the accuracy of the calculation of the overall thermal transmittance of double windows.

Keywords: energy labeling program for windows; double windows; overall thermal transmittance of windows

---

1. Introduction

Since 2012, the Korean government has operated the Energy Efficiency Standards and Labeling Program for windows, which requires window companies to provide the energy ratings for their products prior to sale. The grades, on a scale of 1–5, are determined based on the test results of the thermal transmittance (U-value), airtightness, and thermal resistance of the windows and doors [1,2] according to Korean Standards KS F 2278 and KS F 2292, respectively. The government has suggested a simulation method in the program for the determination of the thermal performance of windows. This method provides an alternative procedure by which window companies can save time and money on laboratory testing, which is necessary for the determination of energy ratings. Following this method, window companies prepare a window product with a determined energy rating and conduct the simulation evaluation to review the validity of the base model. If the difference between the experimental and theoretical values obtained using the base model does not exceed a range specified in the operational regulations [3], the base model can be implemented to develop a series model. The series model is a partial modification of the base model, which generally changes the glazing system or the thermal break in the window frame. This means that the thickness of the glazing system on the base and series models should be the same. Using the regulations, window companies can get the certified thermal transmittance required of their products faster and at a cheaper cost. The Korean government allows window companies to use the calculation method proposed by the International Organization for Standardization (ISO), standard 15099 [4]. Therefore, WINDOW/THERM [5] is commonly used as a simulation program to evaluate the thermal performance of windows. In a previous study [6], we analyzed the origin of the differences in the calculation results of the thermal performance of a window depending on a simulator and suggested a possible solution. However, window companies are reluctant to use the calculation method because the results obtained using the method are different from those obtained experimentally. The thermal performance of a single window can be calculated according to ISO 15099 such that it does not vary much from the test value [7]. However, this method cannot be used for a double window because of the thickness of the air cavity in the direction of heat flow. Double windows are a common window type in Korea [8], and they are mainly used in residential buildings. These windows consist of four windows that open horizontally in one window frame and have an air cavity between the external and internal windows. The thickness of the air cavity is usually 70–120 mm, which means the length of the heat flow direction. If this thickness exceeds 50 mm, ISO 15099 requires that another calculation method should be used to determine the thermal properties of the air cavity, for example, performing laboratory tests. In a previous study [9], to validate the ISO 15099 method, the thermal properties of the air cavity between internal and external windows were calculated based on computational fluid dynamics (CFD) and ISO 15099. It was observed that the ISO 15099 method was inappropriate for calculating the thermal properties of the air cavity under actual experimental conditions. Therefore, it is necessary to use another method to determine the thermal characteristics of the air cavity between the internal and external windows in a double window to indicate the circumstances of an experimental test. Furthermore, when determining the thermal properties of the air cavity using the ISO 15099 method, it is assumed that the double window is part of a glazing system. This method assumes a double window to be a single window with a huge thick glazing system and a window frame. For these reasons, window companies suspect the reliability of the ISO 15099 method and require a more suitable method for calculating the U-values of double windows.

In this study, to determine an appropriate calculation method, ISO 15099 and ISO 10077 were used in the calculation of the thermal transmittance of double windows. Given that it is relevant to select a calculation method that is appropriate for determining the thermal properties of the air cavity between the internal and external windows in a double window, first, the U-values of double windows were calculated using WINDOW/THERM, based on Korean regulations. Thereafter, series ISO 10077-1 [10] and 10077-2 [11] of ISO 10077 were also employed to calculate the U-values of the double windows. Specifically, ISO 10077-1 specifies a method for the calculation of the thermal transmittance of a double window, whereas 10077-2 provides reference input data for the calculation of the thermal transmittance of frame profiles as well as the linear thermal transmittance of their junction with glazing. Seven cases of double windows, including four types of double window products and six types of glazing systems, were considered. The thermal properties of the air cavity in each case were determined using International Standards and were simulated using CFD. Finally, the U-values computed using ISO 15099 and ISO 10077-1 were compared with the experimental results.

2. Methods

2.1. Double Window Types

Table 1 lists the six types of glazing systems that are available for double windows. The glazing systems were selected based on the International Glazing Database (IGDB), which is operated by the National Fenestration Rating Council (NFRC). These consist of two 5 mm glass panes separated by a 12 mm-wide gap filled with air (Air) or argon gas (Ar). In Table 1, LE and CL correspond to glass panes with and without low-emissivity coating, respectively. The numbers before each abbreviation correspond to the thickness of the glass pane or that of the gap between glass panes.

Table 1. Glazing systems for double windows.

Glazing System	Composition	Ug ($W{\cdot}m^{-2}{\cdot}K^{-1}$)
A	5CL + 12Air + 5CL	2.901
B	5CL + 12Air + 5LE	1.704
C	5CL + 0.76PVB + 3CL + 12Air + 5LE	1.664
D	5CL + 12Air + 5LE	1.624
E	5LE + 12Ar + 5LE	1.278
F	6LE + 14Ar + 5CL	1.124

The glazing systems B and D have glass pane coatings of thickness 5 mm, with emissivities of 0.035 and 0.026, respectively. It causes that thermal the performances of the glazing system B and D are different. The glazing system C is 25.76 mm thick because it comprises one laminated glass pane. It consists of 5 and 3 mm thick clear glass panes and a 0.76 mm thick polyvinyl butyral (PVB) coating in-between. Glazing system F is 25 mm thick and consists of a 14 mm-wide gap filled with argon gas between one 6 mm-coating glass pane and one 5 mm glass pane.

Table 2 indicates the seven double window cases according to the product name and the glazing system. In this study, three double window products with polyvinyl chloride (PVC) frame and one double window product with aluminum frame were chosen. These window products, which are the horizontal slide type, are widely available in the Korean market. That with product name VBF250 has an external window that consists of an upper component that slides and a lower component that is fixed. The others have internal and external windows that slide, which are common in Korea. Each type of double window has a different frame profile, thus, they can have different distances between the external and internal windows. Products S3-235 and S5-250 have the same distance (88 mm) between the external and internal windows, while for HS235D, the distance is 94 mm. Product VBF250 has a different upper and lower component in the external window, so the distance between its external and internal windows is 70.6 mm in the upper part and 94.5 mm in the lower part. Cases 1 and 2 are the same type of double window with two different glazing systems and so are Cases 3 and 4. Cases 6 and 7 are based on VBF250, and the upper part of the external window and the internal window have glazing systems A or B in each case, but the lower part of the external window is the same.

Table 2. Specification on the double window cases.

Case	Product Name	Frame Material	Glazing System		Distance between the External and Internal Windows (mm)
			External Window	Internal Window	
1	<S3-235>	PVC	A	D	88
2			E	E	88
3	<S5-250>	PVC	A	D	88
4			A	A	88

Table 2. *Cont.*

Case	Product Name	Frame Material	Glazing System		Distance between the External and Internal Windows (mm)
			External Window	Internal Window	
5	<HS235D>	Aluminum	F	F	94
6	<VBF250>	PVC	Upper B Lower C	B	70.6 94.5
7			Upper A Lower C	A	70.6 94.5

2.2. Laboratory Tests

Laboratory tests were carried out on all the double window products according to KS F 2278. The test equipment consisted of a 2.0 × 2.0 m attachment frame to which the test specimen was attached and cold and hot chambers, which each included a cold wind blower and a heater box, respectively. The attachment frame was fixed between the cold and hot chambers, and the air temperature of the cold chamber was set to 0 °C, while that in the hot chamber and the heater box was set to 20 °C. The equipment was operated until the two chambers reached a steady state after which the temperature and quantity of heat in each chamber and the heater box were measured three times every 30 min. During this test, the steady state implied that the air temperature and surface temperature were kept constant, and the variation in the difference in the air temperature between the heater box and the cold chamber was within 3% per hour. To measure the surface temperature of the hot and cold sides, each specimen was divided into nine areas, and a T-type thermocouple was attached to the center of each of the nine areas [9], represented by the orange dots in Figure 1. In this study, eight additional T-type thermocouples, represented by the green dots in Figure 1, were installed in the corner area of the glazing systems, 3 cm away from the window frame. Figure 1a shows the locations at which the thirteen T-type thermocouples were installed on the one side of a specimen, these were applied to the surfaces of the internal and external windows, as shown in Figure 1b.

2.3. Calculation of the Thermal Resistance of the Air Cavity between the Windows

The thermal resistance of the air cavity between the internal and external windows impacts the calculation of the thermal transmittance of double windows, as the computed U-value should be similar to the experimental result. Table 3 denotes the thermal resistance of the air cavity in each window product. The thermal resistance of the air cavity in each case can be calculated using three methods, i.e., ISO 15099, ISO 10077-1, and CFD. ISO 15099 allows for the calculation of the effective conductivity of the unventilated frame cavity, and is defined according to the thickness or width of the air cavity in the direction of heat flow. In this study, WINDOW/THERM was chosen for computing the thermal properties of the air cavities in the double windows according to ISO 15099. Thus, the thermal resistance of the air cavity was calculated using the effective conductivity and thickness. In this software, the air cavity between the external and internal windows was considered as a wide gap between the glass panes, i.e., the air cavity presumably belongs to a 132 mm thick giant glazing system, consisting of four pane glass and three gaps filled with air or argon.

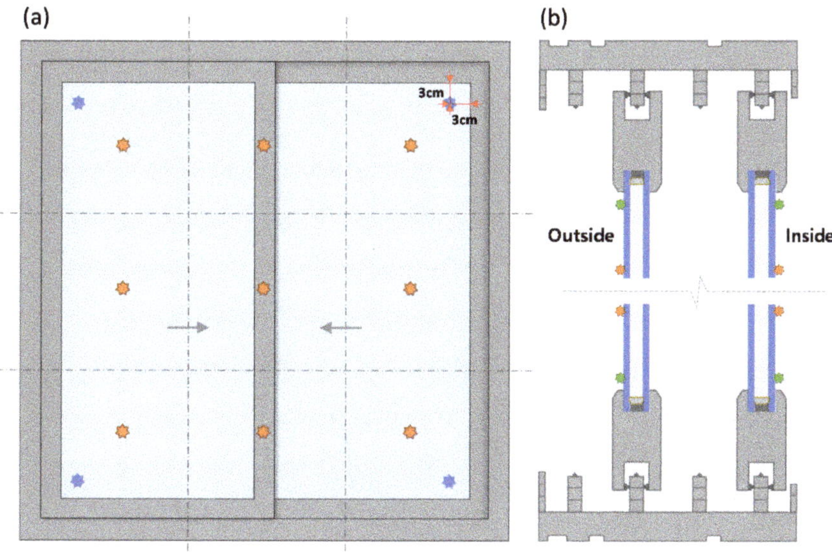

Figure 1. (**a**) Locations of T-type thermocouples on one side of a specimen, and (**b**) sections of specimens with T-type thermocouples on the internal and external windows.

However, in ISO 10077, the calculation of the overall thermal transmittance of the double windows is based on component parts, the elements constituting the glazing systems, the thermal transmittance of the frame, and the linear thermal transmittance of the frame/glazing junction. Specifically, ISO 10077-1 provides the values of the thermal resistance of unventilated air cavities in double windows according to the thickness of the air gap (6, 9, 12, 15, and 50 mm) in the form of a table. These values depend on whether the glazing status on one side has a normal emissivity coating or is uncoated. Unfortunately, there is no exact value for the thermal resistance of air cavities with thicknesses in the range 70.6–94 mm in ISO 10077-1. As previously reported [9], CFD was used to analyze the actual thermal characteristics of the air cavities. The CFD model is not a precise simulation of the air cavity between the external and internal windows. However, it aims to analyze the thermal properties stemming from the actual width and height. Based on CFD, Cases 1–5 imitating the air cavity in a double window were modeled as a closed 1 × 2 m air cavity (width and height, respectively). The thicknesses, measured in the horizontal direction of the heat flow based on the distance between the external and internal windows, are specified in Table 3. Two 1 × 2 m air cavities (width and height, respectively), were formed in a double window because the window was divided. For this reason, the two air cavities exhibited symmetrical air flows and temperature distributions. In this study, it was assumed that the CFD results obtained for one air cavity can be applied to all the air cavities in a double window. The ambient temperature of the CFD air cavity model were 0 and 20 °C, and these were defined by the external and internal glazing systems for the surface emissivity and surface heat transfer coefficient of the CFD model. Cases 6 and 7 had an exterior window divided into four sides. Therefore, the CFD model for these cases had two types of exterior boundary conditions and two different thicknesses. The upper and lower parts of the air cavity had different boundary conditions owing to the different glazing systems in the double windows. However, the air cavities exhibited symmetrical thermal properties. All the values of the thermal resistance computed using CFD were higher than those calculated using ISO 15099. This explains why the laboratory test values of the thermal transmittance of the double windows are lower than those obtained theoretically [9].

Table 3. Thermal characteristics of air cavities for the double window cases.

Case	Glazing System			Distance between the External and Internal Windows (mm)	Effective Thermal Conductivity ($W{\cdot}m^{-1}{\cdot}K^{-1}$)	Thermal Resistance ($m^2{\cdot}K{\cdot}W^{-1}$)		
	External Window		Internal Window			ISO 15099 (R_i)	ISO 10077-1 (R_s)	CFD
1	A		D	88	0.4333	0.203	N/A	0.219
2	E		E	88	0.4252	0.207		0.219
3	A		D	88	0.4333	0.203		0.219
4	A		A	88	0.4577	0.194		0.219
5	F		F	94	0.4489	0.209		0.220
6	Upper	B	B	70.6	0.3475	0.203		0.216
	Lower	C		94.5	0.4649	0.203		
7	Upper	A	A	70.6	0.3607	0.196		0.216
	Lower	C		94.5	0.4867	0.194		

2.4. Calculation of the Thermal Transmittance of the Double Windows

Generally, the overall thermal transmittance of a window can be calculated using WINDOW/THERM according to ISO 15099 and using the thermal resistance values of the air cavity obtained via CFD based on ISO 10077. Unfortunately, the thermal properties of an air cavity according to ISO 10077-1 are not available for all the seven cases used in this study. The ISO 10077-1 methodology includes the calculation of the linear thermal transmittance of the double window. The corresponding values could be obtained directly from a table in ISO 10077-1 or calculated from a formula in ISO 10077-2.

ISO 15099 includes the procedure for calculating thermal transmittance. In this procedure, the effect of three-dimensional heat transfer in frames and glazing units is not considered. Additionally, in this procedure, the linear thermal transmittance and frame thermal transmittance, U_f, was calculated. However, there is an alternative procedure that can be used to calculate these values, which is used in area-based calculations and by WINDOW/THERM. In this case, Equation (1) was used to calculate the total thermal transmittance:

$$U_t = \frac{\sum U_{cg}A_c + \sum U_{fr}A_f + \sum U_{eg}A_e + \sum U_{div}A_{div} + \sum U_{de}A_{de}}{A_t} \tag{1}$$

With this method, it is unnecessary to determine the linear thermal transmittance. Instead, the glass area, A_{gv}, is divided into the center-glass area, A_c, plus the edge-glass area, A_e. Similarly, the thermal transmittance of the glazing system is divided into the center-glass and edge-glass systems, U_{cg} and U_{eg}, respectively, which are used to characterize each glass system area. If dividers are present, then the divider area, A_{div}, and the divider thermal transmittance, U_{div}, were calculated along with the corresponding divider edge area, A_{de}, and thermal transmittance, U_{de}. U_{eg} can be determined from the following equation:

$$U_{eg} = \frac{\Phi_{eg}}{l_{eg}(T_{ni} - T_{ne})} \tag{2}$$

where l_{eg} is the length of the edge of the glass area and is equal to 63.5 mm. These lengths were measured from the internal side. The quantity, Φ_{eg}, represents the heat flow rates through edge-glass areas (internal surfaces), including the effect of glass and spacer, and it is expressed in units, per length of edge-glass. In WINDOW/THERM, the thermal resistance of an air cavity between internal and external windows was used to calculate the thermal transmittance of the glazing system, given that the air cavity was assumed to be a component of the glazing system. The thermal transmittance of the glazing system, according to ISO 15099, can be determined from the following equation:

$$U_g = U_{cg} + U_{eg} = \frac{1}{R_t} \tag{3}$$

where R_t is obtained by adding the thermal resistances at the external and internal boundaries and of the glazing cavities and layers.

$$R_t = \frac{1}{h_{ex}} + \sum\nolimits_{i=2}^{n} R_i + \sum\nolimits_{i=1}^{n} R_{gv,i} + \frac{1}{h_{int}}. \tag{4}$$

Figure 2 shows the numbering scheme of the glazing system. Specifically, the thermal resistance of the ith glazing is given by:

$$R_{gv,i} = \frac{t_{gv,i}}{\lambda_{gv,i}} \tag{5}$$

and the thermal resistance of the ith space is given by:

$$R_i = \frac{T_{f,i} - T_{b,i-1}}{q_i} \tag{6}$$

where $T_{f,it}$, and $T_{b,i-1}$ are the external and internal facing surface temperatures of the ith glazing layer, respectively. It should be noted that the first space corresponds to the external environment, the last space corresponds to the internal environment, and the spaces in between correspond to the glazing cavities. Therefore, Equation (6) gives the thermal resistance of an air cavity between external and internal windows in a double window using ISO 15099.

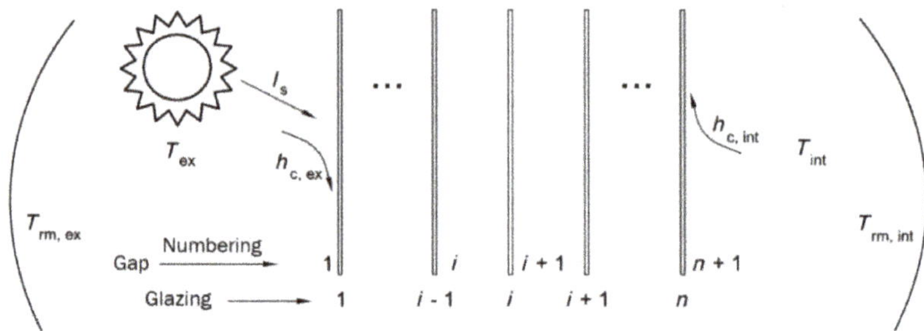

Figure 2. Numbering scheme for glazing system layers.

However, the U-value of a double window according to the ISO 10077 method needs linear thermal transmittance. The preferred method of establishing the values of the linear thermal transmittance was by numerical calculations using the formulas included in ISO 10077-2, Annex F. However, when the results of a detailed calculation are unavailable, ISO 10077-1 provides default values of the linear thermal transmittance for typical combinations of frames, glazing, and spacers.

ISO 10077-1 details the procedure for calculating the thermal transmittance, U_w, of a double window, which is a system consisting of two separate windows, as shown in Figure 3. For this method, it is necessary to calculate the thermal transmittances of the internal and external windows, U_{w1} and U_{w2}, respectively. The thermal transmittance of a single window, U_{w1} or U_{w2}, was calculated using the following formula:

$$U_w = \frac{\sum A_g U_g + \sum A_f U_f + \sum l_g \Psi_g + \sum l_{gb} \Psi_{gb}}{A_f + A_g} \tag{7}$$

where U_g and U_f are the thermal transmittances of the glazing system and frame, respectively, Ψ_g is the linear thermal transmittance due to the combined thermal effects of glazing, spacer and frame, and Ψ_{gb} is the linear thermal transmittance due to the combined thermal effects of glazing and glazing bar. The internal surface resistance, R_{si}, of the external window when used alone. The external surface resistance, R_{se}, of the internal window when used alone, and the thermal resistance, R_s, of the space between the glazing in the two windows were also calculated according to the given equations. In this study, the thermal resistance of each air cavity could not be defined by ISO 10077-1, whereas it could be defined by ISO 15099 and CFD, as shown in Table 3. The U-value of a double window could be calculated based on ISO 10077-1 using the thermal resistance of the air cavity computed using the CFD method. Then, the thermal transmittance of the double window was calculated using the following formula:

$$U_w = \frac{1}{U^{-w1} - R_{si} + R_s - R_{se} + U^{-w2}} \tag{8}$$

In this study, three methods were considered for calculating the thermal transmittance of double glazing, as shown in Table 4. Method A involves calculating the thermal resistance of an air cavity between external and internal windows, according to ISO 15099, and finally calculating the U-value of a double window in WINDOW/THERM software. In this method, the linear thermal transmittance does not need to be calculated. Methods B and C are based on the ISO 10077-1 methodology for determining the U-value of a double window. The thermal resistance of the air cavity computed by CFD, as shown in Table 3, is used as input data for these methods; however, Method B uses the default value of the linear thermal transmittance in ISO 10077-1. Method C uses the linear thermal transmittance determined by numerical calculations using the formulas included in ISO 10077-2, Annex F. The three methods will be evaluated for the validity of whether the U-value of a double window similar to experimental values can be derived.

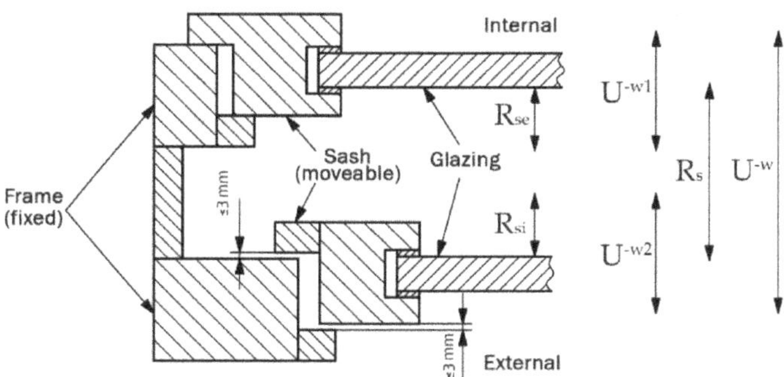

Figure 3. Illustration of a double window defined according to ISO 10077-1.

Table 4. Methods for calculating the U-value of a double window using three different sets of data.

	Calculation Method for the Data	Method A	Method B	Method C
1	U-value of a double window	ISO 15099	ISO 10077-1	ISO 10077-1
2	Thermal resistance of air cavity	ISO 15099	CFD	CFD
3	Linear thermal transmittance	N/A	ISO 10077-1 (table)	ISO 10077-2 (calculation)

3. Results and Discussion

Table 5 summarizes the thermal transmittance of seven double windows according to the three calculation methods and the laboratory tests. All the calculated U-values were higher than those obtained based on the laboratory tests, except for Case 1. According to Korean regulations, for U-values to be valid, the difference between the experimental and theoretically obtained values should not exceed 0.14 $W{\cdot}m^{-2}{\cdot}K^{-1}$. Calculation method A based on ISO 15099 exhibited a difference between experimental and theoretical U-values of 10% to 24%, although the Korean government operates the regulation to ensure the use of this method. Four cases were considered valid in Korean regulations, and three cases were considered invalid. The difference between the theoretical and experimental values was lower when ISO 10077-1 (Methods B and C), where the thermal resistance of the air cavity was calculated via CFD was used than when ISO 15099 was used. The CFD value used in this method was calculated using a simple air cavity model that depended on the actual height and width of the cavity. Nevertheless, the ISO 10077-1 method with CFD implementation (Method B and C) was considered valid based on Korean regulations because the difference between the experimental and calculated U-values was less than 0.14. However, it needs to be revised when determining the calculation method for the U-value of a double window to ensure the effective operation of Korean regulations. It should include the method that can reflect the thermal characteristics of the air cavity between the internal and external windows under experimental conditions, and this method should be able to present numerical results.

Table 5. The thermal transmittance of double windows (U-value) in the laboratory and the calculation methods.

Case	Laboratory Test ($W{\cdot}m^{-2}{\cdot}K^{-1}$)	Method A ($W{\cdot}m^{-2}{\cdot}K^{-1}$)	Method B ($W{\cdot}m^{-2}{\cdot}K^{-1}$)	Method C ($W{\cdot}m^{-2}{\cdot}K^{-1}$)
1	1.220	1.216 (−0.3%)	1.147 (−6.0%)	1.134 (−7.0%)
2	0.737	0.915 (24.2%)	0.872 (18.3%)	0.845 (14.7%)
3	1.113	1.227 (10.2%)	1.142 (2.6%)	1.129 (1.4%)
4	1.314	1.470 (11.9%)	1.355 (3.1%)	1.335 (1.6%)
5	1.006	1.152 (14.5%)	1.030 (2.4%)	0.991 (−1.5%)
6	0.950	1.046 (10.1%)	0.986 (3.8%)	0.988 (4.0%)
7	1.187	1.320 (11.2%)	1.246 (5.0%)	1.233 (3.9%)

The method for calculating the linear thermal transmittance influenced the accuracy of the thermal transmittance of a double window. The U-value of the double window computed using the calculated linear thermal transmittance (Method C) was closer to the experimental value than that computed using the given linear thermal transmittance (Method B). U-values of Case 6 were similar, regardless of whether Method B or C was used, but the results obtained by Method C were closer to the experimental value than those obtained by Method B. This is because the default values of linear thermal transmittance, provided by ISO 10077-1, are considered conservatively. These values are larger than the actual linear thermal transmittance value, according to ISO 10077-2.

Although the same frame is used in Cases 1 and 2, Case 2 had a lower U-value than Case 1 because of the good thermal performance of the applied glazing system. The calculated U-values for Case 2 were most deviated from the experimental value compared to those for the other cases. It is thought that there are factors to be considered when calculating the U-value of a double window with excellent thermal performance of the glazing system. Therefore, in future studies, it would be necessary to identify the reasons behind this analysis, as well as the methods that should be included in the calculation.

4. Conclusions

The Korean government has been operating a simulation system for assessing the thermal performance of windows to allow companies to save time and money in determining window energy ratings. However, the uncertainty in the calculation results with respect to double windows has been discussed steadily. Thus, window companies are reluctant to use this calculating method. It has led window companies attempting to realize experimental U-values, even though the process is costly and time-consuming. According to Korean regulations, the procedure provided in ISO 15099 is used in the calculation of the thermal transmittance of double windows. However, our findings indicate that this method, ISO 15099 resulted in only four out of seven calculated values satisfying the criteria imposed by the Korean regulations. Further, all four valid values differ significantly from the experimental values. In a previous study [9], the importance of adopting the appropriate thermal properties of the air cavity between internal and external windows during the calculation of the thermal performance of double windows was reported. Therefore, the ISO 15099 method is no longer suitable for determining the thermal properties of the air cavity between internal and external windows, which is used to calculate the U-value of double windows. This method should be improved such that it can adopt the thermal resistance of the air cavity under experimental conditions, and the CFD method used in this study is one of several methods that can be used. With the CFD method, it is possible to provide a table that can be used in calculating the U-values of double windows by pre-calculating the thermal resistance according to various glazing systems. In subsequent studies, it would be necessary to consider this alternative method so as to make it easier to use the calculation method for the U-value of a double window. If the U-value of a double window is calculated according to ISO 10077-1, the result approximates the experimental value. This also overcomes the error associated with existing methods, which assume that the air cavity between internal and external windows is part of the glazing system. Therefore, the procedure detailed in ISO 10077 should be considered for the appropriate calculation of the thermal transmittance of double windows.

Author Contributions: Investigation, M.B. and Y.L.; Methodology, Y.L.; Project administration, J.K.; Validation, G.C.; Writing—original draft, M.B.; Writing—review & editing, S.K. All authors have read and agreed to the published version of the manuscript.

Funding: This research was supported by a grant (20RERP-C146906-03) from the Residential Environment Research Program funded by the Ministry of Land, Infrastructure, and Transport of the Korean government.

Conflicts of Interest: The authors declare no conflict of interest.

References

1. Korea Agency for Technology and Standards (KATS). *KS F 2278: Test Method of Thermal Resistance for Windows and Doors*; KATS: Emseong, Korea, 2014.
2. Korea Agency for Technology and Standards (KATS). *KS F 2292: The Method of Air Tightness for Windows and Doors*; KATS: Emseong, Korea, 2013.
3. Ministry of Trade, Industry and Energy (MOTIE). Operational Regulation on Equipment for Efficiency Management. 2016. Available online: https://eep.energy.or.kr/download/Korean%20Energy%20Efficiency%20Policies%20(2015).pdf (accessed on 25 November 2020).
4. ISO 15099:2003(E). Thermal Performance of Windows, Doors and Shading Devices—Detailed Calculations. 2003. Available online: https://www.iso.org/standard/26425.html (accessed on 11 December 2020).
5. NFRC. *THERM 7/WINDOW 7 NFRC Simulation Manual*; NFRC: Berkeley, CA, USA, 2017.
6. Bae, M.J.; Choi, H.J.; Choi, G.S.; Kang, J.S. A study on the evaluation methods of window simulation for the reliability and reproducibility of result. In Proceedings of the Summer Conference of the Society of Air-conditioning and Refrigerating Engineers of Korea, Pyeongchang, Korea, 22–25 June 2016; pp. 108–109.
7. Lee, Y.J.; Oh, E.J.; Kim, S.K.; Choi, H.J.; Kim, Y.M. A comparative analysis of the simulation results of total window thermal transmittance (Uw) according to the evaluation method—Focused on comparison of the single window simulation results. *Int. J. Korea Inst. Ecol. Archit. Environ.* **2016**, *16*, 77–82.
8. Bae, M.J.; Cho, S.H.; Choi, G.S. The study on windows registered as energy standards and labeling Program based on the frame materials and opening types. *Int. J. Korea Inst. Ecol. Archit. Environ.* **2018**, *18*, 81–87.
9. Kang, J.S.; Oh, E.J.; Bae, M.J.; Song, D.S. A numerical study of the thermal characteristics of an air cavity formed by window sashes in a double window. *Int. J. Thermophys.* **2017**, *38*, 180. [CrossRef]
10. ISO 10077-1:2006(E). Thermal Performance of Windows, Doors and Shutters—Calculation of Thermal Transmittance—Part 1: General. 2006. Available online: https://www.iso.org/standard/40360.html (accessed on 11 December 2020).
11. ISO 10077-2:2012(E). Thermal Performance of Windows, Doors and Shutters—Calculation of Thermal Transmittance—Part 2: Numerical Method for Frames. 2012. Available online: https://www.iso.org/obp/ui/#iso:std:iso:10077:-2:ed-2:v1:en (accessed on 11 December 2020).

Article

Optical Sensing Approach to the Recognition of Different Types of Particulate Matters for Sustainable Indoor Environment Management

Hosang Ahn [1,*], Jae Sik Kang [2], Gyeong-Seok Choi [1] and Hyun-Jung Choi [1]

[1] Building Information Research Center, Korea Institute of Civil Engineering and Building Technology, 283 Goyangdaero, Ilsanseogu, Goyang, Gyeonggi 10223, Korea; bear717@kict.re.kr (G.-S.C.); mingineu@kict.re.kr (H.-J.C.)

[2] Living Environment Research Center, Korea Institute of Civil Engineering and Building Technology, 283 Goyangdaero, Ilsanseogu, Goyang, Gyeonggi 10223, Korea; jskang@kict.re.kr

* Correspondence: hahn@kict.re.kr; Tel.: +82-31-910-0744

Received: 19 October 2020; Accepted: 16 December 2020; Published: 17 December 2020

Abstract: The indoor environment is a crucial part of the built environment where our daily time is mostly spent. It is governed not only by indoor activities, but also affected by interconnected activities such as door opening, walking and routine tasks throughout the inside and outside of buildings and houses. Pollutant control is one of the major concerns for maintaining a sustainable indoor environment, and finding the source of pollutants is a relatively hard part of that task. Pollutants are emitted from various sources, transformed by sunlight, react with vapor in ozone and are transported into cities and from country to country. Due to these reasons, there has been high demand to monitor the transportation of particulate matters and improve air quality. The monitoring of pollutants and identification of their type and concentration enables us to track and control their generation and consequently discover reliable suitable mitigation measures to control air quality at regulated levels by contaminant source removal. However, the monitoring of pollutants, especially particulate matter generation and its transportation, is still not fully operated in atmospheric air due to its open nature and meteorological factors. Even though indoor air is relatively easier to monitor and control than outdoor air in the aspect of specific volume and contaminant source, meteorological parameters still need to be considered because indoor air is not fully separated from outdoor air flow and contaminants' transportation. In this study, an optical approach using a spectral sensor was attempted to reveal the feasibility of wavelength and chromaticity values of reflected light from specific particles. From the analysis of reflected light of various particulate matters according to different liquid additives, parameter studies were performed to investigate which experimental conditions can contribute to the enhanced selective sensing of particulate matter. Five different particulate matters such as household dust, soil, talc powder, gypsum powder and yellow pine tree pollen were utilized. White samples were selectively identified by the peak at 720 nm for talc and 433 nm and 690 nm in wavelength for gypsum under chemical additives. Other grey household dust and yellowish soil and pine tree pollen revealed a distinct chromaticity x, y coordinates shift in vector within the maximum range from (0.22, 0.19) to (0.55, 0.48). Applicable approaches to assist current particle matter sensors and improve the selective sensing were suggested.

Keywords: optical sensing; particulate matter; sustainable indoor environment; contaminant control

1. Introduction

It has been continuously necessary to control indoor air quality more precisely and with more detailed information. However, indoor air is quite different from atmospheric air in the aspects of air flow characteristics and type of contaminants [1]. Depending on the type of building and purpose of usage, the contaminant type and its level vary due to different human activities and emission sources. For a residential house, cooking is reported as a primary factor for the emission of gaseous pollutants such as formaldehyde, CO and Total Volatile Organic Compounds (TVOC). Particulate pollutant $PM_{2.5}$ is also one of the contaminants highly detected in indoor air during cooking [2]. In addition, human activities such as ironing, vacuum cleaning, lighting candles and smoking are also known to increase the level of pollutants, and even walking can increase the PM level by resuspension [3]. Relatively large particulate matters such as soil dust, flower pollen and PM_{10} are well-known to be transported into the indoors by air flow from the outside and their generation and behavior have been reported quite differently [4].

Even though various technologies to detect both gaseous and particulate contaminants have been developed and widely applied to practical fields, any sensing data to inform us with both the contaminant source and its concentration simultaneously does not exist, and even its accuracy remains low [5]. Most commercial sensors to detect particulate matters are generally used as dust sensors and are mostly based on the light scattering principle. As many particles exist in a specific volume of the sensor when used inside, more light is scattered and reflected to the detector and represented as particle levels. For this reason, it is necessary to introduce sufficient air containing contaminants that can represent a statistically mean concentration per volume into the sensor inside by fan or air compressor for reliable accuracy. The other factor to govern the dust level is the interaction between the light source and particulate matters. In previous research, light sources such as laser diode, infrared and LED photodiode were used to examine how light source can influence the sensing of particulate matters [6–8].

Depending on the light source, single point detection, uniformity issue and brightness difference were reported to limit the sensitivity of dust sensors [9]. For more accurate concentration, particle counters utilizing a beta ray absorption method were tested and authorized to report daily data of particulate matters that have an aerodynamic diameter of less than 10 and 2.5 μm in Korea [10]. According to the purpose of measurement, both optical sensing and beta attenuation monitoring (BAM) were adopted to research the area or air pollution forecast, but simple light-scattering-based sensors were mostly utilized in daily life measurement for a single household's air quality monitoring, including a dust sensor, air conditioner and air purifier. As recognized in the above explanations, the concentration of particulate matter is primary information for sensors in monitoring particulate matter contaminants and is provided relatively sufficiently with various methods. However, other information such as the type of particle, and the chemical composition to inform us of its origin and where it is generated and transported from, is still under laboratory level observation [11].

Nowadays, characterization to determine the origin of contaminants, especially for particulate matters, is a major concern in Korea. This is because daily concentration of particulate matters $PM_{2.5}$ and PM_{10} have caused a noticeable increase in the reported number of patients with respiratory disease, and personal protective equipment (PPE) including air pollution masks, filters and air purifiers are selling significantly above production amounts [12]. In several reports, particulate matters are characterized and chemical compositions have reported that PM_{10} and $PM_{2.5}$ contain organic compound and heavy metal ions, which may cause health issues [13]. Furthermore, it is necessary to analyze particulate matters at the laboratory level to know the source of particulate matters and their chemical properties that can potentially be harmful to respiratory health. However, chemical analysis is expensive and it takes a long time to reach to the desired results. As a result, there is at least demand to identify the types of contaminants using a simple dust sensor at an economic cost as a prescreening level test.

In this study, two approaches were tested. A small-scale spectral sensor was utilized to find the feasibility of light wavelength in terms of position and intensity to discriminate the type of particulate matter. The other approach was to use a chromameter to reveal the color data of particulate matter in a chromaticity diagram. Five different particles, household dust, soil, pine tree pollen, talc and gypsum powder were chosen and tested to find the feasibility of optical approaches using color and reflected light to distinguish different particulate matters. It is our expectation that the intrinsic color of particles can be a key parameter to identify particulate matters having unique colors. Particulate matters which havea tendency to react easily with water and refractive index liquid can be selectively detected by observing reflected light and characterizing its spectrum. Our study can assist current light-scattering-based sensors to identify the type of particulate matter contaminants and concentration with higher accuracy for reliable indoor environment management.

2. Materials and Methods

Five different particulate matters were collected in Korea and prepared for the characterization as they were. Household dust was collected by a regular vacuum cleaner from a living room in a typical apartment complex in Goyang city. Korean pine tree pollen was collected during spring season by washing a glass plate located under a pine tree bush in Jeongbal mountain, located in Goyang city for one day. Illite powder, a commonly found yellow soil in Korea, was used for the representative soil sample. It was purchased from Yong Gung Illite®Inc., and the average size of illite powder was characterized to be less than 200 μm. Talc powder, a raw material widely used as a construction material and usually suspended in indoor air during the construction process was purchased from a chemical company to have the chemical formula $Mg_3H_2(SiO_3)_4$; $H_2Mg_3O_{12}Si_4$. Gypsum powder was prepared by grinding gypsum insulation board manufactured by KCC Inc., Korea, which has a 9.5 mm thickness, 900 mm width and 1800 mm length in general grade. A total of 20 samples for five different particulate matters were ground and filtered with Whatman®qualitative paper filter having 20 μm particle retention by flushing with distilled water to exclude the size-induced difference. After drying at room temperature, the collected powders were used for the experiment. All samples were prepared by cutting them into pieces small enough to grind and sieve to make a desired powder size of 20 μm. Those powders were denoted "as prepared" to distinguish between untreated powders and other powders treated by chemical additives.

Filters and liquid additives to modulate the reflected light of particle samples were tested. Cellophane filters ranged from red, orange, yellow, green, blue, pink and violet in a visible light range as shown in Figure 1. Three color filters, dark blue, green and yellow were utilized, having 400–450 nm, 500–550 nm, and 550–600 nm in wavelength, respectively. Two liquid additives, refractive index liquid (n = 1550, Cargille Inc. Cedar Grove, NJ, USA) and distilled water, were tested.

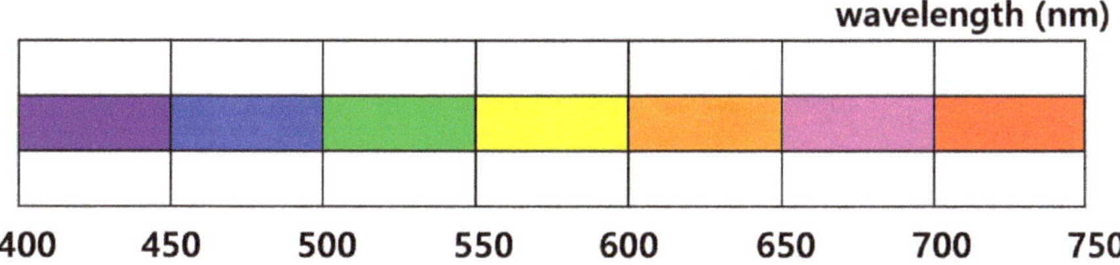

Figure 1. Color chart of seven cellophane filters in wavelength ranging in visible light from 400 nm to 750 nm. All colors represent the prepared colors of the cellophane filters in a specific wavelength range.

Reflected light was observed in the same 10 cm distance from the sample surface to the detectors, spectral sensor and chromameter. An 80 W–6500 K white LED light bulb was used for the light source to provide sufficient light in the visible light range and avoid a light color effect. In addition to this, a UV light with 365 nm in wavelength was used.

As shown in Figure 2, a schematic (a) and a picture (b) of the experimental apparatus, chromameter (c) and spectral sensor (d) were prepared. A spectral sensor, Apollo™, developed by NanoLambda in Korea, was used to differentiate reflected light into the light spectrum in a small chamber and to examine the applicability to a small-scale sensor. The configuration of the chamber and detailed experimental method was described in our previous study [14]. A chromameter CR-400 by Konica Minolta was used to acquire color data in terms of chromaticity values.

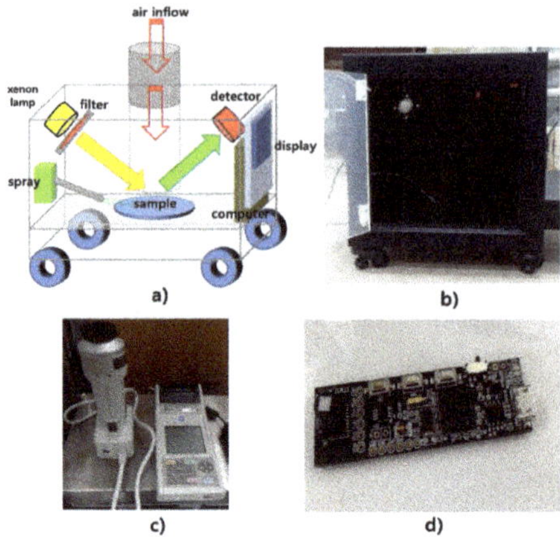

Figure 2. Schematics and pictures of experimental apparatus. (**a**) A schematic of chamber, (**b**) Black-coated closed chamber, (**c**) Chromameter, and (**d**) Spectral sensor (20 mm width and 60 mm length).

3. Results

As described in the Introduction, the main purpose of this study was to find the feasibility of optical approaches in identifying the specific types of particulate matters among whole particulate mixtures in the air and their influence on other parameters, filters and liquid additives on their selectivity in terms of light intensity, wavelength and chromaticity value. A spectral sensor and chromameter were tested, respectively, under the same conditions by liquid additives. Chromaticity values in a chromaticity diagram are shown in Figures 3–5. Details of the conditions and results are denoted in Tables 1–5.

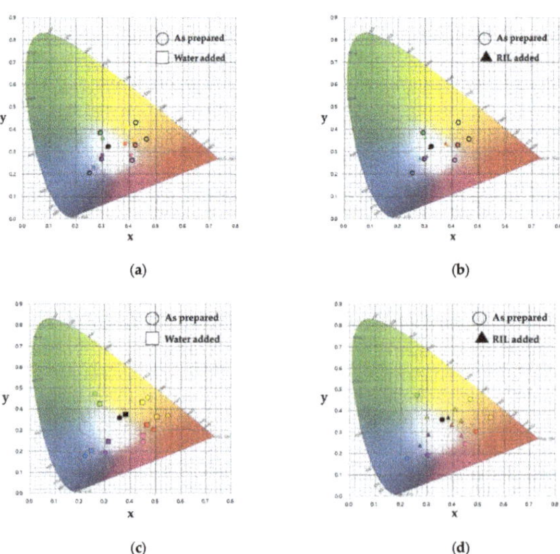

Figure 3. Chromaticity diagram of household dust, and soil dust. All dots are chromaticity values (Y, x, y) and only x, y coordinates are denoted. Each figure is listed as: (**a**) as prepared (○) and water added (□) household dust; (**b**) as prepared (○) and refractive index liquid (RIL) added (▲) household dust; (**c**) as prepared (○) and water added (□) soil dust; (**d**) as prepared (○) and RIL added (▲) soil dust.

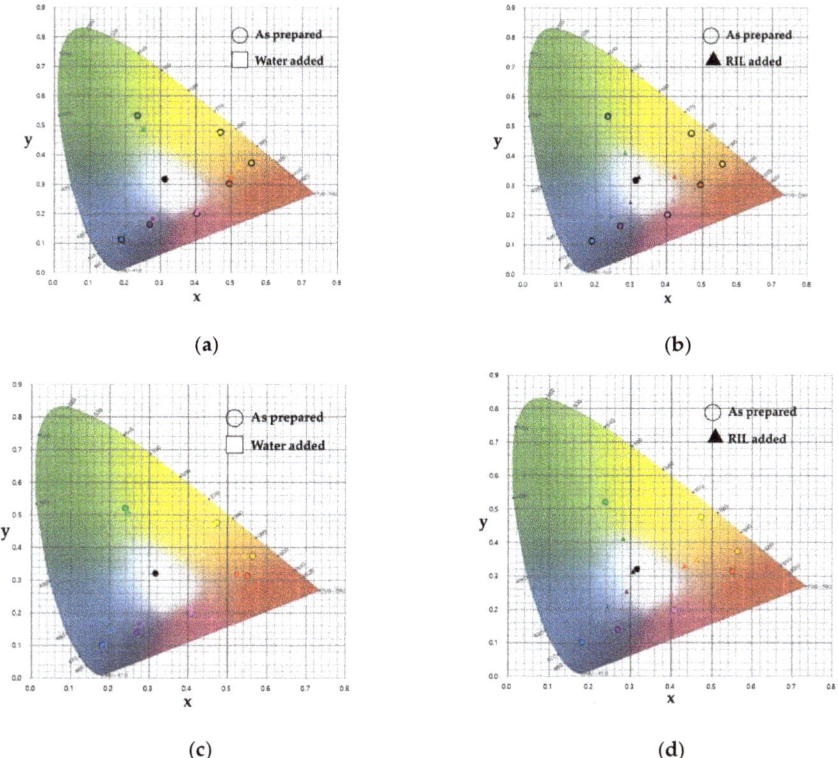

Figure 4. Chromaticity diagrams of samples under pink cellophane filter; (**a**) as prepared (○) and water added (□) talc powder; (**b**) as prepared (○) and RIL added (▲) talc powder; (**c**) as prepared (○) and water added (□) gypsum powder; (**d**) as prepared (○) and RIL added (▲) gypsum powder.

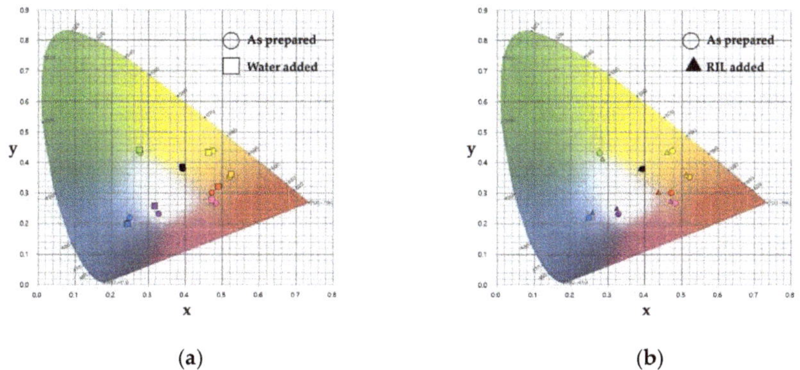

Figure 5. Chromaticity diagram of pine tree pollen; (**a**) as prepared (○) and water added (□) pine tree pollen; (**b**) as prepared (○) and RIL added (▲) pine tree pollen.

Table 1. Chromaticity values of household dust and soil powder samples measured under cellophane filter conditions.

			Chromaticity Values						Chromaticity Values		
	Sample	Cellophane	Y	x	y		Sample	Cellophane	Y	x	y
Household Dust	As prepared	-	13.40	0.3229	0.3244	Soil Powder	As prepared	-	49.15	0.3576	0.3581
		Red	4.87	0.4216	0.3307			Red	7.63	0.4914	0.3031
		Orange	8.01	0.4625	0.3576			Orange	18.21	0.5472	0.3697
		Yellow	13.47	0.4230	0.4305			Yellow	34.44	0.4687	0.4546
		Green	5.08	0.2923	0.3853			Green	8.84	0.2617	0.4733
		Blue	3.16	0.2516	0.2058			Blue	4.71	0.2227	0.1778
		Pink	9.91	0.4098	0.2624			Pink	16.94	0.4496	0.2458
		Violet	4.40	0.2970	0.2686			Violet	3.46	0.3027	0.1933
	Water Added	-	5.18	0.3190	0.3208		Water Added	-	6.25	0.3824	0.3670
		Red	4.43	0.3869	0.3357			Red	4.56	0.3970	0.3341
		Orange	6.16	0.4197	0.3508			Orange	6.67	0.4332	0.3524
		Yellow	9.02	0.3955	0.4066			Yellow	9.91	0.4103	0.4069
		Green	4.41	0.3025	0.3586			Green	4.17	0.3002	0.3711
		Blue	3.11	0.2692	0.2320			Blue	3.09	0.2732	0.2384
		Pink	8.52	0.4081	0.2856			Pink	8.09	0.4349	0.2858
		Violet	4.35	0.3008	0.2851			Violet	4.32	0.3054	0.2890
	Refractive Index Liquid Added	-	3.48	0.3163	0.3177		Refractive Index Liquid Added	-	22.23	0.3818	0.3749
		Red	4.37	0.3768	0.3369			Red	5.57	0.4647	0.3254
		Orange	5.87	0.3859	0.3305			Orange	11.02	0.5044	0.3627
		Yellow	8.17	0.3756	0.3864			Yellow	19.05	0.4484	0.4310
		Green	3.92	0.3021	0.3635			Green	5.67	0.2814	0.4245
		Blue	4.29	0.2800	0.2699			Blue	3.24	0.2475	0.2012
		Pink	7.38	0.4170	0.2834			Pink	12.04	0.4501	0.2742
		Violet	3.39	0.3080	0.2797			Violet	3.34	0.3145	0.2454

Table 2. Chromaticity values of talc powder and gypsum powder samples measured under cellophane filter conditions.

			Chromaticity Values						Chromaticity Values		
	Sample	Cellophane	Y	x	y		Sample	Cellophane	Y	x	y
Talc Powder	As prepared	-	74.79	0.3127	0.3191	Gypsum Powder	As prepared	-	79.97	0.3157	0.3213
		Red	7.53	0.4932	0.3040			Red	8.52	0.5498	0.3143
		Orange	22.17	0.5568	0.3732			Orange	24.57	0.5630	0.3738
		Yellow	50.31	0.4682	0.4770			Yellow	53.23	0.4719	0.4763
		Green	12.40	0.2337	0.5337			Green	12.90	0.2382	0.5213
		Blue	3.84	0.1888	0.1134			Blue	3.99	0.1816	0.1021
		Pink	20.73	0.4025	0.2020			Pink	20.89	0.4078	0.1989
		Violet	4.57	0.2687	0.1634			Violet	3.74	0.2681	0.1400
	Water Added	-	47.39	0.3136	0.3205		Water Added	-	19.09	0.3065	0.3134
		Red	6.69	0.4999	0.3211			Red	5.07	0.4332	0.3300
		Orange	15.50	0.5189	0.3610			Orange	9.48	0.4626	0.3495
		Yellow	32.18	0.4565	0.4669			Yellow	16.81	0.4224	0.4363
		Green	8.73	0.2529	0.4859			Green	6.00	0.2818	0.4111
		Blue	3.61	0.2012	0.1323			Blue	4.44	0.2431	0.2094
		Pink	14.18	0.4069	0.2127			Pink	10.71	0.4011	0.2453
		Violet	3.57	0.2783	0.1835			Violet	4.35	0.2906	0.2543
	Refractive Index Liquid Added	-	15.30	0.3225	0.3297		Refractive Index Liquid Added	-	56.70	0.3159	0.3217
		Red	4.91	0.4221	0.3310			Red	7.34	0.5239	0.3181
		Orange	8.70	0.4525	0.3457			Orange	18.51	0.5450	0.3724
		Yellow	14.58	0.4276	0.4370			Yellow	37.61	0.4629	0.4703
		Green	5.23	0.2828	0.4115			Green	9.90	0.2457	0.5043
		Blue	3.21	0.2443	0.1962			Blue	4.91	0.2043	0.1460
		Pink	7.38	0.4170	0.2834			Pink	17.27	0.4064	0.2135
		Violet	3.39	0.3080	0.2797			Violet	3.54	0.2757	0.1664

Table 3. Chromaticity values of pine tree pollen measured samples under cellophane filter conditions.

			Chromaticity Values						Chromaticity Values		
	Sample	Cellophane	Y	x	y		Sample	Cellophane	Y	x	y
Pine tree Pollen	As prepared	-	32.77	0.3938	0.3796	Pine tree Pollen	Water Added	-	28.76	0.3926	0.3846
		Red	6.79	0.4714	0.3017			Red	6.28	0.4887	0.3205
		Orange	14.02	0.5200	0.3524			Orange	12.88	0.5247	0.3614
		Yellow	24.74	0.4745	0.4402			Yellow	23.08	0.4626	0.4336
		Green	6.84	0.2764	0.433			Green	6.30	0.2746	0.4422
		Blue	4.52	0.2479	0.2204			Blue	3.25	0.2432	0.1983
		Pink	13.51	0.4823	0.2669			Pink	13.54	0.4722	0.2769
		Violet	3.40	0.3286	0.2329			Violet	4.42	0.3174	0.2579
	Refractive Index Liquid Added	-	22.62	0.3892	0.3804		Refractive Index Liquid Added	Green	5.88	0.2844	0.4117
		Red	6.06	0.4374	0.3039			Blue	4.39	0.2589	0.2374
		Orange	11.17	0.5106	0.3595			Pink	11.28	0.4708	0.2735
		Yellow	19.14	0.4601	0.4351			Violet	3.35	0.3240	0.2506

Table 4. Peak positions of detected reflected light of samples as a function of wavelength by spectral sensor under experimental conditions.

Sample Conditions			As Prepared			Water			Refractive Index Liquid		
Filter	Measurement	Materials									
No filter	Peak Intensity		Low	High	Other	Low	High	Other	Low	High	Other
	Peak Positions (nm)	Pine tree pollen	421	681	601	421	681	591	421	676	721
		Soil	421	681	597	421	677	721	421	681	-
		Household Dust	420	678	-	420	677	720	420	677	-
		Talc	420	677	720	420	680	720	420	677	720
		Gypsum	420	679	-	433	691	-	433	690	-
Pink filter	Peak Intensity		Low	High	Other	Low	High	Other	Low	High	Other
	Peak Positions (nm)	Household Dust	440	-	-	440	-	-	440	-	-
		Talc	430	-	-	430	490	-	430	-	-
		Gypsum	439	820	-	453	-	-	453	-	-

Table 5. Peak intensity ratio of detected reflected light of samples as a function of wavelength by spectral sensor under experimental conditions.

Sample Conditions			As Prepared			Water			Refractive Index Liquid		
Filter	Measurement	Materials									
No filter	Peak Intensity		Low	High	Other	Low	High	Other	Low	High	Other
	Peak Intensity Ratio	Pine tree pollen	1	10.46	0.98	1	8.75	0.14	1	5.86	0.81
		Soil	1	8.51	0.14	1	4.61	0.41	1	8.12	-
		Household Dust	1	9.31	-	1	6.31	0.44	1	9.56	-
		Talc	1	6.32	0.45	1	4.51	1.41	1	6.21	0.68
		Gypsum	1	9.63	-	1	9.65	-	1	9.71	-
Pink filter	Peak in Intensity		Low	High	Other	Low	High	Other	Low	High	Other
	Peak Intensity Ratio	Household Dust	0.48	-	-	0.30	-	-	0.39	-	-
		Talc	0.21	-	-	0.08	0.06	-	0.21	-	-
		Gypsum	0.36	0.13	-	0.40	-	-	0.44	-	-

3.1. Color Detection

Chromaticity diagrams of five samples are shown in Figures 3–5. As denoted in these Figures, all samples were prepared as dried, water added and refractive index liquid added. Water and refractive index liquid were utilized to investigate the additional effect of additives on the reflected light by also predicting changes to the refractive indices and associated colors. A cellophane filter was used to block a specific range of wavelength depending on its color. Details of the wavelength and color of each cellophane filter were drawn in consecutive color bars as shown in Figure 1. The prepared samples, denoted as circles in the diagram, were located with the coordinates x, y, Y in the diagram using matching color circles according to the color of the cellophane filter.

Here, coordinates; x, y, and Y represent the color space and luminance value, respectively, of the reflected light [15]. Details of the chromaticity values for household dust, soil, talc, gypsum and pine tree pollen are summarized in Tables 1–3, which correspond to the chromaticity diagram in Figures 3–5. Water-added samples and refractive-index-liquid-added samples were depicted as a square and filled triangle, respectively, as in the above manner. Prior to a detailed explanation, the five samples can be simply divided into two groups, a white group and non-white group (grey and yellowish samples) after the bare-eye observation. Talc and gypsum powders belong to the white group and household dust, soil and pine tree pollen are regarded as part of the non-white group.

All chromaticity values in the tables represent the positions in the diagram. Seven colors of cellophane filters and no filter case were denoted. For household dust, the initial coordinates of x and y were recorded, ranging from (0.3163, 0.3177) to (0.3229, 0.3244) for the prepared sample. Most cases were observed in x and y ranges between (0.25, 0.2) and (0.45, 0.45), close to the center white regions regardless of the color of the filters. Indices in the range represent the boundary limit of the x and y coordinates. Even with the additions of water and refractive index liquid, no significant shift in coordinates value were measured. For soil dust, the initial coordinates of x and y were observed to be slightly close to the yellowish region as (0.3157, 0.3213), and coordinates corresponding to the seven filters were found to spread out to locate at each color region, which range from (0.22, 0.19) to (0.55, 0.48). For additive cases, the movements of the coordinates in the vector scale from the initial point were different depending on water and refractive index liquid and an increased shift was found in the case of the refractive index liquid. This might be attributed to the intrinsic yellowish color of soil dust and deep yellowish color of refractive index liquid.

In Table 2, the white color group, talc and gypsum powders are characterized. The intensity of color, Y values for the as prepared and additive cases under the yellow filter and no filter were found to be highest and similar to the previous soil dust case. This is also due to the intrinsic color of the two powders. White can easily reflect the colors of the yellow filter and refractive index liquid. Starting points (black dots) of white color powders were noticed to similarly position at (0.3127, 0.3191) and (0.3157, 0.3213), which are closer to the center white region than in previous cases. Seven color points were also found to locate at their own color region in the range of (0.18, 0.11) to (0.56, 0.54) for talc, and (0.18, 0.10) to (0.57, 0.53) for gypsum. Interesting results were observed for the additive cases. In the case of water, not much movement in the coordinates in the vector was seen in both cases, but a noticeable amount of shift was found for the refractive-index-added cases. As noted in the Figures of the talc and gypsum powders, the coordinates of the refractive-index-added cases ranged from (0.24, 0.19) and (0.46, 0.44) for talc and (0.20, 0.14) and (0.55, 0.48) for gypsum, respectively. The only exception was for the yellow filter in used condition and this was due to the similar color of the filter and refractive index liquid. Pine tree pollen was also characterized and it has a yellowish natural color. As expected, the initial coordinates were found to be around the yellow region in the diagram as (0.3938, 0.3796). Its intrinsic color is strong yellow. Thus, it appears that the influence of color and additives did not have an impact on the modulating of the coordinates in the diagram. The indices varied in a relatively narrow band from (0.24, 0.22) to (0.52, 0.45) for the as prepared, from (0.24, 0.19) to (0.53, 0.45) for water and (0.25, 0.44) for refractive index liquid.

Based on the coordinate value analysis, a noticeable difference was measured for the white-colored powders such as the talc and gypsum samples. For the white-colored powders, more distinct shifts to each color region were measured than for the yellowish pollen, soil and grey household dust. This may be attributed to the intrinsic color of the powder being close to white, more light reflected to the detector and induced to increased intensity to the spectral sensor. Meanwhile, other non-white, yellowish and grey powders were detected at lower intensity.

In addition, overall chromaticity values for the yellowish powders were observed to shift into the yellow region in the diagram, in an upper left direction from the central white region.

Similar experiments were studied by Dang et al. In their report, the chromaticity value of five different colored inorganic pigments of drawing points revealed corresponding measured chromaticity value according to their color [16]. For water- and refractive-index-liquid-added cases, obvious differences in chromaticity values were observed. In Figure 3a,b, household dust revealed to shift more in the yellow and red regions, which correspond to a long wavelength range in the light spectrum. However, soil samples as shown in Figure 3c,d were detected to have more movement in the red and green regions. In the case of the pollen sample, no significant change in chromaticity values were observed under additive conditions. For the talc and gypsum powders, an obvious shift for the talc was observed only for the refractive-index-liquid case and chromaticity values were centered in the white region more than any other samples as shown in Figure 4a,b. This is well described in the previous study and in accordance with results [17]. This means that more white light is reflected to the chromameter detector. For the comparison with talc, the same experiments were executed for the gypsum powder as shown in Figure 4c,d. Under the chromameter measurement, no noticeable difference was observed. This means that similar intrinsic colors and particle shapes can be hardly differentiated under the chromameter study.

3.2. Light Spectrum Detection

A spectral sensor was used to characterize the light spectrum of samples under experimental conditions. Figure 6a,b shows the spectrum of reflected light for household dust and talc powder as a function of wavelength. The same measurement was performed under different conditions. Details of the measurements are summarized in Tables 4 and 5 according to peak position and peak intensity ratio. As shown in Tables 4 and 5, five samples revealed obvious difference in terms of peak position and peak intensity after the additive treatment and cellophane filter usage.

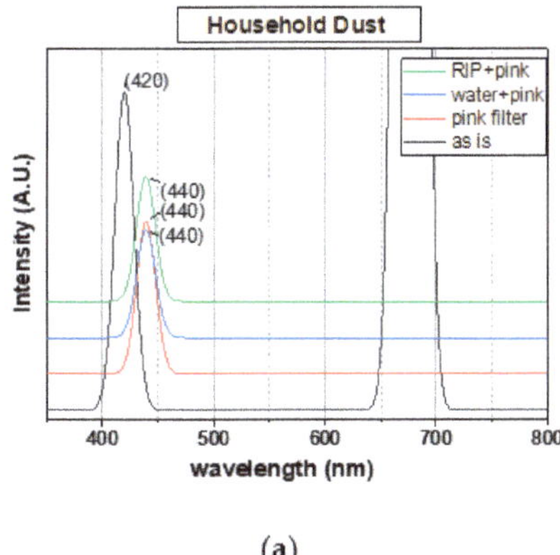

(a)

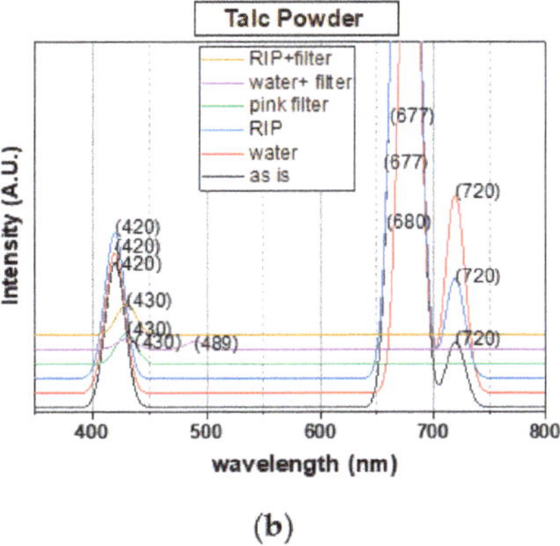

(b)

Figure 6. (**a**) Spectrum of reflected light of household dust according to the sample as prepared (black), pink filter(red), water + pink filter (blue), and RIL + pink filter (green); (**b**) Spectrum of reflected light of talc powder according to sample status as: as prepared (black), water added (red), RIL added (blue), pink filter (green), water + pink filter (violet), and RIL + pink filter (brown).

In Table 5, peak intensity at each wavelength was calculated in a ratio. The peak intensity values at low wavelength for samples are regarded to "1" as a base, then, peak intensities at other higher wavelengths were divided by base peak intensity. After pink filter usage, the overall light intensity decreased and was calculated with the same method for three household dust, talc and gypsum samples. Therefore, a higher ratio value means a relatively strong peak and vice versa. Two representative samples, household dust and talc powder, were graphed to scrutinize the light spectrum changes by additives and filter and compared by peak position in wavelength and peak intensity as well. In the case of household dust as shown in Figure 6, two peaks at 420 and 678 nm in wavelength were observed.

Under the pink filter, the peak at 678 nm was observed to be removed and a peak shift from 420 to 440 nm was also observed. This is due to the light filtering at a long wavelength range by the pink cellophane. Even when water and refractive index liquid are added, no significant shift was observed. These results correspond well with the measured chromaticity values under additives. Chromaticity values are previously discussed in Figure 3,b; overall values are centered to the white region with relatively more shifts for the blue, red, and yellow regions observed after refractive index liquid was added. Meanwhile, the talc powder revealed slightly different results than that of the household dust. The light spectrum for as prepared talc powder shows three peaks at 420, 677, and 720 nm, respectively. Two peaks at 420 and 677 nm showed a relatively low ratio value of 6.32, but no noticeable changes were observed for both additives, which show ratio values of 4.51 and 6.21 in Table 5. Considering the combined conditions of the pink filter and water addition, a peak shift from 420 to 430 nm and an additional peak was observed at 490 nm. Other peaks at a long wavelength range were filtered the same as previously. For comparison, similar gypsum power was also characterized under pink filter and additives conditions. The peak at a short wavelength region shifted from 420 to 453 nm, but this was a big difference from the additional peak observed for the talc case at approximately 490 nm, which was not detected. Instead, an additional peak at approximately 820 nm was detected for the water-added gypsum sample. These results appear to be correlated with the absolute amount of shift in the chromaticity values which are larger for gypsum than for the talc powder.

4. Discussion

From the analysis of color and reflected light of five different particulate matters according to different liquid additives, chromaticity coordinate values in the diagram were effective by measuring the shift amount in the vector under the refractive-index-liquid-added case to identify the non-white color group particulate matters including yellowish pine tree pollen, soil dust and grey household dust from the white color group particulate matters. Soil dust under refractive index liquid and cellophane filters revealed the largest movement in coordinates. Talc and gypsum powder belonging to the intrinsic white color group were identified by comparing the peak at around the reflected light spectrum. For the light spectrum cases, peaks of approximately 420 nm can be regarded as the guideline peak to determine the influence of the filter and additives according to the type of sample. Relative peak intensities of the other peaks were compared with a ratio of the peak intensity at 420 nm. The peak at approximately 720 nm in wavelength was solely observed for talc powder. Water and refractive index liquid could selectively shift the peak at approximately 420 nm and 679 nm to 433 and 690 nm, respectively, only for gypsum powder. More distinct dependency on filter color was observed for gypsum powder rather than household dust, pollen, soil and even similar talc powder. This was in good agreement with the method called the "browning index" used in other color and light characterization methods using a spectrophotometer [18].

In this study, we were concerned with indoor particulate matters, which possibly can be generated by daily activities such as cooking, vacuum cleaning and installed construction materials, or transported from outdoors by daily activity. It was our intention to find the feasibility of an optical approach to assist in the identification of the source of particulate matters. In cases where a large portion of particulate matters are observed to inflow from outdoor sources, people can be air showered at the front door and contaminants are pre-removed before transporting these into the indoors. In other cases, more indoor sources are found to contribute to indoor air contamination, and mechanical ventilation can be operated with a priority to outflow the indoor air to maintain the air quality level. Our study can at least contribute to this decision making as the first stage at a pre-screening level, but results showed that it has limitations in distinguishing the similar intrinsic color of powders. Light intensity was also vulnerable to the light bulb's source (>6500 k) and power. In addition, other particulate matters such as organic compounds and black carbon, generated from combustion, should be studied to assist in supporting these methods. In future studies, other chemical additives should be tested to reveal different colors after their reaction with white particulate matters. Another approach can be attempted to combine the presorting method by using a camera according to its color and shape to improve the identification rate by a simple method.

5. Conclusions

Color- and light-spectrum-assisted optical sensing of particulate matters was operated on a laboratory scale and under controlled experimental conditions. The chromaticity values and reflected light spectrum in terms of wavelength and intensity revealed the difference between samples according to color and water or refractive-index-liquid additives. Noticeable results can be summarized as below.

Five different particulate matters such as household dust, soil dust, pine tree pollen, talc and gypsum powders were optically characterized under as prepared, water-added and refractive-index-liquid-added cases combined with seven colors of light filter.

Depending on the intrinsic color of the as prepared sample, the non-white color group particulate matters including grey household dust, yellowish soil dust and pine tree pollen were selectively detected by observing the amount of chromaticity value shift in the vector. The white color group samples such as the talc and gypsum powders were distinguished by observing the unique peak positions in wavelength using a spectral sensor.

The intrinsic color of particulate matters can be the distinct point for identifying the types of non-white group samples such as household dust, soil dust and pine tree pollen under chromameter characterization rather than for the white color samples. Under the combined conditions of a cellophane filter and refractive index liquid, yellowish soil dust and pine tree pollen showed relatively large shifts in the chromaticity diagram in the vector. However, white color samples including talc and gypsum powers did not reveal a noticeable change. This was attributed to the intrinsic white color of powders under the xenon lamp.

In light spectrum measurements, the white color group samples showed more meaningful results than previous color measurements. All samples had consistent peaks at approximately 420 nm and 680 nm in wavelength. At other positions, a unique peak was observed depending on the type of particulate matter. Talc powder was the only sample that did not show any change in spectrum regardless of the additives and under no light filter conditions. An additional peak was observed at approximately 720 nm with water.

In the case of the pink filter, the gypsum sample revealed an obviously unique result in that a distinct peak at approximately 820 nm was detected for the as prepared case and peak position shifts from 439 nm to 453 nm were observed. When both water and the pink filter were used, the peak at approximately 490 nm was the index peak for talc powder. It can be concluded that the light spectrum study was more effective to identify white color powders and distinguish them from other non-white group powders in terms of distinct peak position and peak shift.

Two approaches using a chromameter and spectral sensor were attempted and meaningful results were observed. Simple color and reflectance change by water or refractive index liquid were able to draw the noticeable differences in observation. However, these still have limitations in intuitive observation, such as the way of the camera. Post analysis was inevitable to perform chromaticity value sorting and calculation.

It was our observation that there is a noticeable relationship between the intrinsic color of the particulate matter and the appropriate approach, whether it is chromaticity value, shift or the peak of the reflectance spectrum. In addition, water or refractive index liquid were found to be effective to enhance the relative deviation in color or reflected light depending on the type of particulate matter. An appropriate combination of chromameter and spectral sensor can be an alternative approach to detect particulate matters with higher selectivity. In future studies, various colors of particulate matters are required to be characterized to determine the relationship between the intrinsic color of particles and optical identification. It is also our hope that particulate matters less than 10 and 2.5 μm in aerodynamic diameter should be separated to investigate their size dependence under the above approaches.

Author Contributions: Conceptualization, methodology, validation, formal analysis, investigation, writing—original draft preparation, H.A.; data acquisition, experiments, data processing, H.-J.C.; writing—review and editing, G.-S.C.; resources, project administration, J.S.K. All authors have read and agreed to the published version of the manuscript.

Funding: This research was supported by a grant (20AUDP-B151639-02) from KAIA Urban Construction Research Program funded by Ministry of Land, Infrastructure and Transport of Korean government.

Acknowledgments: Authors would like to thank KICT supporting division members for their sincere assistance regarding this research related experiments, data acquisition, and sample preparation from the field and laboratories.

Conflicts of Interest: To the best of our knowledge, authors have no conflict of interest, finance or otherwise.

References

1. Biliaiev, M.M.; Kharytonov, M.M. Numerical Simulation of Indoor Air Pollution and Atmosphere Pollution for Regions Having Complex Topography. In *Air Pollution Modeling and Its Application XXI*; Springer: Berlin/Heidelberg, Germany, 2011; pp. 87–91.
2. Chafe, Z.; Brauer, M.; Klimont, Z.; Dingenen, R.V.; Mehta, S.; Rao, S.; Riahi, K.; Dentener, F.; Smith, K.R. Household cooking with solid fuels contributes to ambient $PM_{2.5}$ air pollution and the burden of disease. *Environ. Health Perspect.* **2014**, *12*, 1314–1320. [CrossRef]
3. Afshari, A.; Matson, U.; Ekberg, L.E. Characterization of indoor sources of fine and ultrafine particles: A study conducted in a full-scale chamber. *Indoor Air* **2005**, *15*, 141–150. [CrossRef]
4. Colome, S.; Kado, N.; Jaques, P.; Kleinman, M. Indoor-outdoor air pollution relations: Particulate matter less than 10 µm in aerodynamic diameter (PM_{10}) in homes of asthmatics. *Atmos. Environ. Part A Gen. Top.* **1992**, *26*, 2173–2178. [CrossRef]
5. Tasić, V.; Jovašević-Stojanović, M.; Vardoulakis, S.; Milošević, N.; Kovačević, R.; Petrović, J. Comparative assessment of a real-time particle monitor against the reference gravimetric method for PM_{10} and $PM_{2.5}$ in indoor air. *Atmos. Environ.* **2012**, *54*, 358–364. [CrossRef]
6. Kelly, K.E.; Whitaker, J.; Petty, A.; Widmer, C.; Dybwad, A.; Sleeth, D.; Martin, R.; Butterfield, A. Ambient and laboratory evaluation of a low-cost particulate matter sensor. *Environ. Pollut.* **2017**, *211*, 491–500. [CrossRef]
7. Vázquez, L.; Zorzano, M.; Jimenez, S. Spectral information retrieval from integrated broadband photodiode martian ultraviolet measurements. *Opt. Lett.* **2007**, *32*, 2596–2598. [CrossRef]
8. Li, H.; Sang, X. LED array light source illuminance distribution and photoelectric detection performance analysis in dust concentration testing system. *Sens. Actuators A Phys.* **2018**, *271*, 111–117. [CrossRef]
9. Vaca-Oyola, L.S.; Marín, E.; Rojas-Trigos, J.B.; Cifuentes, A.; Cabrera, H.; Alvarado, S.; Cedeño, E.; Calderón, A.; Delgado-Vasallo, O.A. Liquids refractive index spectrometer. *Sens. Actuators B Chem.* **2016**, *229*, 249–256. [CrossRef]
10. Shin, S.; Jung, C.; Kim, Y. Analysis of the measurement difference for the PM_{10} concentrations between beta-ray absorption and gravimetric methods at gosan. *Aerosol Air Qual. Res.* **2011**, *11*, 846–853. [CrossRef]
11. Sówka, I.; Chlebowska-Sty's, A.; Pachurka, Ł.; Rogula-Kozłowska, W.; Mathews, B. Analysis of particulate matter concentration variability and origin in selected urban areas in Poland. *Sustainability* **2019**, *11*, 5735. [CrossRef]
12. Jo, E.J.; Lee, W.S.; Jo, H.Y.; Kim, C.H.; Eom, J.S.; Mok, J.H.; Kim, M.H.; Lee, K.; Kim, K.U.; Lee, M.K.; et al. Effects of particulate matter on respiratory disease and the impact of meteorological factors in Busan, Korea. *Respir. Med.* **2017**, *124*, 79–87. [CrossRef]
13. Zhang, K.; Shang, X.; Herrmann, H.; Meng, F.; Mo, Z.; Chen, J.; Lv, W. Approaches for identifying $PM_{2.5}$ source types and source areas at a remote background site of South China in spring. *Sci. Total Environ.* **2019**, *691*, 1320–1327. [CrossRef]
14. Ahn, H.; Jung, B.K.; Joo, J.C.; Park, J.R. Spectral sensing of asbestos according to concentration in various asbestos containing materials. *Appl. Mech. Mater.* **2014**, *627*, 7–11. [CrossRef]
15. Mortimer, R.J.; Varley, T.S. Quantification of colour stimuli through the calculation of CIE chromaticity coordinates and luminance data for application to in situ colorimetry studies of electrochromic materials. *Displays* **2011**, *32*, 35–44. [CrossRef]
16. Dang, R.Y.; Liu, Y.; Liu, G. Chromaticity changes of inorganic pigments in Chinese traditional paintings due to the illumination of frequently-used light sources in museum. *Color Res. Appl.* **2018**, *43*, 596–605. [CrossRef]

17. Steuber, F.; Staudigel, J.; Stössel, M.; Simmerer, J.; Winnacker, A.; Spreitzer, H.; Weissörtel, F.; Salbeck, J. White light emission from organic LEDs utilizing spiro compounds with high-temperature stability. *Adv. Mater.* **2000**, *12*, 130–133. [CrossRef]
18. Ferrer, E.; Alegría, A.; Farré, R.; Clemente, G.; Calvo, C. Fluorescence, browning index and color in infant formulas during storage. *J. Agric. Food Chem.* **2005**, *53*, 4911–4917. [CrossRef]

Article

Dynamic Energy Performance Gap Analysis of a University Building: Case Studies at UAE University Campus, UAE

Young Ki Kim [1,*], Lindita Bande [1], Kheira Anissa Tabet Aoul [1] and Hasim Altan [2]

[1] Department of Architectural Engineering, United Arab Emirates University, Al Ain 15551, UAE; lindita.bande@uaeu.ac.ae (L.B.); kheira.anissa@uaeu.ac.ae (K.A.T.A.)

[2] Department of Architecture, Faculty of Design, Arkin University of Creative Arts and Design, Kyrenia 99300, Cyprus; hasimaltan@gmail.com

* Correspondence: kim9519021@gmail.com; Tel.: +971-3-713-5330

Abstract: As a result of an increasing demand for energy-efficient buildings with a better experience of user comfort, the built environment sector needs to consider the prediction of building energy performance, which during the design phase, is achieved when a building is handed over and used. There is, however, significant evidence that shows that buildings do not perform as anticipated. This discrepancy is commonly described as the 'energy performance gap'. Building energy audit and post occupancy evaluation (POE) are among the most efficient processes to identify and reduce the energy performance gap and improve indoor environmental quality by observing, monitoring, and the documentation of in-use buildings' operating performance. In this study, a case study of UAE university buildings' energy audit, POE, and dynamic simulation were carried out to first, identify factors of the dynamic energy performance gap, and then to identify the utility of the strategy for reducing the gap. Furthermore, the building energy audit data and POE were applied in order to validate and calibrate a dynamic simulation model. This research demonstrated that the case study building's systems were not operating as designed and almost a quarter of the cooling energy was wasted due to the fault of the building facility management of the mechanical systems. The more research findings were discussed in the paper.

Keywords: energy performance gap; dynamic energy performance gap; building energy audit; POE study; dynamic building simulation; simulation model validation and calibration

Citation: Kim, Y.K.; Bande, L.; Tabet Aoul, K.A.; Altan, H. Dynamic Energy Performance Gap Analysis of a University Building: Case Studies at UAE University Campus, UAE. *Sustainability* **2021**, *13*, 120. https://dx.doi.org/10.3390/su13010120

Received: 1 December 2020
Accepted: 22 December 2020
Published: 24 December 2020

Publisher's Note: MDPI stays neutral with regard to jurisdictional claims in published maps and institutional affiliations.

1. Introduction

More than a third of global energy consumption and CO_2 emission are associated with the building sector [1]; and awareness of the importance of buildings' energy performance related with CO_2 emissions has increased worldwide. The International Energy Agency (IEA) describes energy efficiency as the "first fuel" [2] and may be more important than any energy-generating technologies. Furthermore, Cullen et al. stated that one of the greatest potentials for improving energy efficiency and reducing CO_2 emission lies in the building sector [3]. However, as the demand for buildings has become increasingly complex, with design criteria such as multi-purpose plans, larger sizes and higher standards for services and user comfort in terms of Heating, Ventilation, Air Conditioning (HVAC), lighting, acoustic, and data processing [4], the analysis and evaluation of the energy performance of buildings have become more intricate than ever before. With the introduction of the first building regulations in the 1970s, energy conservation in general and later the interest in building energy consumption and its energy efficiency have increased multifold into the late 1990s and 2000s [5]. This strong interest and necessity have led to the development of a wide range of methodologies for predicting, analyzing, evaluating, and validating the energy performance of buildings. To date, numerous studies have shown that while in use, buildings usually do not perform as predicted [6–13]. This discrepancy is referred to as the 'Energy Performance Gap' and is typically demonstrated through an energy audit exercise. This is even more critical, given the global call to reach Zero Energy Buildings, i.e., a building that combines passive building designs and highly energy efficient systems to minimize the heating, cooling, ventilation, lighting and electricity demand and consumption with on-site energy generation system, such as PVs, combined heat and power (CHP), wind turbines to recover the energy demand and consumption from the building to achieve the energy balance [14]. Regardless of how energy-efficient a building is designed to be and an efficient system being applied, zero energy buildings cannot be achieved if there is an energy performance that is different from the plan in the operation stage of the building, which further heightens the need to underhand the reason behind the energy performance gap in order to better control it.

The main purpose of an energy audit is to identify opportunities to improve the energy efficiency of a building and it typically takes a whole building assessment approach from building envelope to mechanical systems, operations and maintenance, and building schedules. Therefore, wholesale building audits provide the most accurate picture of existing conditions as well as an understanding of energy savings' opportunities in buildings. The American Society of Heating, Refrigerating and Air-Conditioning Engineers (ASHRAE) defines three levels of audit to assess the building inside and out. They are: Level 1: Site assessment or preliminary audits, Level 2: Energy survey and engineering analysis audits, and Level 3: Detailed analysis of capital-intensive modification audits [15]. The aim of post occupancy evaluation (POE) is to assess the performance of a building after it is occupied and to address any performance and energy gap issues. It will also help the designers with valuable feedback on its actual performance as opposed to its assumed one and to improve the building mechanical design process. Moreover, by using the POE study data to predict the energy performance of the in-use building, it will help calibrate a building energy model to actual operating conditions [11]. Detailed dynamic simulation models (DSMs) are able to obtain more accurate predictions of energy performance/consumption in buildings. The DSMs are more often used in the study of commercial buildings' performance because they allow the collection of extensive input data, and a database for materials and systems [16]. Nonetheless, it must be recognized that despite these attributes and many other additional features, there are still significant differences between the expected and actual energy consumption of non-domestic buildings.

In this case study, a university building's energy audit and a POE assessment were carried out. The results were analyzed to identify the underlying factors of the energy performance gap. In addition, the energy audit data and POE analysis were used to calibrate and validate a dynamic simulation model. Finally, a gap reduction achievement after the energy audit and POE analysis was evaluated using a dynamic simulation. Through this research, it shows, first, that the energy audit is an effective way to identify the underlying causes of the dynamic energy performance gap, and second, that the POE is able to produce more accurate dynamic energy simulation models with an indication for the potential increase in the indoor environmental quality, especially users' thermal comfort. The research findings will be implemented to reduce the energy consumption while improving user comfort for case study building under study as well as in other university buildings.

2. The Energy Performance Gap

2.1. The Energy Performance Gap Review Status

A number of studies show that in-use buildings do not necessarily perform as designed [6–13]. Findings from the PROBE studies (post occupancy review of buildings and their engineering) addressed that actual energy consumption when in-use could be twice as much as predicted [9]. Low Carbon Building Programme and Carbon Trust's Low Building Accelerator have studied that in-use energy consumption could be up to five times higher than prediction [11].

Figure 1 shows that the energy performance gaps in between prediction and actual electricity consumption in three building sectors in the UK: schools, general offices and university buildings. The value of energy consumption was averaged for each sector.

As it can be seen in Figure 1, the actual electricity consumption could be 60–70% higher than predicted in schools and general offices, and even over 85% higher in university campuses [17]. Another study by Chris et al. [18] is summarized in Table 1, which gathered the reported case studies of the discrepancies in the energy usage in different types of buildings which are located in different climates, different building types with an average performance gap between the predicted and actual energy consumption. It also confirms that gaps between the design/prediction and actual buildings energy consumption have become a matter of fact.

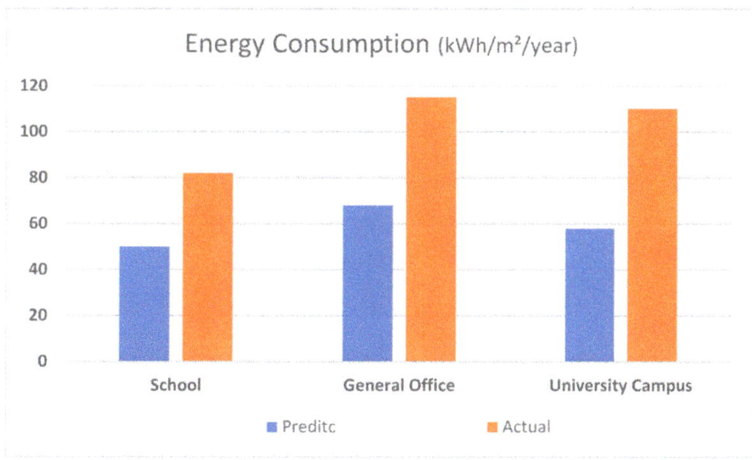

Figure 1. CarbonBuzz median electrical consumption per-sector: predicted vs. actual (Source: [11]).

Table 1. Discrepancy in different buildings.

Building Type	Gap Average	Total Number of Reported Cases
Office	16%	25
School	67%	11
Multipurpose	45%	8
University	67%	3
Laboratory	32%	2
Restaurant	31%	2
Retail	37%	2
Supermarket	−10%	2
Library	8%	2

Again, Figure 1 and Table 1 show the energy performance gap average from different studies and building types. In order to satisfy Zero Energy Building/Net Zero Energy Building (ZEB/NZEB), it seems really important to reduce the energy performance gap that occurs in the actual operation of the building. Without reducing this gap, well planned buildings to reduce energy consumption and renewable energy technologies planned to cover the reduced energy would be meaningless. Therefore, reducing the gap between the design/prediction and actual buildings' energy consumption needs to be considered.

2.2. Classification of the Gap

Figure 2 shows an overview of the common root causes of performance discrepancies that exist at each stage of a building's life cycle. Based on the S-curve visualization of a building performance proposed by Bunn and Burman [19], the performance gap can be classified in three categories: 'Perceived gap', 'Static gap', and 'Dynamic gap' [18]:

- Perceived gap: compares predictions from compliance modeling to performance modeling energy consumption;
- Static gap: compares predictions from performance modeling to measured energy use; and
- Dynamic performance gap: utilizes calibrated predictions from performance modeling with measured energy use taking a longitudinal perspective to diagnose underlying issues and their impact on the performance gap.

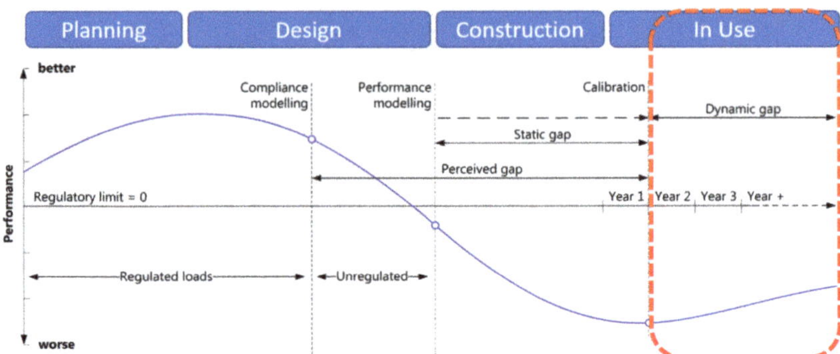

Figure 2. Three possible performance gaps throughout the life cycle of a building (s-curve visualization was adapted from [19]).

These classifications can help identify the underlying causes of the energy performance gaps throughout the life cycle of a building and highlights specifically in each phase of the building's life the potential needs to reduce the performance gaps. In this research, the 'Dynamic performance gap' has been adopted to identify the underlying cause of the gaps as described below.

2.3. Sources of the Daymic Performance Gap

The dynamic performance gap between the predicted and actual energy performance in buildings results from several causes. Causal factors for prediction and actual performance aside, the current predictions tend to be unrealistically low, while the actual energy performance is generally unnecessarily high. This can ultimately be associated with the lack of feedback for the actual use and operation of the building and its associated energy consumption. Currently, there is a great lack of information on the actual energy performance of existing building sectors [20]. Such a great lack of information leads to an increase in the gap between prediction and in-use, and an evident failure to achieve the reducing energy performance gaps in the built environment [21].

The study by the Lawrence Berkeley National Laboratory studied 85 commercial buildings in the US and listed over 3500 deficiencies from the case study buildings and 53% of them were related with the dynamic performance gap and 32% to the overall HVAC system (Figure 3) [22]. The most common deficiencies in the case study buildings were modifications of set-points, scheduling, control sequences, calibration, mechanical fixes and equipment replacements. This shows that operation and control, and management and maintenance, need focusing to reduce the dynamic performance gap during the in-use phase of life cycle of the buildings.

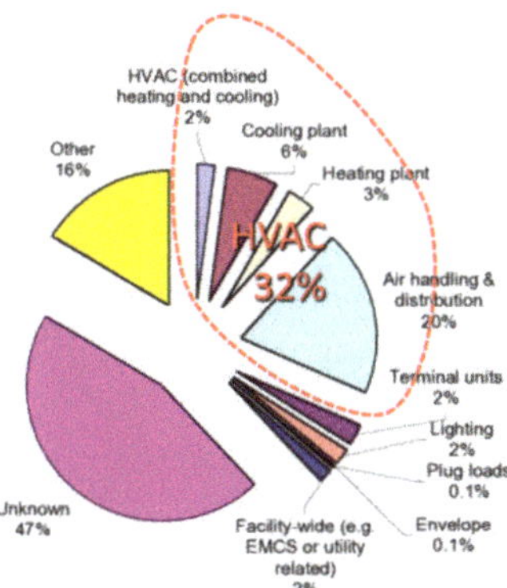

Figure 3. Number (%) of deficiencies identified by in-use building system (Source: [22]). EMCS: Energy Management Control System.

3. Materials and Methods

Taking a case study approach, this research analyzes the energy performance of a university building in the United Arab Emirates University (UAEU), located in Al Ain, UAE. As discussed earlier, this research was guided by ASHRAE Building Energy Audit Level 1 methodology, followed by POE monitoring study. Results from the energy audit and POE monitoring data were used to calibrate and validate the dynamic energy simulation model, aiming to produce more accurate predictions of energy consumption and finding the source of discrepancy of dynamic energy performance gap to reduce the gap with the objective to propose recommendations for an improved indoor environmental quality.

3.1. Case Study Building Description

The case study building is known as the 'F1 Building' (Figure 4) and is located in the male side campus of UAEU. The current building houses three colleges and 13 departments located in three wings of three floors each (e.g., College of Engineering, College of Science, and College of Food & Agriculture) with over 600 occupants and a total building area of 21,360 m^2.

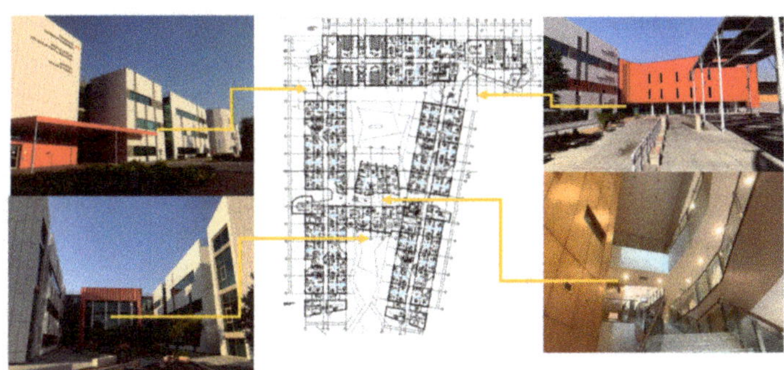

Figure 4. Floor plan and a view of the case study building, UAEU Campus (Source: Author).

It was planned and constructed as an energy efficient building and was completed in 2011. The building is fully air-conditioned by 13 rooftop Air Handling Units (AHUs) which provide cooling and fresh air to all floors and atrium. Variable Air Volume systems (VAVs) were part of the design strategy to save energy when the building is not fully occupied during weekends, holidays and vacations (Table 2).

Table 2. Detail of the case study building.

	Name	Total Number of Reported Cases
Architecture	Site	Al-Ain, UAE
	Programs	Offices, laps and lecture rooms
	Building area	7120 m^2
	GFA (Gross Floor Area)	21,360 m^2
Mechanical	Cooling plant	Campus District cooling
	Cooling system	13 AHUs on rooftop
	System control	VAVs
Electrical	Lighting	T5 flounce lamp (offices, laps, lecture rooms), energy efficient light bulbs (circulation)
	Target illumination	400–500 Lux (Offices), 200–300 Lux (circulation)
	Renewable	N/A

3.2. Energy Audit and POE Monitoring

ASHRAE Energy Audit Level 1, which is 'Site Assessment or Preliminary Audits, was carried out to assess the case study building. Typically, energy audits take a whole building approach by examining the building envelope, building systems, operations and energy consumption, especially the electricity usage. The energy data were provided by the Facility Management department but only monitored total electricity consumption.

The monitoring of POE data was conducted for one calendar year from January to December 2019 via walkthrough inspections with hand-carrying devices to measure the Lux and noise level, and the fixed unit measuring data logger for measuring temperature, Relative Humidity (RH), noise (Db), lighting (lux), CO_2, Particulate Matter (PM) 2.5, PM 10 and Total Volatile Organic Compounds (TVOCs) on every floor (Figure 5). Table 3 shows the details of the energy audit and POE study in this research. The energy audit data were used for making an accurate energy simulation model that reflects the actual state of the building, such as the building envelope information (U-value and construction details), HVAC systems, level of airtightness, and building operation and schedule, and it shows in Table 4. In the case of POE data, it was being used to implement the indoor environment of the actual building in the simulation model. Internal temperature, RH, and lighting data especially will be used to define the input data for the simulation model, such as set the indoor target temperature and lighting power density according to the operating schedule. Both the energy audit and POE were used for identifying the dynamic performance gap, management problems, and other aspects that may have negative impacts on indoor environmental quality.

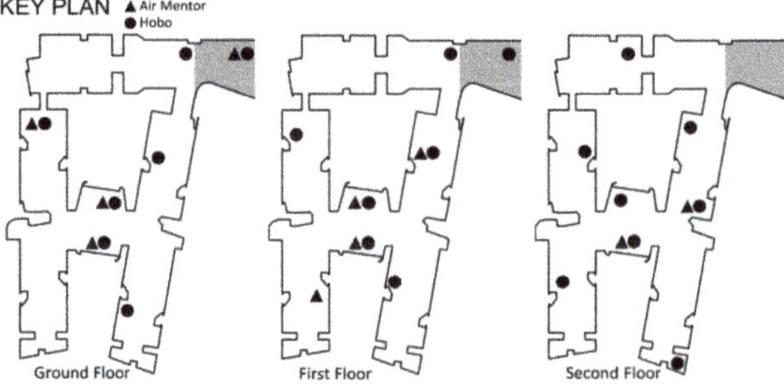

Figure 5. Floor plan and a view of the case study building, UAEU Campus (Source: author).

Table 3. Detail of the energy audit and post occupancy evaluation (POE) study.

	Measurements	Devises	Monitoring	Image
Energy Audit	Building envelops HVAC system and operation Energy consumption	Testo 872 Testo 440 dp - -	Surface temperature Airleakage MEP CAD files with FM team interviews 2019 energy consumption was provided by FM team	
POE Study	Thermal comfort	HOBO	Temperature (°C) RH (%) Lighting (lux) Every 15 min	
	Acoustic and lighting	PRECISION GOLD Environment Meter	Acoustic (dB) Lighting (lux) Spot measurement	
	Indoor Air Quality (IAQ)	Air Mentor Pro	Temperature (°C) RH (%) PM 2.5 ($\mu g/m^3$) PM 10 ($\mu g/m^3$) CO_2 (ppm) TVOC (ppb) Every 15 min	

MEP CAD: Mecahnica, Electrical, Pluming Comupter-aided Design; FM: Facility Management; PM: Particulate Matter; TVOC: Total Volatile Organic Compounds

Table 4. Simulation models' specifications and input data.

	Baseline Model (ASHRAE 90.1)	Abu Dhabi Code (National Code)	As-Designed Model (Prediction)	In-Use Model (Current)
Wall Roof Floor Window	0.705 ($W/m^2 \cdot K$) 0.360 ($W/m^2 \cdot K$) 1.986 ($W/m^2 \cdot K$) 6.81 (SHGC 0.25)	0.329 ($W/m^2 \cdot K$) 0.329 ($W/m^2 \cdot K$) 1.823 ($W/m^2 \cdot K$) 2.2 (SHGC 0.25)	0.537 ($W/m^2 \cdot K$) 0.403 ($W/m^2 \cdot K$) 1.423 ($W/m^2 \cdot K$) 2.2 (SHGC 0.25)	0.537 ($W/m^2 \cdot K$) 0.403 ($W/m^2 \cdot K$) 1.423 ($W/m^2 \cdot K$) 2.2 (SHGC 0.25)
HVAC	Package Rooftop DX, CAV (System 3)	Package Rooftop DX, CAV	Package Rooftop DX, VAV	Package Rooftop DX, CAV **
Cooling Set Temp	24 °C (28 °C *)	24 °C (28 °C *)	24 °C (28 °C *)	21 °C **(23 °C *)
Airtightness	0.6 ACH	0.6 ACH	0.6 ACH	1.5 ACH **
Lighting	9.7 W/m^2	9.7 W/m^2	9.7 W/m^2	9.7 W/m^2 ***

* cooling setback temperature (ASHREAE Standard 55); ** input data from the energy audit and POE studies; *** lighting power density remains the same for all cases. However, the in-use model's schedule is set as '24 h on'; CAV: Constant Air Volume; AHC: Air Change Hour.

3.3. Dynamic Energy Simulation

A dynamic simulation energy model was developed for the case study building based on the findings generated by the energy audit and POE study. The simulation model was calibrated with the actual performance including indoor environmental condition, HVAC operation, airtightness and lighting schedule. To define the energy performance gap in case study buildings, four different simulation models were developed and each of the models was named as the baseline model as based on ASHRAE Standard 90.1–2007, Abu Dhabi code, the as-designed model for prediction and in-use model for current building. Each model's specifications and input data are shown in Table 4. The input data and building specifications for in-use model were based on the energy audit and POE data analysis results.

In this research, the dynamic performance gap due to the operation method of the building system and management was investigated. In order to investigate the dynamic energy performance gap, the Scenario C and D's basic building external envelopes' thermal performance, lighting power density, and HVAC system were modeled in the same way (Table 4). Basically, the energy consumption of the building due to changes in the operation method of the HVAC system from VAV to Constant Air Volume (CAV), with the internal temperature setting from 24 to 21 °C during occupied hours—with the operation schedule of the lighting equipment in the corridor and the remaining entrance doors open—was investigated. All these operational and managing changes were based on the energy audit and POE studies.

To perform the dynamic energy simulation, this study used the Trane TRACE 700 v 6.1 which was designed to simulate the building energy performance check, especially the detailed HVAC systems. The TRACE 700 complies with ASHRAE Standard 90.1–2007 and 2010 Appendix G for the performance rating method for Leadership in Energy and Environmental Design (LEED) analysis. The Trace 700 is validated by ASHRAE Standard 140–2011 and 2014 for using the dynamic simulation to study the energy consumption of whole building with a detailed energy consumption breakdown of HVAC systems and plants [23]. A computer model for dynamic energy simulation was created using REVIT 2017, which was converted to a gbXML file and imported into TRACE700. The computational model and the interface of TRACE 700 is shown in Figure 6.

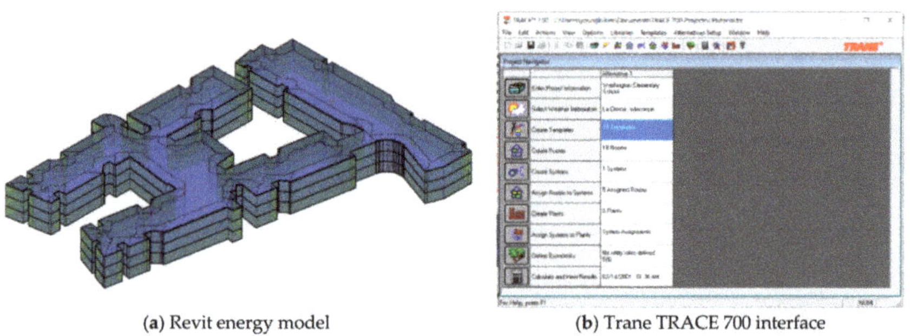

(**a**) Revit energy model (**b**) Trane TRACE 700 interface

Figure 6. Energy simulation model and TRACE 700 interface.

In the case of the simulation model, it was largely divided into the conditioned zone and unconditioned zone. In the case of the conditioned zone, it was defined as the area cooled/conditioned by HVAC systems such as offices, corridors and stairs, storages and toilets. For the unconditioned zoned, it was defined as the area not conditioned by HVAC systems such as the emergency stairs and emergency exits in this paper. As can be seen in Figures 4 and 5, the simulation study was considered as a whole building simulation. A baseline model was developed based on ASHRAE Standard 90.1–2007 Appendix G method and this baseline was used to evaluate the expected energy performance of the case study building in international standards. The American Society of Heating, Refrigerating and Air-Conditioning Engineers (ASHERAE) standard, which was in effect when the building was designed, was the latest version in 2007. Therefore, the ASHRAE version used to create the baseline in this study was 2007. The benchmark model is the 'as-designed model' which was not a calibrated model, by adapting the energy audit and POE data for calibration. After calibrating with the energy audit and POE data, which is the 'in-use model', it shows the actual building energy performance. To compare the 'as-designed model' against the 'in-use model', it is able to identify the dynamic performance gap in the case study building during the in-use life cycle phase (Table 5).

Table 5. Dynamic simulation scenario.

	Scenarios	Purpose
Simulation	A: Baseline B: National Code C: Benchmark D: Actual Building	Define the level of energy consumption at international level. Define the level of energy consumption at current national level. Define the level of energy consumption at as-designed level. To identify the dynamic performance gap in actual building.

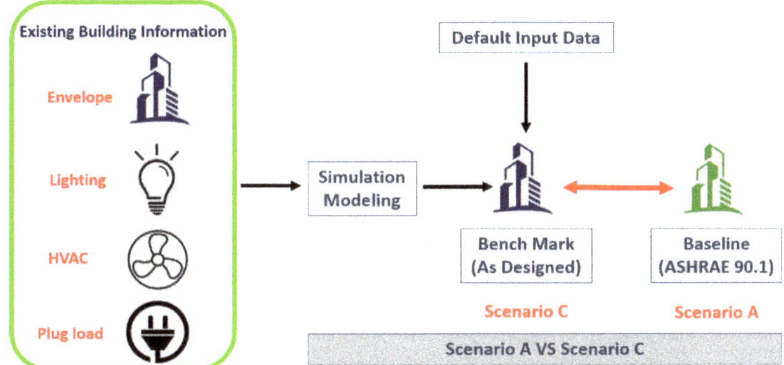

(a) Defining the level of energy consumption in international level

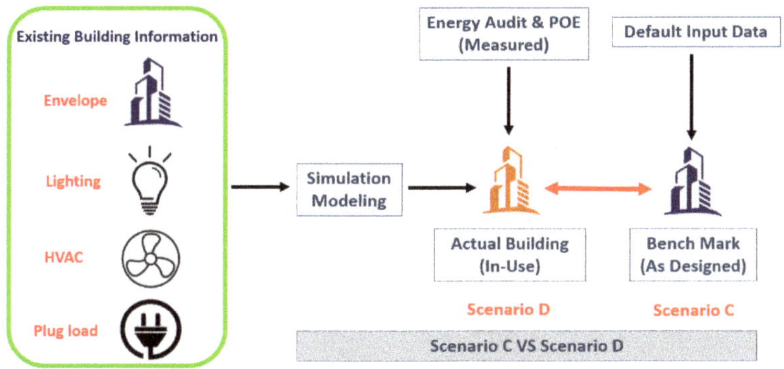

(b) Defining the dynamic energy performance gap

4. Results and Discussion

The energy audit and POE were conducted for one calendar year from January to December 2019, and a dynamic simulation was carried out for an annual energy consumption study. From these studies, several deficiencies of energy performance gap were found in the case study building and are reviewed below.

4.1. Energy Audit Analysis

Through several stages of energy audit, four major deficiencies were found. One was during HVAC operation and performance check, the AHU fans were operated as CAV (constant air volume) rather than VAV (variable air volume), which would increase the operation time and the energy consumption from a supply air fan. These fans continuously work with full power and supply air sets air temperature between 11.5 and 13.2 °C, which was also quite low due to satisfying the internal setting temperature during occupied hours (measured at 21.5 °C), and even during unoccupied hours (measured at 23.8 °C). Similar problems were indicated in lighting controls and scheduling. The lighting remained turned on during the daytime in the corridors.

The light fixtures are located next to the curtainwall in the case study building (Figure 7d) and it is almost 13% of the circulation lighting. Another similar operation and maintenance problem was the exit doors located in the exterior envelope. These doors remained widely open during occupied hours (Figure 7a,b), with the expected loss of airtightness and return air to AHUs for maintaining mixed air temperature with indoor target temperature. Figure 5 shows some examples of IR images of the current situation.

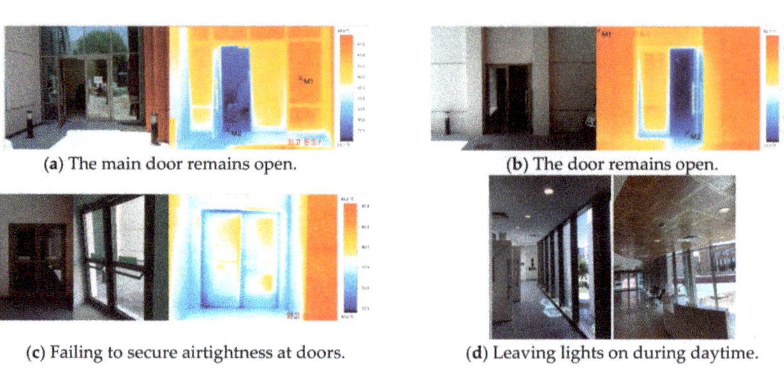

(**a**) The main door remains open. (**b**) The door remains open.

(**c**) Failing to secure airtightness at doors. (**d**) Leaving lights on during daytime.

Figure 7. Deficiencies from energy audit.

Table 6 shows that the location of the open doors and images taken by the IR camera and how much conditioned air escaped from the exterior building doors. This building has a total thirteen air conditioners installed on the roof of the building and supply 203,557 cubic feet per minute (CFM), about 30% (59,350 CFM) of which is supplied from the outside air, and the remaining 70% (144,200 CFM) is reused. To measure the escaped air from the doors, TESTO 440 dp was used and every 15 min an average CFM was measured at three of the doors which, all year round, remained open during the occupied hours. Based on the measurement, it is noticed that 10% (21,072 CFM) of supply air could escape from the building and 15 and 36% of the return air and the outside air, respectively. This direct air leakage from the doors affects the cooling energy consumption of the building and it can also affect on operating hours and energy consumption for AHU fans. These measurement data were used to calibrate the simulation model (Scenario C) to increase the model accuracy for validating the simulation model's (Scenario D) airtightness rate from 0.6 to 1.5 ACH which corresponds to the actual building.

4.2. POE Study Anlysis

Table 7 summarizes the results collected by the mobile and fixed monitoring data logger and devices. During the occupied hours, the indoor temperature was around 21.5 °C in both enclosed and open-plan offices. The monitored indoor temperatures were significantly lower than the ASHRAE Standard 55-recommended temperatures, which are 24 °C for occupied hours and 28 °C for unoccupied hours. This indoor setting temperature could have an effect on the energy performance gap in the case study building. To define the level of performance gap by setting the different indoor temperatures between as-designed model and in-used model, the monitored indoor temperatures were applied as indoor setting temperatures for in-use simulation model and ASHRAE-recommended temperatures for the as-designed simulation model. Except for the indoor temperature, the rest of the air quality measures such as noise, lighting, CO_2, TVOCs, PM 2.5 and PM 10 are within the comfort and safe ranges based on the WELL Building Standard recommendation.

Table 6. Infiltration of air from the opened doors.

	Supply Air	Return Air	Outside Air	Escape Air
CFM	203,557	144,200	59,350	21,072

6301 CFM

6363 CFM

8408 CFM

Table 7. Summary of Indoor Environment Quality (IEQ) data from the POE study.

	Temperature (°C)			RH (%)		
	Range	**Whole Average**	**Occupied Average**	**Range**	**Whole Average**	**Occupied Average**
East enclosed/private? Office	19.9–23.1	20.8	20.9	44.9–9.9	52.6	53.5
East open-plan offices	21.2–23.3	22.3	22.3	39.9–53.5	47.5	48.4
West private? Office	21.2–23.8	22.2	21.9	39.7–55.9	49.1	50.9
West open-plan offices	20.2–23.2	21.5	21.1	40.9–59.3	51.1	53.3
Corridor	20.9–23.7	21.6	21.5	45.6–58.8	52.7	53.8

4.3. Energy Performance Gap in the Case Study Building

The overall energy consumption with end-use results for each simulation model is illustrated in Figure 8. This indicates that the baseline model, which used the ASHRAE 90.1 Appendix G method, was the most energy consumed and the best case/benchmark was the as-designed model. To compare the energy consumption between the baseline model and the Abu Dhabi code, which showed a 7.4% lesser consumption per the Abu Dhabi code, where the reduction was mainly related to enhanced building envelopes only.

To improve the accuracy of the energy simulation models, the in-use model was calibrated with energy audit analysis and POE data such as internal target temperature, HVAC supply air fans operation method, corridor lighting schedule, and airtightness. After calibrating the model, in-use building consumed almost 25% more energy than the as-designed model (Figure 8). From this comparison study, it was noted that the dynamic performance gap could not be ignored as part of the life cycle of the buildings. The main discrepancies from this study were mainly related to operation and management issues such as doors remaining open, lights kept on during daytime, VAVs working as CAVs, and these issues can be solved without increasing the cost or extra investment to update the existing mechanical and operating systems.

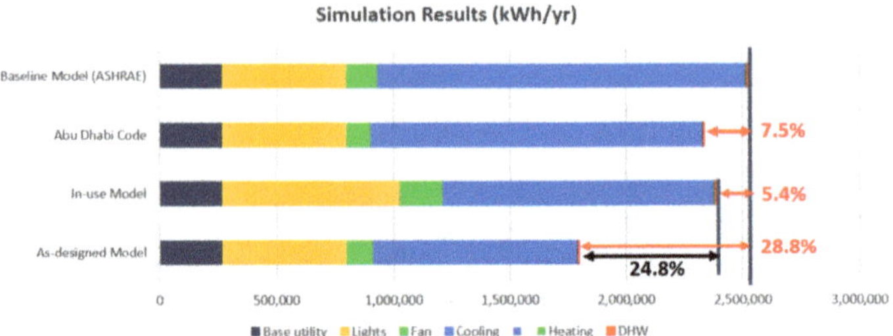

Figure 8. Dynamic energy simulation results.

After finding the dynamic energy performance gaps from the case study building through the series of simulation scenarios, it was simulated with a detailed dynamic simulation to investigate the degree of dynamic energy performance gaps of each end-use in the case study building. The energy consumption was increased by 29.7% for lighting use due to failing to operate the correct lighting schedules in the corridors, by 39% for fan power due to operating the VAV systems as CAV systems, and by 24.8% increased by cooling energy consumption and it was associated with an incorrect internal temperature setting, an incorrect operation of the HVAC systems and the failure to secure the airtightness of the case study building (Figure 9). A previous study from Papadopoulos et al. also found one major operational issue in the university building which was related to the thermostat setpoints [24] and showed the same recommendation found in this study.

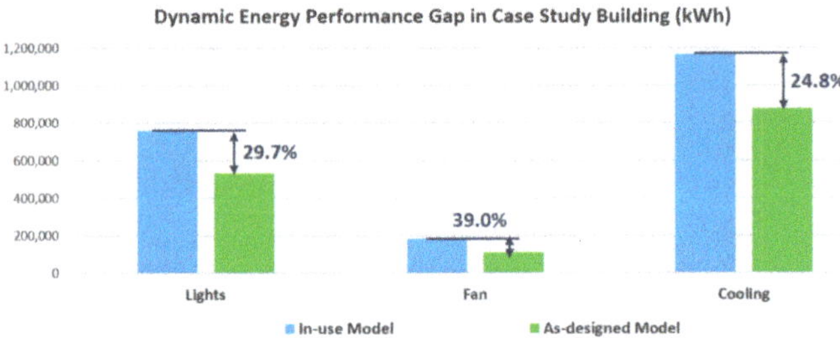

Figure 9. Energy performance gap by deficiencies.

The comparison between the total dynamic energy performance gap in the as-designed model and the in-use model was confirmed by the fact there was almost 25% of the gap. This shows that a building system with the correct operation with schedule, and careful management scheme would certainly reduce the energy consumption to bring it closer to the as-design level, and at the same time, it could improve the indoor environmental quality of the building.

5. Conclusions

More than a third of global energy consumption and CO_2 emission is generated by the building sector and the awareness of the importance of buildings' energy performance related to CO_2 emissions has increased worldwide. Improving energy efficiency is one of the key policy directions for the UAE government to tackle energy security, energy conservation, and climate change. As a direct response, to increase energy efficiency in the building sector in the UAE, substantial steps have been taken in recent decades such as setting up the minimum levels of U-values for building envelopes, introducing building rating systems, and improving HVAC systems' efficiency. However, there is still an extensive evidence which shows that the buildings usually do not perform as well as predicted whilst in-use and this discrepancy is commonly referred to as the "energy performance gap".

The focus area of this research is using the building energy audit data and the POE and to reduce the dynamic energy performance gap in one of the university buildings in the UAEU, UAE. To achieve the main objective of this study, the energy audit data and the POE, followed by a detailed dynamic simulation method, were applied to several simulation scenarios. Identifying the cause of discrepancies between the as-designed and actual building, and to increase the accuracy of the simulation model, POE and energy audit data were used for calibrating the simulation model (Scenario C) to validate (Scenario D).

The research clearly indicated that the case study building in-use condition was not operated as designed and almost a quarter of the cooling-related energy was wasted by mismanaged and poorly understood building's active system operations and management. This type of performance gap is commonly found in the UAE and can be easily solved without increasing the cost or extra investment for systems upgrade. This method is able to use the calibration of the actual conditions of the building with a computer-generated energy simulation model to improve the accuracy of the simulation model, and efforts to identify the underlying causes of the gap between numerical/computational predictions and actual usage as well. This research method is very cost effective, and less than one year of the return of investment could be achievable. It also shows that the energy audit and POE study are possible to reduce the dynamic energy performance gaps in the building sector by improving the indoor environment quality, especially after the buildings are occupied as in-use phase.

Author Contributions: Y.K.K. conceived the presented idea and developed the theory and performed the computations. Y.K.K. and L.B. verified the analytical methods. K.A.T.A. and H.A. encouraged Y.K.K. to investigate the POE study and supervised the findings of this work. All authors have read and agreed to the published version of the manuscript.

Funding: This research was funded by the UAEU, grant name and numbers: 2018 Start-UP (31N380).

Institutional Review Board Statement: Not applicable.

Informed Consent Statement: Not applicable.

Data Availability Statement: The data presented in this study are available on request from the corresponding author. The data are not publicly available due to the case study building is governmental facility.

Acknowledgments: The authors would like to thank the UAEU for supporting the study and allowing for the measurements to take place.

Conflicts of Interest: The authors declare no conflict of interest and the funders had no role in the design of the study; in the collection, analyses, or interpretation of data; in the writing of the manuscript, or in the decision to publish the results.

References

1. International Energy Agency. Total Energy Use in Buildings Analysis and Evaluation Methods. International Energy Agency, 2016. Available online: https://www.iea-ebc.org/Data/publications/EBC_PSR_Annex53.pdf (accessed on 24 December 2020).
2. International Energy Agency. Capturing the Multiple Benefits of Energy Efficiency. International Energy Agency, 2014. Available online: https://webstore.iea.org/capturing-the-multiple-benefits-of-energy-efficiency (accessed on 24 December 2020).
3. Cullen, J.M.; Allwood, J.M.; Borgstein, E.H. Reducing energy demand-What are the practical limits. *Environ. Sci. Technol.* **2011**, *45*, 1711–1718. [CrossRef] [PubMed]
4. Pérez-Lombard, L.; Ortiz, J.; Pout, C. A review on buildings energy consumption information. *Energy Build.* **2008**, *40*, 394–398. [CrossRef]
5. Davies, H. Tracing the Continuing Development of Part L. Modern Building Services, 2013. Available online: http://www.modbs.co.uk/news/fullstory.php/aid/12062/TracingthecontinuingdevelopmentofPartL.html (accessed on 24 December 2020).
6. Demanuele, C.; Tweddell, T.; Davies, M. Bridging the gap between predicted and actual energy performance in schools. In Proceedings of the World Renewable Energy Congress XI, Abu Dhabi, UAE, 25–30 September 2010.
7. Bordass, B.; Cohen, R.; Field, J. Proceedings of the International Conference on Improving Energy Efficiency in Commercial Buildings, Frankfurt, Germany, 19–20 April 2004.
8. Probe Archive Held by the Usable Buildings Trust (UBT). Available online: http://www.usablebuildings.co.uk/Pages/UBProbePublications1.html (accessed on 15 December 2020).

9. Bordass, B.; Cohen, R.; Standeven, M.; Leaman, A. Assessing building performance in use 3: Energy performance of probe buildings. *Build. Res. Inform.* **2001**, *29*, 114–128. [CrossRef]
10. Carbon Trust. *Closing the Gap-Lessons Learned on Realising the Potential of Low Carbon Building Design CTG047*; Carbon Trust: London, UK, 2011.
11. Menezes, A.C.; Cripps, A.; Bouchlaghem, D.; Buswell, R. Predicted vs actual energy performance of non-domestic buildings: Using post-occupancy elvaluation date to reduce the performance gap. *Appl. Energy* **2012**, *97*, 355–364. [CrossRef]
12. Burman, E.; Mumovic, D.; Kimpain, J. Towards measurement and verification of energy performance under the framework of the European directive for energy performance of buildings. *Energy* **2014**, *77*, 153–163. [CrossRef]
13. Cohen, R.; Bordass, B. Mandating transparency about building energy performance in use. *Build. Res. Inform.* **2015**, *43*, 534–552. [CrossRef]
14. Igor, S.; Assunta, N.; Karsten, V. Net zero energy buildings: A consistent definition framework. *Energy Build.* **2012**, *48*, 220–232.
15. U.S. Department of Energy. Energy Efficiency & Renewable Energy Building Technologies Program: A Guide to Energy Audits. Pacific Northwest National Laboratory, 2011. Available online: https://www.pnnl.gov/main/publications/external/technical_reports/PNNL-20956.pdf (accessed on 15 December 2020).
16. Raslan, R.; Davies, M.; Doylend, N. An analysis of results variability in energy performance compliance verification tools. In Proceedings of the Eleventh International IBPSA Conference, Glasgow, Scotland, UK, 27–30 July 2009.
17. Hamilton, I.; Steadman, P.; Bruhns, H. *CarbonBuzz-Energy Data Audit*; UCL Energy Institute: London, UK, 2011.
18. van Dronkelaar, C.; Dowson, M.; Burman, E.; Spataru, C.; Mumovic, D. Review of the Energy Performance Gap and Its Underlying Causes in Non-Domestic Buildings. *Front. Mech. Eng.* **2016**, *1*, 17. [CrossRef]
19. Bunn, R.; Burman, E. S-curves to model and visualize the energy performance gap between design and reality-first steps to a practical tool. In Proceedings of the CIBSE Technical Symposium, London, UK, 17 April 2015.
20. Lowe, R.; Oreszczyn, T. Regulatory standards and barriers to improved performance for housing. *Energy Policy* **2008**, *36*, 4475–4481. [CrossRef]
21. Oreszczyn, T.; Lowe, R. Challenges for energy and buildings research: Objectives, methods and funding mechanisms. *Build. Res. Inform.* **2010**, *38*, 107–122. [CrossRef]
22. Evan., M.; Hannah., F.; Tehesia., P.; Norman., B.; David., C.; Tudi., H.; Mary., A.P. *The Cost-Effectiveness of Commercial-Building Commissioning: A Meta-Analysis of Energy and Non-Energy Impacts in Existing Buildings and New Construction in the United States*; Lawrence Berkeley National Laboratory: Berkeley, CA, USA, 2004.
23. Trane TRACE 700. Available online: https://www.trane.com/commercial/north-america/us/en/products-systems/design-and-analysis-tools/trace-700.html (accessed on 15 December 2020).
24. Sokratis., P.; Constantine., E.K.; Alex., V.; Elie., A. Rethinking HVAC temperature setpoints in commercial buildings: The potential for zero-cost energy savings and comfort improvement in different climates. *Build. Environ.* **2019**, *155*, 350–359.

Article

Analysis on Operation Modes of Residential BESS with Balcony-PV for Apartment Houses in Korea

Jiyoung Eum and Yongki Kim *

Green Building Research Center, Korea Institute of Civil Engineering and Building Technology, Goyang 10223, Korea; eumjiyoung@kict.re.kr
* Correspondence: kimyk@kict.re.kr; Tel.: +82-31-910-0490

Abstract: The integration of battery energy storage systems (BESS) with renewable energy is a potential solution to address the disadvantages of renewable energy systems, which is irregular and intermittent power. In particular, residential BESS is advancing in numerous countries. The residential BESS connected to the photovoltaic system (PV) can store the PV power in the battery through charging, and supply the PV power, which was stored in the battery, to the load through discharging when there is no PV power. Therefore, the utilization of residential BESS with PV reduces the daily electric power consumption and the electricity bills that households have to charge. However, it is understood that there is no case of installing and using residential BESS in Korea yet. Most residential houses in Korea are apartment houses, and thus residential BESS can be used with balcony PV. This paper presents operation modes of residential BESS with balcony PV for apartment houses. The BESS capacity was estimated by considering the balcony PV capacity, which can be installed in households, and power consumption. The applicability of the residential BESS was analyzed through performance and economics evaluation under current and various conditions. The operation modes of BESS were divided into four types according to PV power supply priority and battery charging source, and a test took place in a demonstration house. The risk of fully discharging the battery has been discovered when PV power is first charged to the battery or when only PV power is charged with the battery. As a result, preferential charging of the battery with PV power and then with PV and grid power was found to be the most optimal operation mode. In addition, additional functions were proposed for residential BESS in apartment households. The results will contribute to effective application of residential BESS with balcony PV in the near future.

Keywords: BESS (battery energy storage system); balcony photovoltaic system; apartment houses; operation modes; zero-energy houses

Citation: Eum, J.; Kim, Y. Analysis on Operation Modes of Residential BESS with Balcony-PV for Apartment Houses in Korea. *Sustainability* **2021**, *13*, 311. https://doi.org/10.3390/su13010311

Received: 25 November 2020
Accepted: 29 December 2020
Published: 31 December 2020

1. Introduction

With increasing interest in zero-energy buildings, which minimize energy consumption in buildings, a battery energy storage system (BESS) along with a renewable energy system is also attracting attention for their efficient energy management [1,2]. Among renewable energy systems, photovoltaic systems (PVs) are most commonly installed in buildings to reduce energy consumption. However, PVs have a disadvantage in that they supply power only during solar radiation time. By connecting BESS, PV power can be stored and supplied at other times [3,4]. In particular, the market for residential BESS is expanding internationally in regards to zero-energy houses. The operation of the residential BESS with PV can show the effect of reducing the daily electric power consumption and the electricity bills. In addition, it can increase the rate of self-sufficiency electric power in the household.

North America, Europe, and Japan are encouraging dissemination by offering a variety of benefits, such as subsidies and tax reductions for the residential BESS. However, it was investigated that there is no market for residential BESS in Korea yet because of the high price of residential BESS, low electric rate, progressive rate system, and limited PV capacity [5]. The residential BESS used overseas is for detached houses and has a capacity of 3–10 kWh, so at least 1–3 kW of PV capacity is required. In contrast, most residential houses in Korea are apartment houses. Individual households of apartment houses use a small capacity balcony PV of around 1 kW, and not a rooftop PV. Further, the surplus PV power remaining after it has been supplied to the load cannot be sold in households. Therefore, BESS implemented overseas cannot be readily used in Korea.

Research topics for using the residential BESS with PV are typically operation scheduling, capacity calculation, economic analysis, etc. Research related to this was conducted under variable electric power rate conditions such as Time-of-Use (TOU) pricing and Real-time Price (RTP), which are demand-managed options. Yoon et al. performed simulations for the control of a residential BESS using energy generation and consumption data for 64 residences with the Pecan Street Project in the United States and a range of seasonal dynamic price tables [6]. Hassan et al. developed a model to optimize FiT (PV generation and export tariff) revenue streams of PVs with BESS and simulated it as residential data [7]. Vieira et al. modeled and simulated BESS with PV with real data and MATLAB/Simulink for residential buildings in Portugal, showing a reduction on the energy sent and consumed from the grid, as well as the energy bill [8]. Jung et al. developed an optimal scheduling model of BESS with PV in residential buildings using the electric power and electricity price variables in Korea through simulation using a Python and Cplex solver [9]. Ratnam et al. organized the optimization approach methods for the scheduling of BESS with residential PVs to assess the customer benefit under incentives, such as time-of-use (TOU) pricing, feed-in-tariffs, and net metering [10]. Cucchiella et al. proposed the economic feasibility of residential lead-acid BESS combined with PV panels in Italy and the assumptions at which these systems become economically viable combined of electric power prices, investment costs, tax deduction and etc. [11]. Stelt et al. assessed and compared the technical and economic feasibility of both Household Energy Storage (HES) and Community Energy Storage (CES) in Netherlands using a mathematically optimized Home Energy Management System (HEMS) schedules scenario [12]. Koskela et al. analyzed the profitability and sizing of a photovoltaic system with an associated BESS from an economic perspective for an apartment building and detached houses in Finland [13]. Mulleriyawage et al. tried to calculate the optimal capacity for the fiscal benefits based on the TOU (Time-of-Use) tariff scheme because it is difficult to use residential BESS due to its expensive price in Australia [14]. These studies were based on the entire building, not the unit of apartment households.

This paper presents operation modes of residential BESS with balcony PV for individual households of apartment houses in Korea. An experiment on various operation modes was conducted in a demonstration house. The results of this experiment show that some functions need to be added for residential BESS to be made applicable to individual households of apartment houses in Korea.

2. Experiment Method

2.1. System Configuration

BESS connected with a PV consists of a battery pack, Power Conversion System (PCS; included DC/DC converter and AC/DC Inverter), Battery Management System (BMS), and Power Management System (PMS). The BMS is installed inside the battery pack to protect and control the battery, and the PMS manages the PCS and the BMS. The residential BESS installed in a residential building are typically connected to the PV module, with the load and grid as shown in Figure 1 [15]. In particular, as this load is a battery load separate from the grid, the battery discharge power is supplied only to the home appliance connected to the load line. The load is supplied with the PV or battery power, and the remaining PV power can be stored in the battery and supplied at the required time.

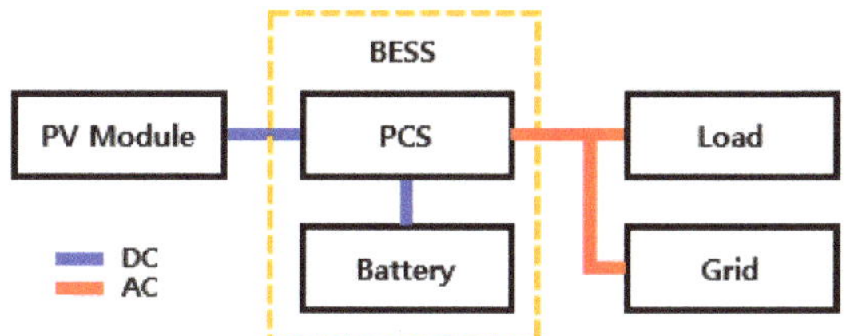

Figure 1. Schematic diagram of battery energy storage system (BESS) with photovoltaic system (PV).

As mentioned earlier, Korea has many apartment houses and the PV capacity that can be installed in individual households is limited. The capacity of the balcony PV is about 300 W (PV module 1ea) per household and it can be larger by connecting the PV modules in series [16]. Therefore, BESS should be designed and operated in consideration of the environment used. The operation mode of the residential BESS can basically charge the battery with PV power and the PV power can be set preferentially to either load or battery. The battery is charged with grid power when the battery is in system check or the SOC (state of charge) level of the battery is at an emergency level.

2.2. Experiment Apparatus and Method

In order to analyze the operation modes of the residential BESS connected with the balcony PV, an experiment was conducted in one household of a demonstration house in Goyang City. The experimental apparatus included a 2.016 kWh BESS prototype (48 V 42 AH LiFePO4 Battery, 1 kW PCS, All-in-one type, efficiency 93%), a 1.2 kW balcony PV (300 W module 4 series connection, efficiency 18%, south facing, installation tilt angle 70°), and an electric fan (power consumption 260 W). The BESS capacity was estimated by considering the capacity of the balcony PV, which can be installed in individual households, and the total household power consumption. For data collection, a PCS monitoring software (SolarPower) was used for the PV, battery, and load side. On the grid side, data were collected using a power meter (Wattman Power Meter) and a power monitoring software (Wattman Viewer). The inclined solar radiation and temperature were measured using a pyranometer (EKO MS-602), a thermocouple (TC-T), and a data logger (GRAPHTEC 260-16CH). The items data that were measured are PV generation power (kW), battery voltage (V) and current (A), load power (kW), grid power (kW), solar radiation (W/m^2), PV module temperature (°C), and outdoor temperature (°C). The data sampling/recording time was 1 s/1 min (average 60 s). Figure 2 shows the apparatus used in the experiment.

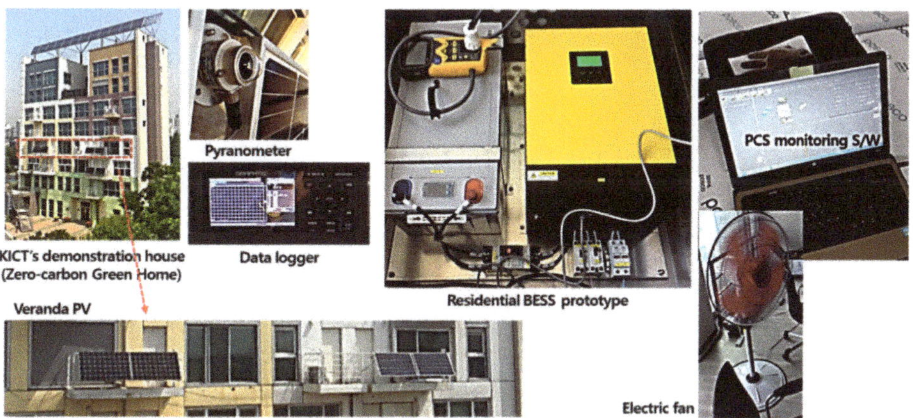

Figure 2. Photographs of the building and experiment apparatus.

The basic settings for BESS operation were set to battery bulk charge voltage 58.4 V, battery discharge cut-off voltage 41 V, and battery re-charge voltage 44 V. The experimental procedure was to set four operation modes for 3 days and to connect the electric fan as a load at 13:00 on the first day of changing the operation mode.

According to the PV power supply priority and battery charge source, the operation mode of BESS was divided into four, as shown in Table 1. Mode 1 preferentially supplies PV power to the load and charges the battery only with PV power. Mode 2 supplies PV power to the load first and charges the battery with PV and grid power. In mode 3, the battery is preferentially charged with PV power, and only the PV power is supplied. Mode 4 preferentially charges the battery with PV power and then with PV and grid power. Figure 3 is a schematic diagram of PV power supply priority and battery charge source.

Table 1. Operation modes of BESS.

Operation Mode	PV Power Supply Priority	Battery Charge Source
Mode 1	Load	PV only
Mode 2	Load	PV and Grid
Mode 3	Battery	PV only
Mode 4	Battery	PV and Grid

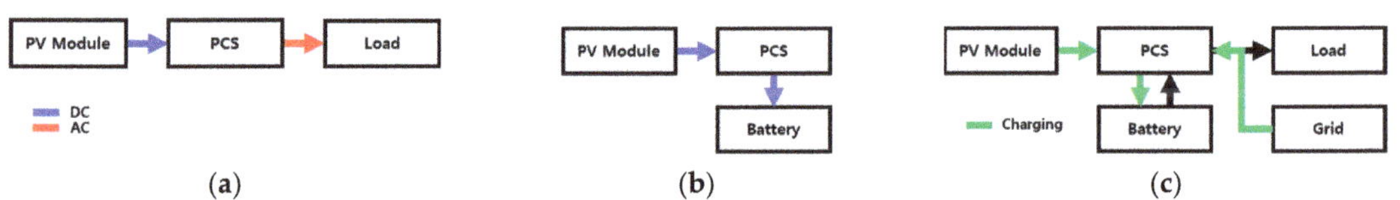

Figure 3. Schematic diagram of BESS operation: (**a**) PV power supply to load first; (**b**) PV power supply to battery first; (**c**) battery charge source.

3. Experiment Results

The power data per day were analyzed for each operation mode of BESS. The experiment lasted about 3 months from 21 August 2019 to 8 November 2019. In order to check the charging and discharging results, the experiment was conducted on all days except rainy days and days that the apparatus were checked. This paper describes by selecting an experimental date that representatively showed the characteristics of the modes. In this experiment, the connection load was an electric fan with a power consumption of 260 W, but it actually consumed 200 W. For reference, the PV power differed between day to day due to weather effects. Table 2 summarizes the experiment date, weather, load, and BESS states for each mode.

Table 2. Experiment schedule, weather, and state by BESS operation mode.

Mode	Date	Weather	BESS State
Mode 1	2019.08.23 (Fri.) 2019.08.24 (Sat.) 2019.08.25 (Sun.)	Sunny Cloudy Sunny	Load connection at 13:00 Full discharge
Mode 2	2019.08.27 (Tue.) 2019.08.28 (Wed.) 2019.08.29 (Thu.)	Rainy Cloudy Rainy	Load connection at 13:00 - -
Mode 3	2019.11.05 (Tue.) 2019.11.06 (Wed.) 2019.11.07 (Thu.)	Sunny Partly Cloudy Sunny	Load connection at 13:00 Full discharge
Mode 4	2019.09.17 (Tue.) 2019.09.18 (Wed.) 2019.09.19 (Thu.)	Cloudy Sunny Sunny	Load connection at 13:00 - -

Under mode 1, the experiment was conducted for a total of 3 days from 23 to 25 August 2019. The weather was cloudy. Figure 4 shows the graph of time variations of the daily PV, battery, grid and load power, battery voltage, and battery capacity (SOC) in mode 1. For reference, the grid power in the graph was expressed without distinction between supply and demand based on households. Charging started at 06:18 and was completed at 09:54 on 23 August 2019. The power was fully discharged about 10 h after the battery began to discharge. The discharge power was 1.854 kWh (discharge peak power of 0.199 kW), and the load power at this time was 2.037 kWh. An intermittent discharge occurred between 15:00 and 17:00, but the start and end times of discharge were at 17:33 on 23 August 2019 and at 03:36 on 24 August 2019. About 92% of the battery capacity was discharged and about 91% of the load was supplied to the battery. In this mode, the battery voltage reached about 33 V, which is the full discharge voltage range. Therefore, the battery could not be used because the battery voltage could not be maintained until the next PV charge.

Mode 2 was designed to charge with grid power when the battery voltage drops below the reference voltage while PV charge is not possible. The experiment was conducted for a total of 3 days from 27 to 29 August 2019. The weather was rainy and cloudy. Figure 5 shows the graph of time variations of the daily PV, battery, grid and load power, battery voltage, and battery capacity (SOC) in mode 2. The battery began to discharge at 17:41 on 27 August 2019 and ended at 03:38 on 28 August 2019. It remained discharged for about 9 h and 45 min. The total discharge power was 3.733 kWh (discharge peak power of 0.249 kW), and the load power at discharge was 3.875 kWh. Similar to mode 1, intermittent discharge occurred between 16:00 and 17:00, before the discharge start time. On 28 August 2019, the battery voltage fell at 4:00, but the battery voltage was maintained without further dropping due to the charging of the grid power. The battery was charged with 0.365 kWh of grid power. Since PV power is supplied to the load with priority, it is impossible to charge the battery if there is no remaining PV after being supplied to the load. Therefore, it is necessary to prevent the discharge by reducing the battery idle time and charging the grid power.

Under mode 3, the experiment was conducted for a total of 3 days from 5 to 7 November 2019. The weather was sunny and partly cloudy. Figure 6 shows the graph of time variations of the daily PV, battery, grid and load power, battery voltage, and battery capacity (SOC) in mode 3. Charging started at 07:38 on 5 November 2019 and full charge was reached at 11:31 on 5 November 2019. The battery was fully discharged about 9 h and 30 min after the battery discharge began at 17:38 on 5 November 2019. The discharge power was 1.739 kWh (discharge peak power of 0.199 kW), and the load power at that time was 1.867 kWh. The battery was completely discharged and battery voltage was not maintained until the charge of PV power the next day. The next day's PV power was sent to the load and grid.

Under mode 4, the experiment was conducted for a total of 3 days from 17 to 19 September 2019. The weather was cloudy and sunny. Figure 7 shows the graph of time variations of the daily PV, battery, grid and load power, battery voltage, and battery capacity (SOC) in mode 4. The battery is preferentially charged by PV power, and then by grid power when there is no PV power and the battery voltage reaches the full discharge protection voltage. In other words, if the battery is likely to be discharged, the battery is charged with grid power. The battery started to recharge at 04:03 on 18 September 2019, about 9 h and 30 min after it was first discharged at 18:31 on 17 September 2019. 18 On September 2019, the battery voltage fell at 4:00, but it was not fully discharged due to the charging of the grid power. Based on the 2-day data, the total discharge power was 2.894 kWh (discharge peak power of 0.199 kW), the load power at this time was 2.916 kWh, and the total power charged from the grid was 0.214 kWh.

In the modes where PV power was supplied to the load first (mode 1 and 2), the charge/discharge of the battery is determined by the load. Therefore, charge/discharge may occur frequently and the battery may not be sufficiently charged. As the battery voltage drops rapidly after complete discharge, it is necessary to reduce the battery idle time between discharge and charge. On the other hand, in the modes where PV power was supplied to the battery first (mode 3 and 4), PV power is supplied to the load after the battery is fully charged. The graph of mode 4 (18 September 2019) representatively shows this characteristic. The modes, in which the battery was charged only with PV (mode 1 and 3), risk full discharge in the absence of PV. Therefore, the ability to maintain battery voltage is required to use this mode. If the battery charge sources are PV and grid (mode 2 and 4), the battery is charged by grid power to maintain the battery voltage. The graphs of mode 2 and mode 4 show the charge of the grid power before and after the PV time. Further, in this mode, full discharge did not occur even on the third day of the experiment period.

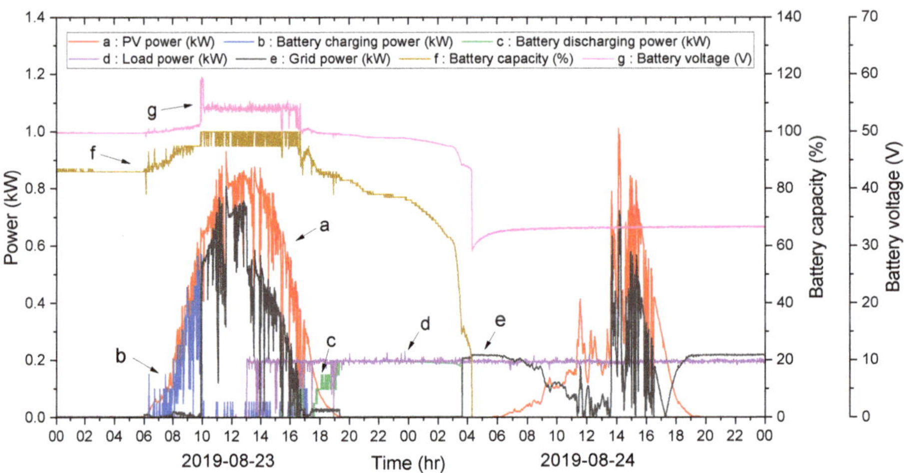

Figure 4. Performance characteristics of BESS prototype with balcony PV based on mode 1 (23–24 August 2019).

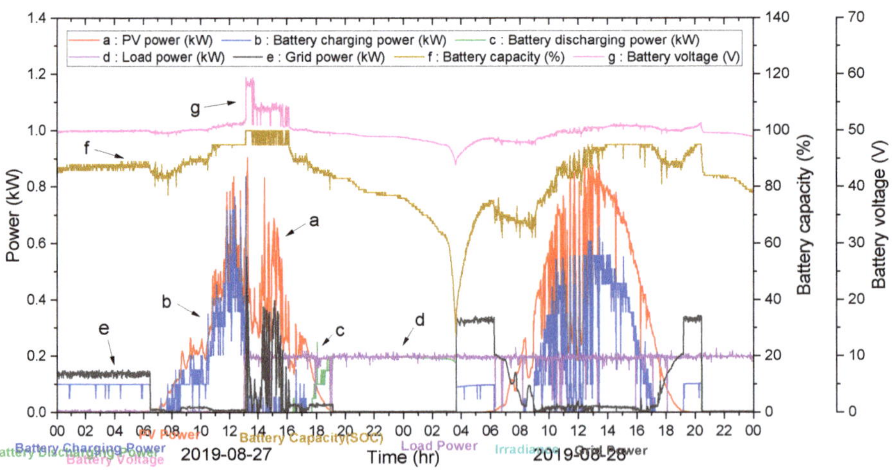

Figure 5. Performance characteristics of BESS prototype with balcony PV based on mode 2 (27–28 August 2019).

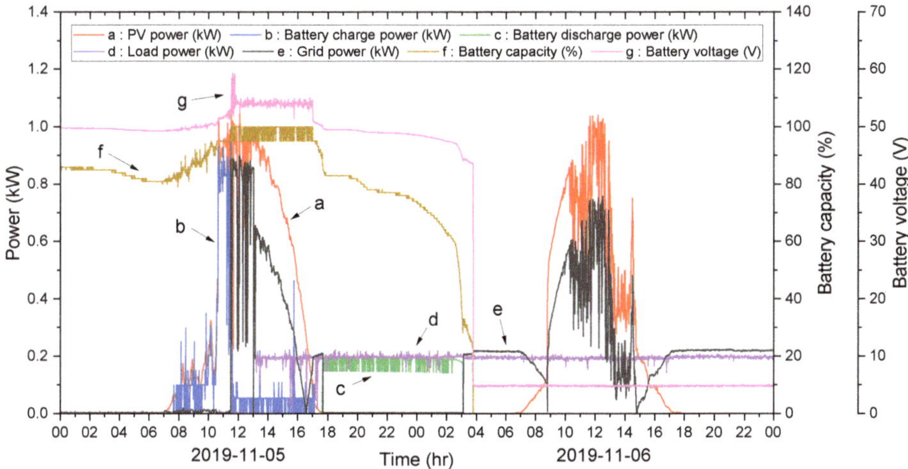

Figure 6. Performance characteristics of BESS prototype with balcony PV based on mode 3 (5–6 November 2019).

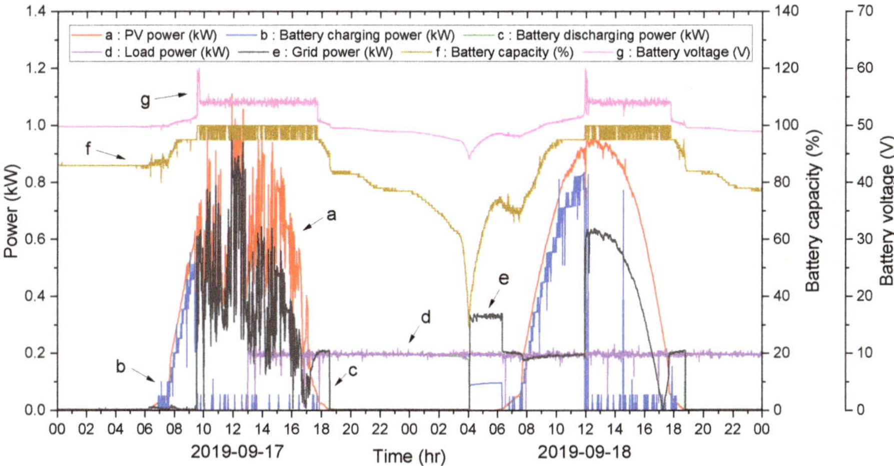

Figure 7. Performance characteristics of BESS prototype with balcony PV based on mode 4 (17–18 September 2019).

As an additional experiment, mode 4 was applied by changing the load to an air conditioner (power consumption of 2280 W). The experimental results are shown in Figure 8. The set temperature of the air conditioner was 22 °C, and the actual power consumption was about 1400 W. Based on the PV generation time, it was confirmed that the battery was charged by PV power and discharged to the load.

The operation modes of the residential BESS with the balcony PV were confirmed in individual households of apartment houses through experiments for each operation mode of the residential BESS. The results suggest that mode 4 is the most appropriate among the four operation modes of BESS. Nevertheless, some functions are still required to apply the scheme to individual households of apartments. First, the load connected with BESS should be the total power consumption in the household, not the power consumption of the home appliances connected to a separate load line. When only the balcony PV is installed, the PV power is supplied to the household through the plug of the inverter to reduce the total power consumption. In order to use the residential BESS to reduce total power consumption, such as the balcony PV, it is necessary to integrate the grid and load lines that are currently separated. Second, the battery must maintain the minimum voltage using the grid power before reaching an unusable battery condition. The charge of grid power is a concept that maintains the battery voltage at the minimum current rather than the normal charge. Third, it should be possible to set the voltage range or time for the PV charge. The BESS stops the grid power charge because it recognizes that PV can charge the battery when PV generation starts. However, there is a risk that the battery will reach a full discharge state because the PV power is unstable and cannot maintain a constant charge.

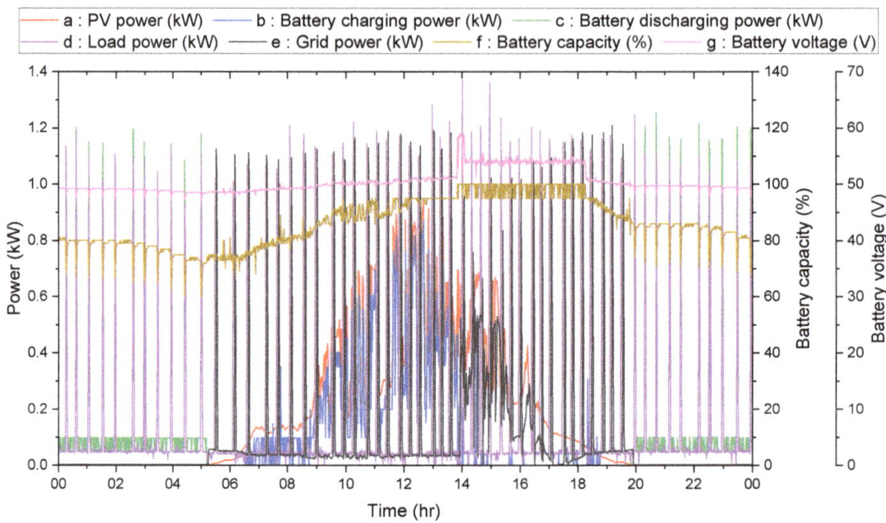

Figure 8. Performance characteristics of BESS prototype with balcony PV applied the air conditioner's load (7 July 2019).

4. Conclusions

In this study, several experiments were conducted with different operation modes to suggest optimal operation modes of residential BESS with balcony PV in Korea. The experiment apparatuses were a 2.016 kWh BESS, a 1.2 kW balcony PV, and an electric fan. The operation mode of BESS was divided into four types according to PV power supply priority and battery charge source.

The results show that if PV power was supplied to the load first, the charge/discharge of the battery was determined by the load (mode 1 and 2). However, when PV power was supplied to the battery first (mode 3 and 4), PV power was supplied to the load after the battery was fully charged. Furthermore, if the battery was only charged with PV, there was a risk of full discharge in the absence of PV (mode 1 and 3), but charging with the grid prevented this (mode 2 and 4).

Based on the characteristics of each operation mode, it was determined that mode 4, in which PV power preferentially charges the battery and then charges the battery with PV and grid power, is appropriate for the BESS operation mode of individual households of apartment houses. However, some functions need to be added to ensure applicability of residential BESS to individual households of apartment houses in Korea. First, the load connected with BESS should be the total power consumption in the household, not the power consumption of home appliances connected to a separate load line. Second, the battery must maintain the minimum voltage using the grid power before reaching an unusable battery condition. Third, it should be possible to set the voltage range or time for PV charge. By satisfying the above conditions, BESS with balcony PV is expected to be used efficiently for individual households of apartment houses in Korea. Furthermore, it is believed that it will contribute to zero energy in houses by improving the energy independence rate of households.

Author Contributions: Methodology, data collection and analysis, visualization, validation, writing—original draft preparation, J.E.; conceptualization, methodology, writing—review and editing, supervision, Y.K. All authors have read and agreed to the published version of the manuscript.

Funding: This work was supported by the Korea Institute of Energy Technology Evaluation and Planning (KETEP) and the Ministry of Trade, Industry and Energy (MOTIE) of the Republic of Korea (No. 20172410104720).

Data Availability Statement: The data presented in this study are available on request from the corresponding author.

Conflicts of Interest: To the best of our knowledge, the named authors have no conflict of interest to declare, financial or otherwise.

References

1. Xu, Z.B.; Guan, X.H.; Jia, Q.S.; Wu, J.; Wang, D.; Chen, S.Y. Performance Analysis and Comparison on Energy Storage Devices for Smart Building Energy Management. *IEEE Trans. Smart Grid.* **2012**, *3*, 2136–2147. [CrossRef]
2. Liu, J.; Chen, X.; Yang, H.; Li, Y. Energy storage and management system design optimization for a photovoltaic integrated low-energy building. *Energy* **2020**, *190*, 116424. [CrossRef]
3. O'Shaughnessy, E.; Cutler, D.; Ardani, K.; Margolis, R. Solar plus: Optimization of distributed solar PV through battery storage and dispatchable load in residential buildings. *Appl. Energy* **2018**, *213*, 11–21. [CrossRef]
4. Bingham, R.D.; Agelin-Chaab, M.; Rosen, M.A. Whole building optimization of a residential home with PV and batter storage in the Bahamas. *Renew. Energy* **2019**, *132*, 1088–1103. [CrossRef]
5. Eum, J.Y.; Kim, Y.K. Economical analysis of the PV-linked residential ESS using homer in Korea. *J. Korea Acad. Ind. Coop. Soc.* **2019**, *20*, 36–42.
6. Yoon, Y.; Kim, Y.-H. Effective scheduling of residential energy storage systems under dynamic pricing. *Renew. Energy* **2016**, *87*, 936–945. [CrossRef]
7. Hassan, A.S.; Cipcigan, L.; Jenkins, N. Optimal battery storage operation for PV systems with tariff incentives. *Appl. Energy* **2017**, *203*, 422–441. [CrossRef]
8. Vieira, F.M.; Moura, P.S.; De Almeida, A.T. Energy storage system for self-consumption of photovoltaic energy in residential zero energy buildings. *Renew. Energy* **2017**, *103*, 308–320. [CrossRef]
9. Jung, S.; Kang, H.; Lee, M.; Hong, T. An optimal scheduling model of an energy storage system with a photovoltaic system in residential buildings considering the economic and environmental aspects. *Energy Build.* **2020**, *209*, 109701. [CrossRef]
10. Ratnam, E.L.; Weller, S.R.; Kellett, C.M. An optimization-based approach to scheduling residential battery storage with solar PV: Assessing customer benefit. *Renew. Energy* **2015**, *75*, 123–134. [CrossRef]
11. Cucchiella, F.; D'Adamo, I.; Gastaldi, M.; Stornelli, V. Solar Photovoltaic Panels Combined with Energy Storage in a Residential Building: An Economic Analysis. *Sustainability* **2018**, *10*, 3117. [CrossRef]
12. Stelt, S.; Alskaif, T.; Sark, W. Techno-economic analysis of household and community energy storage for residential prosumers with smart appliances. *Appl. Energy* **2018**, *209*, 266–276. [CrossRef]
13. Koskela, J.; Rautiainen, A.; Järventausta, P. Using electrical energy storage in residential buildings—Sizing of battery and photovoltaic panels based on electricity cost optimization. *Appl. Energy* **2019**, *239*, 1175–1189. [CrossRef]
14. Mulleriyawage, U.; Shen, W. Optimally sizing of battery energy storage capacity by operational optimization of residential PV-Battery systems: An Australian household case study. *Renew. Energy* **2020**, *160*, 852–864. [CrossRef]
15. Eum, J.Y.; Kim, Y.K. Estimation methods for the optimal capacity of veranda PV-linked ESS in apartment house. In Proceedings of the SAREK (the Society of Air-conditioning and Refrigeration Engineers of Korea) Summer Annual Conference, Seoul, Korea, 19–21 June 2019; pp. 171–174.
16. Seoul Metropolitan Government. Mini Photovoltaic Power Plant Application/Supplier. Available online: http://solarmap.seoul.go.kr/mini/minisolarRequest12.do (accessed on 20 May 2019).

Article

Experimental Performance of an Advanced Air-Type Photovoltaic/Thermal (PVT) Collector with Direct Expansion Air Handling Unit (AHU)

Jin-Hee Kim [1], Sang-Myung Kim [2] and Jun-Tae Kim [3,*]

[1] Green Energy Technology Research Center, Kongju National University, Cheonan 314701, Korea; jiny@kongju.ac.kr
[2] Graduate School of Energy System Engineering, Kongju National University, Cheonan 314701, Korea; mtanzania@smail.kongju.ac.kr
[3] Department of Architectural Engineering, Kongju National University, Cheonan 314701, Korea
* Correspondence: jtkim@kongju.ac.kr; Tel.: +82-41-521-9333

Abstract: In addition to electrical energy generation, photovoltaic/thermal (PVT) systems utilize heat from building-integrated photovoltaic (BIPV) modules for domestic hot water and space heating. In other words, a PVT system can improve the electricity efficiency of BIPVs while using the waste heat of BIPVs as a source of thermal energy for the building. By generating thermal and electrical energies simultaneously, PVT systems can improve the utilization of solar energy while enhancing the energy performance of buildings. To optimize the performance of an air-type PVT collector, it is necessary for the system to extract more heat from the PV module. Consequently, this approach decreases PV temperature to improve PV electrical energy generation. The thermal and electrical performance of an air-type PVT collector depends on its design, which affects airflow and heat transfer. Moreover, the performances of the PVT collector can differ according to the coupled facility in the building. In this study, the thermal and electrical performances of an advanced air-type PVT collector with a direct expansion air handling unit (AHU) were analyzed experimentally. For this purpose, six prototypes of an advanced air-type PVT collector were developed. Furthermore, a direct expansion AHU with a heat recovery exchanger (HRX) was designed and built. The advanced PVT collectors with a total capacity of 740 Wp were installed in an experimental house and were coupled to the direct expansion AHU system with a maximum airflow of 700 CMH. The performance of PVT collectors was analyzed and compared with the BIPV system. Results showed that building-integrated photovoltaic/thermal (BIPVT) collectors produced 30 W more power than the BIPV system. When operating the AHU system, the temperature of the BIPVT collector was generally lower than the BIPV. The maximum difference in temperature between BIPVT and BIPV was about 22 °C. During winter season, the BIPVT collector supplied preheated air to the AHU. The supplied air temperature from the BIPVT collector reached 32 °C, which was 15 °C higher than outdoor air temperature.

Keywords: BIPVT (building-integrated photovoltaic/thermal); air-type PVT collector; AHU (air handling unit); mock-up experiment; thermal and electrical efficiency

Citation: Kim, J.-H.; Kim, S.-M.; Kim, J.-T. Experimental Performance of an Advanced Air-Type Photovoltaic/Thermal (PVT) Collector with Direct Expansion Air Handling Unit (AHU). *Sustainability* **2021**, *13*, 888. https://doi.org/su13020888

Received: 2 December 2020
Accepted: 14 January 2021
Published: 17 January 2021

1. Introduction

One of the biggest problems of a building-integrated photovoltaic (BIPV) system is the degradation of photovoltaic (PV) module efficiency. The efficiency of the PV module decreases by about 0.4–0.5% when the temperature of the PV module rises by 1 °C at a PV module temperature of 25 °C [1]. To solve BIPV power loss due to PV temperature rise, a building-integrated photovoltaic/thermal (BIPVT) system has been developed.

The BIPVT system is one of the solar energy systems that produce electricity and heat simultaneously. The PV module of the PVT system produces electricity, and the heat generated from the PV module can be used as a heat source for heating and domestic hot water (DHW) in buildings. By using heat from the PV module as a heat source for the building, the PVT system can prevent the rise of PV module temperature and the related power degradation [2,3]. Kazem et al. [4] conducted an outdoor test of three water cooling PVT systems that had variable flow channels (web type, direct type, spiral type) and compared them with conventional PVs in terms of electrical performance. In their results, the proposed PVT systems reduced the PV cell temperature by an average of 3 °C. The electrical efficiencies of the PVT systems were higher than conventional PVs. The conventional PV module had an electrical efficiency of about 7.8%, while the highest achieved PVT electrical efficiency was found to be 9.1% for a spiral flow collector.

BIPVT systems can be categorized into air type, liquid type, and hybrid type, depending on the heat transfer medium used. BIPVT, which uses air as a heat medium, has an advantage of being the easiest to apply and maintain in building systems. The performance of the air-type PVT system is affected by a number of factors such as airflow, flowrate, absorber configuration, baffle shape and arrangement, and so on. Bakari [5] conducted an experiment to analyze the effect of different numbers of baffles. In the study, air solar collectors that were integrated with 2, 3, 4, and 8 baffles were evaluated and compared with a conventional flat plate collector. The results showed that the collector with 2, 3, 4, and 8 baffles had an efficiency of 29.2%, 31.3%, 33.1%, and 33.7% respectively, whereas the efficiency of the air collector without baffle was 28.9%. Furthermore, the impact of the shape and arrangement of baffles and absorption plates on air-type PVT performance has been evaluated through experimental and numerical analyses [6–9].

In addition to baffles and absorber plates, some studies have presented various factors that affect the performance of the PVT system such as nanoparticles, PCM, and coolant. Many researchers have shown that the efficiency of the PVT system can be improved by using metallic nanoparticles together with a PCM. In addition, some results showed that the thermal performance of the coolant is one of the key elements that has potential to improve the PVT system [10,11]. Various related studies have focused on the application of PVT systems together with building facilities. Boutina et al. [12] studied an air-type PVT collector integrated with a chimney tower, and the effects were analyzed through CFD simulation. The proposed PVT collector improved heat transfer rate by approximately 78% over conventional air-type PVT collectors. Tiwari et al. [13] designed and tested a greenhouse dryer using heat sources from an air-type PVT collector. The thermal efficiency, electrical efficiency, and overall thermal efficiency of the air-type PVT collector were found to be 26.68%, 11.26%, and 56.30%, respectively. In addition, the air-powered PVT collectors were approximately twice as cheap as electric dryers. Fan et al. [14] proposed an air-conditioned PVT-SAH (solar air heater) model with heat pipes. The payback period of the proposed system was evaluated at be 5.7 to 16.8 years. The thermal efficiency of the system was 12% higher, and it was found that the temperature of the PV module decreased effectively. In a previous study, Kim et al. [15] studied the energy performance of a building's heating system combined with BIPVT collectors that used water as a heat medium. For their study, a water-type unglazed BIPVT collector was developed and installed on the roof of an experimental house; it was then combined with a heating system consisting of a thermal storage tank, an auxiliary boiler, an inverter, and a fan-coil unit in order to use the thermal energy of PVT collectors. Results confirmed the thermal and electrical efficiencies of BIPVT collectors to be 30% and 17%, respectively. In particular, the electrical efficiency showed a high performance of more than 16% when the heating system was running. The BIPVT increased the water temperature in the thermal storage tank by 40 °C, and thus it can be utilized as a heat source for heating.

Aside from the energy saving potential, connecting an air handling unit (AHU) system to an air-type PVT collector can help improve the performance of the AHU systems. Especially during the winter season when very cold outdoor air enters the AHU system directly, damages to the AHU equipment can occur. However, the connection between the BIPVT system and the AHU can protect the AHU system by using warmer air through

PVT collectors instead of the outdoor cold air directly. Moreover, the use of the warmed air can reduce the energy used to heat the cold outdoor air.

In this study, the performance of air-based BIPVT collectors that are connected to an AHU system in the experimental building is evaluated. Firstly, the electrical performance and the temperature characteristics of the developed BIPVT collector are evaluated by an outdoor test. To compare the performance of the BIPVT collector, the BIPV system is manufactured and tested together. The electrical and thermal characteristics of the BIPVT collector and the BIPV system are analyzed. After this, the performance of the AHU system with BIPVT is evaluated. The temperature characteristics of the BIPVT collector is analyzed through operating the AHU system. In addition, with the AHU system connected to the air-type BIPVT collector, the energy saving potential of the building can also be analyzed.

2. Experimental House of the AHU System with BIPVT Collector

For this study, an air-type PVT collector was designed, as shown in Figure 1. In previous studies, the PVT collectors were developed using conventional PV modules with little gap between PV cells [1–3,11,12]. For this study, the developed PVT collector was integrated with a glass-to-glass PV module, and the PV module was designed to keep a constant gap between PV cells to improve the thermal performance of the PVT collector. The spacing between PV cells allows more solar heat to enter the collector than conventional collectors. Moreover, below the gaps, the absorber plates which have high thermal conductivity were installed. Absorber plates that have high thermal conductivity is one of the main parameters affecting the thermal performance of PVT system [16,17]. By placing the absorber plate below the space between PV cells, the temperature inside the collector can be increased by directly transferring the solar heat to the absorber plate. The installed absorber plates have a round bending shape, and they works as internal baffles that can improve heat collecting efficiency [18]. When air passes through the PVT collector from the inlet to the outlet, the absorber plates lengthen the airflow pathway and create turbulence inside the collector. The lengthened air pathway and turbulence both help to increase heat transfer and improve the thermal performance of the PVT collectors [19].

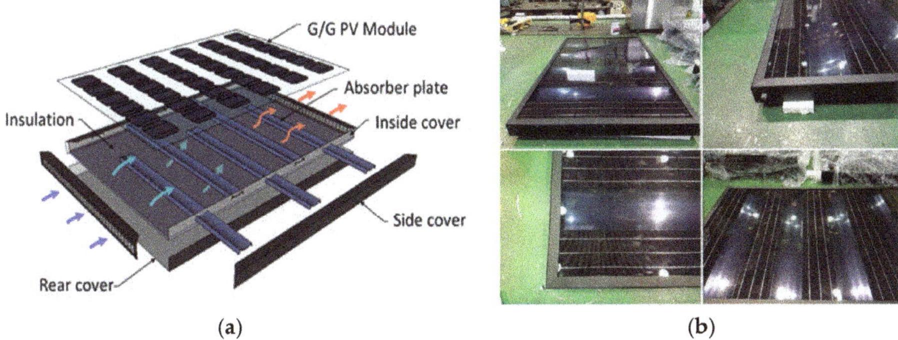

Figure 1. Designed air-type photovoltaic/thermal (PVT) collector: (**a**) Schematic diagram; (**b**) Prototype of the PVT collector.

The designed BIPVT collectors were connected to a direct expansion air handling unit (AHU) and installed in an experimental house. Figure 2 shows the experimental house and the connection of BIPVT collectors and the AHU system. Six BIPVT collectors were installed on the south side; the BIPVT collectors, top and bottom, were connected as one set. Outdoor air enters the inlet below the BIPVT collectors and then is heated by passing through the BIPVT collectors.

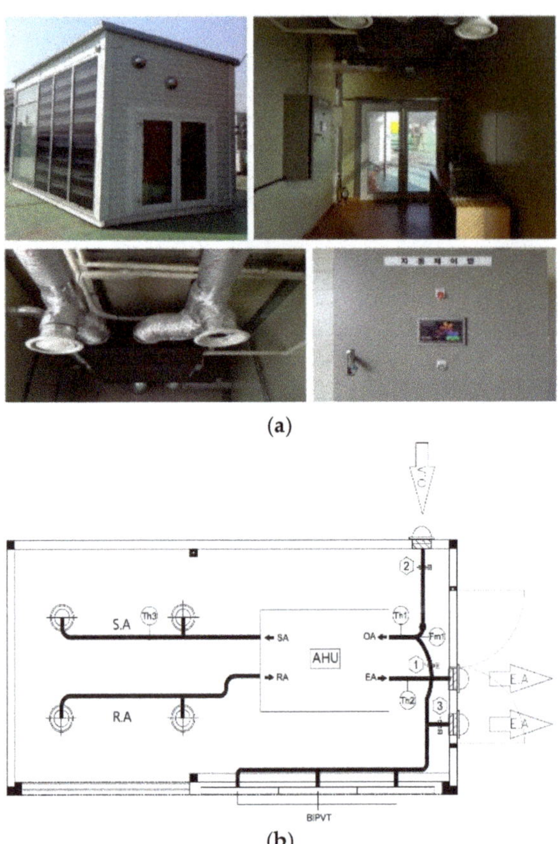

Figure 2. Experimental house: (**a**) Experimental house and equipment system; (**b**) Schematic diagram of the air handling unit (AHU) system with building-integrated photovoltaic/thermal (BIPVT) collectors.

3. Experimental Setup

For the analysis, the BIPVT collectors were evaluated on two bases. One focus is on the BIPVT collector side, and the other one concerns the overall system of the AHU with BIPVT collectors. First, the performance of the BIPVT collector was compared with the BIPV system. Then, the energy savings of the AHU system with the BIPVT collector was investigated. The experiment was evaluated in Cheonan, Republic of Korea (36.815° N, 127.114° E).

3.1. Comparison of BIPVT and BIPV System

In order to analyze the performance of the BIPVT collector, a BIPV system was made as shown in Figure 3. The BIPV system was composed of an insulation behind the PV module to satisfy the performance of the building exterior wall. The temperature of the PV module was measured by thermocouples, and the temperature characteristics of the BIPVT and BIPV system were analyzed.

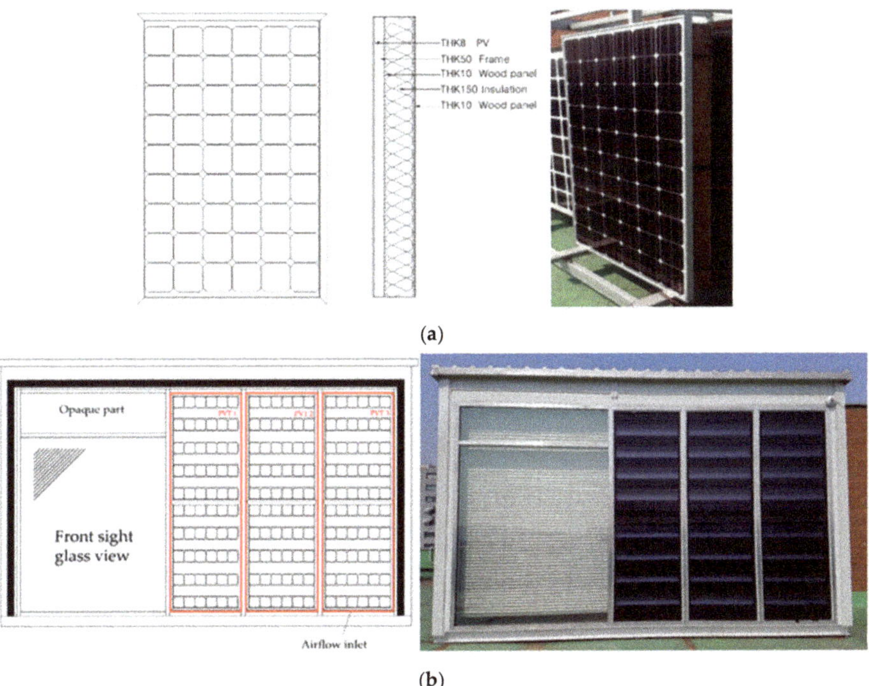

Figure 3. Experimental house: (**a**) BIPV system; (**b**) Experimental house with BIPVT system.

The specification of the PV modules are summarized in Table 1. For the experimental measurements, the associated sensitivity and inaccuracy of measuring equipment should be considered. In this experiment, the uncertainties of the measuring devices were considered. The detailed specification of measuring instruments including uncertainties is presented in Table 2.

Table 1. Specification of the PV module for the BIPVT collector and the BIPV system.

Specification	BIPVT Module	BIPV Module
PV cell type	Mono-crystalline silicon	
PV module efficiency	7.6%	16.2%
Maximum power	123.3 W	265.08 W
Maximum voltage	15.08 V	31.01 V
Maximum current	8.18 A	8.55 A
Open circuit voltage (V_{oc})	19.05 V	38.53 V
Short circuit current (I_{sc})	8.61 A	9.05 A
Collector size	1584 × 1031 × 84.5 mm	1084 × 1031 × 84.35 mm

Table 2. The range and accuracy of the measuring instrument.

Description	Measurement Range	Accuracy
Humidity and temperature transmitter	−50 to 100 °C 0 to 100% RH	±0.8%RH at 23 °C ±0.1 K at 23 °C
Pyranometer (nonlinearity)	0 to 2000 W/m^2	±1.2% at <1000 W/m^2
Thermocouple	−250 to 500 °C	±0.5 °C
Power meter	15–600 V 0.5–20 A	0.1% of the reading 0.1% of the range
Data logger	−10 V to 10 V, 20 mA	0.003% DCV

3.2. AHU System with BIPVT Collectors

BIPVT collectors that were installed in the experimental building were connected to the AHU system. The heated air through the BIPVT collectors passed through the outlet and entered the supply air channel of the AHU. It was controlled with a fan and a damper, and there are two operation modes as shown in Figure 4. When the heating for space is not required, such as during summer season, the damper is closed in order to block the channel between the PVT collector and the AHU, and the heated air is exited outdoors. This prevents overheating of the PV module, which can also improve the electrical efficiency of the PV module by decreasing the temperature of the PV module in summer. During winter when heating is required, the damper is opened to allow the heated air to enter into the AHU as supply air. By using the warmed air as supply air for the AHU, the energy needed to heat the cold outdoor air in the AHU system can be reduced. In order to analyze the energy savings for the AHU system, a test was conducted in the heating season. For the test, the temperature of the PVT collector outlet, the supply air temperature for the room, and the outdoor air temperature were measured.

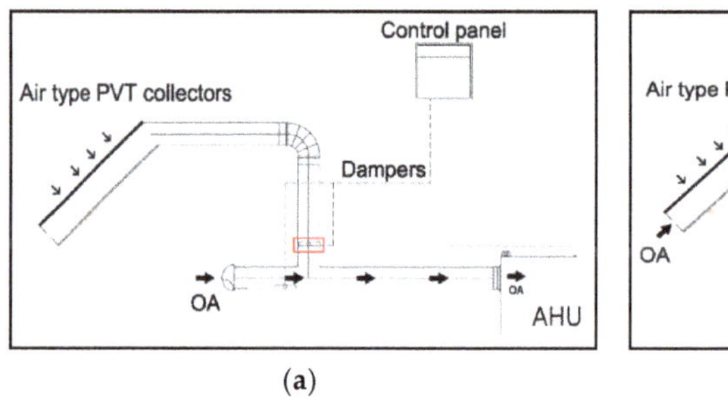

(a)

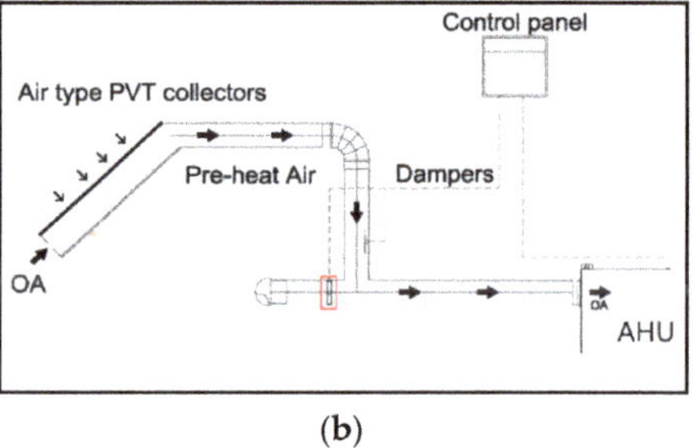

(b)

Figure 4. Operation mode of the AHU system with PVT collectors: (**a**) Nonheating season; (**b**) Heating season.

4. Results and Discussions

The experimental data were collected in the heating period from February to March. When the BIPVT and BIPV system were compared, the heated air of the BIPVT collector was extracted outdoors. During the experiment of the AHU system with the BIPVT collector, the AHU system with the BIPVT collector was run in heating mode. The air heated by BIPVT collectors was fed into the outdoor air (OA) channel of the AHU unit instead of the cold outdoor air.

4.1. Comparison of Performance with the BIPVT and BIPV Systems

The daily electricity yield of the BIPVT collector and the BIPV is shown in Figure 5. When solar radiation was lower than 600 W/m^2, the PV power of the BIPV was higher than the BIPVT collector. However, when solar radiation increased to more than 600 W/m^2, the PV power of the PVT increased more than that of the BIPV. In particular, it was found that during the highest solar radiation around midday (12–2 pm), the difference in PV power generation between BIPVT and BIPV was the largest.

The reason for the difference in PV power between the BIPV system and the BIPVT collectors can be explained with reference to Figure 6. Figure 6 graphically compares the average temperature of the PV module in the BIPV and BIPVT systems. The graph shows that the PV module temperature fluctuated according to changes in solar radiation. The PV module temperatures increased as the solar radiation increased. The PV module temperature of both the BIPVT and BIPV system increased according to increasing solar radiation, but the PV module temperature of the BIPVT collector was lower than that of the BIPV. The temperature of the BIPV was within a range of 15–58 °C for one day, and that for the BIPVT collectors ranged between 12 and 41.8 °C. The lower PV temperature of the BIPVT system was attributed to the exiting heat with air; this can help to prevent degradation of PV power generation caused by temperature increase of the PV module.

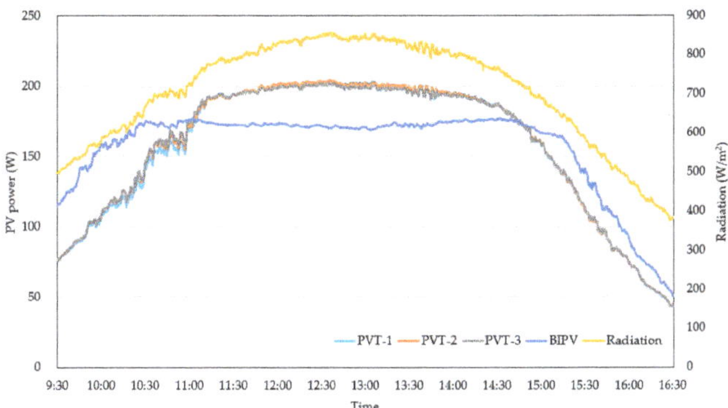

Figure 5. Daily PV power generation of the BIPV and BIPVT collector.

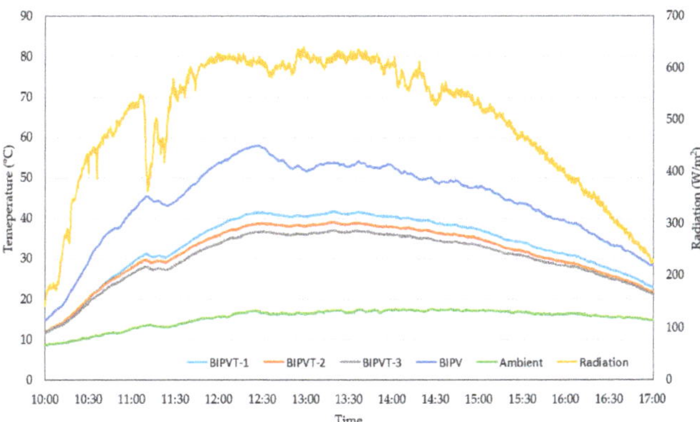

Figure 6. Effect of time on PV temperature and solar radiation on the BIPVT and BIPV systems.

Figure 7 compares the electrical efficiency of the PV module in the BIPV and BIPVT systems. The efficiency was calculated by Equation (1) [20]:

$$\eta_{eff} = \frac{V \times I}{A_{pv} \times G} \tag{1}$$

where η_{eff}, V, I, A_{pv}, and G are the electricity efficiency, the maximum voltage (V), the maximum current (A), the PV area (m^2), and the global solar radiation (W/m^2), respectively. In the case of the BIPV system, when solar radiation increased to more than 600 W/m^2, the electrical efficiency decreased. This was due to increasing the PV module temperature concurrently with high solar radiation. However, for the PVT system, even though solar radiation increased, the electrical efficiency was kept steady without degradation.

4.2. AHU System with BIPVT Collector

Figure 8a shows the temperature of the BIPVT collector and the BIPVT outlet. The outlet temperature of the BIPVT outlet was similar to the midrange temperature of the BIPVT collector. The PV module temperature of the PVT collector was about 25–40 °C, depending on the changes in solar radiation. In addition, the extraction temperature preheated through the BIPVT collector was 22–38 °C, which was 10–20 °C higher than the outside temperature. In Figure 8b, when the temperature of the supply air delivered to the room through the AHU system was about 40 °C, the temperature of outlet air from the BIPVT collector was about 32 °C. When the outlet temperature was 32 °C, the outdoor temperature was 17 °C, which was lower by 15 °C compared to the supplied air from the BIPVT. In the existing AHU system, the cold outdoor air (OA) entered the AHU unit, then it was heated to supply the warm air for the room. By connecting the BIPVT collector with the AHU system, the preheated air from the BIPVT collector can be supplied. It was found that the heating energy of the AHU system was saved by using the BIPVT collector.

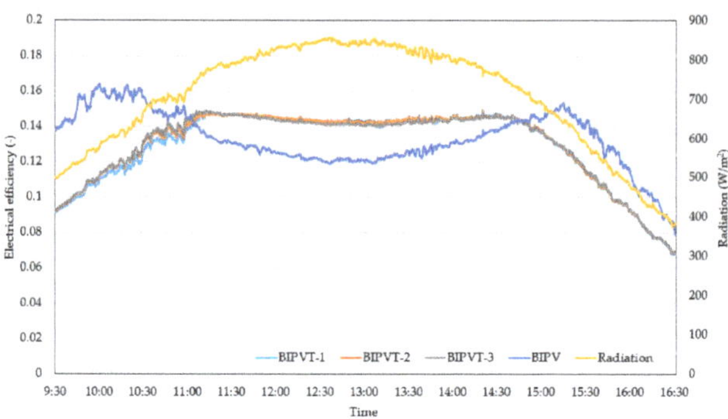

Figure 7. Effect of time on PV temperature and solar radiation on the BIPVT and BIPV systems.

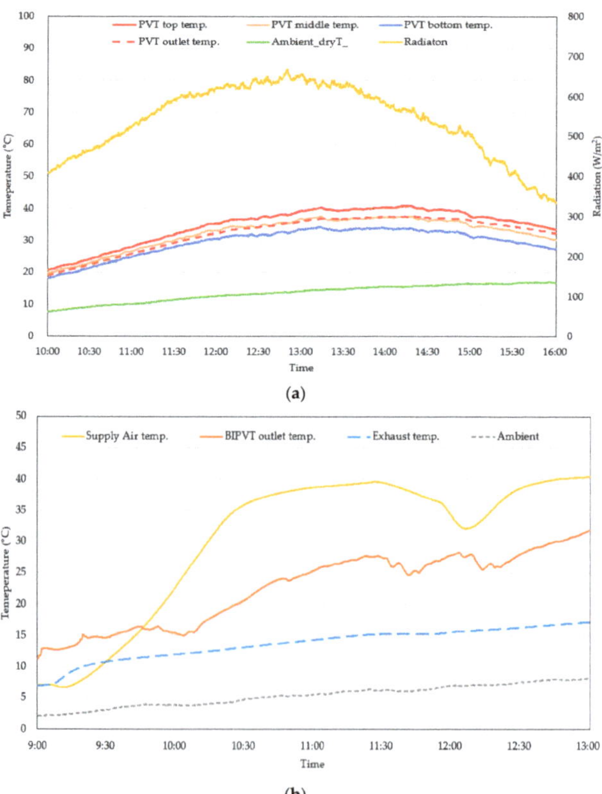

Figure 8. Operation mode of the AHU system: (**a**) Nonheating season; (**b**) Heating season.

5. Conclusions

In this study, an air-type BIPVT collector was manufactured and installed in a real-scale experimental house, and its electrical performance and temperature characteristics were investigated and compared with a BIPV system through experiments. In addition, the effect of energy savings was analyzed when the BIPVT was connected to an AHU system. The key findings are as follows:

(1) The air-type BIPVT collector can prevent the degradation of PV power generation that is often caused by increasing PV temperature. During experimental tests, the BIPVT collector produced electrical energy of about 200 W when the solar radiation was more than 800 W/m^2, which was 30 W more than the total electrical energy produced by the BIPV system. Moreover, the BIPVT collector was kept at a lower PV temperature than the BIPV, where the maximum difference of PV module temperature was about 22 °C.

(2) Air-type BIPVT collectors were found to maintain electrical efficiency even when solar radiation increased. During midday when solar radiation was the highest, PV electrical efficiency of the BIPV decreased up to 12% due to an increased PV module temperature. However, the electrical efficiency of BIPVT was steady at 14%.

(3) Through tests in the experimental building, it was found that the connection of the BIPVT collector with the AHU system can save energy for heating. In the heating period, BIPVT collectors can supply preheated air to the AHU unit, and the AHU system can save energy to heat the cold outdoor air. The temperature of the preheated air from the BIPVT collector was 32 °C, which was 15 °C higher than the outdoor air.

Based on these results, the performance of the BIPVT collector and related energy-saving effect by connecting to an AHU system in the building can be seen. It is expected that this study can be used as a foundation for further study on building systems integrated with BIPVT collectors.

Author Contributions: Conceptualization, J.-H.K., and S.-M.K.; methodology, J.-H.K.; software, S.-M.K.; validation, J.-H.K. and J.-T.K.; formal analysis, J.-H.K. and S.-M.K.; investigation, J.-H.K.; resources, S.-M.K.; data curation, J.-H.K.; writing—original draft preparation, J.-H.K. and S.-M.K.; writing—review and editing, J.-H.K., J.-T.K. and S.-M.K.; visualization, J.-H.K. and S.-M.K.; supervision, J.-T.K.; project administration, J.-H.K.; funding acquisition, J.-T.K. and J.-H.K. All authors have read and agreed to the published version of the manuscript.

Funding: This research was funded by Basic Science Research Program through the National Research Foundation of Korea (NRF) funded by the Ministry of Education (NRF-2018R1D1A1A09083870) and Korea Institute of Energy Technology Evaluation and Planning (KETEP) and the Ministry of Trade, Industry and Energy (MOTIE) of the Republic of Korea, grant number 20188550000480.

Institutional Review Board Statement: Not applicable.

Informed Consent Statement: Not applicable.

Data Availability Statement: Not applicable.

Acknowledgments: The authors would like to thank Natural Resources Canada for their contribution to preliminary stages of the project.

Conflicts of Interest: The authors declare no conflict of interest.

References

1. Sathe, T.M.; Dhoble, A.S. A review on recent advancements in photovoltaic thermal techniques. *Renew. Sustain. Energy Rev.* **2017**, *76*, 645–672. [CrossRef]
2. Kazem, H.A. Evaluation and analysis of water-based photovoltaic/thermal (PV/T) system. *Case Stud. Therm. Eng.* **2019**, *13*, 100401. [CrossRef]
3. Buonomano, A.; Calise, F.; Vicidomini, M. Design, simulation and experimental investigation of a solar system based on PV panels and PVT collectors. *Energies* **2016**, *9*, 497. [CrossRef]
4. Kazem, H.A.; Al-Waeli, A.H.; Chaichan, M.T.; Al-Waeli, K.H.; Al-Aasam, A.B.; Sopian, K. Evaluation and comparison of different flow configurations PVT systems in Oman: A numerical and experimental investigation. *Sol. Energy* **2020**, *208*, 58–88. [CrossRef]

5. Bakari, R. Heat transfer optimization in air flat plate solar collectors integrated with baffles. *J. Power Energy Eng.* **2018**, *6*, 70–84. [CrossRef]
6. Hegazy, A.A. Comparative study of the performances of four photovoltaic/thermal solar air collectors. *Energy Convers. Manag.* **2000**, *41*, 861–881. [CrossRef]
7. Pottler, K.; Sippel, C.M.; Beck, A.; Fricke, J. Optimized finned absorber geometries for solar air heating collectors. *Sol. Energy* **1999**, *67*, 35–52. [CrossRef]
8. Riffat, S.B.; Cuce, E. A review on hybrid photovoltaic/thermal collectors and systems. *Int. J. Low-Carbon Technol.* **2011**, *6*, 212–241. [CrossRef]
9. Tyagi, V.V.; Kaushik, S.C.; Tyagi, S.K. Advancement in solar photovoltaic/thermal (PV/T) hybrid collector technology. *Renew. Sustain. Energy Rev.* **2012**, *16*, 1383–1398. [CrossRef]
10. Sarafraz, M.M.; Safaei, M.R.; Leon, A.S.; Tlili, I.; Khan, I.; Tian, Z.; Goodarzil, M.; Arjomandi, M. Experimental investigation on thermal performance of a PV/T-PCM (Photovoltaic/Thermal) system cooling with a PCM and nanofluid. *Energies* **2019**, *12*, 2572. [CrossRef]
11. Sarafraz, M.M.; Goodarzi, M.; Tlili, I.; Alkanhal, T.A.; Arjomandi, M. Thermodynamic potential of a high-concentration hybrid photovoltaic/thermal plant for co-production of steam and electricity. *J. Therm. Anal. Calorim.* **2020**, 1–10. [CrossRef]
12. Boutina, L.A.; Khelifa, A.; Touafek, K.; Lebbi, M.; Baissi, M.T. Improvement of PVT air-cooling by the integration of a chimney tower (CT/PVT). *Appl. Therm. Eng.* **2018**, *129*, 1181–1188. [CrossRef]
13. Tiwari, S.; Agrawal, S.; Tiwari, G.N. PVT air collector integrated greenhouse dryers. *Renew. Sustain. Energy Rev.* **2018**, *90*, 142–159. [CrossRef]
14. Fan, W.; Kokogiannakis, G.; Ma, Z. Optimisation of life cycle performance of a double-pass photovoltaic thermal-solar air heater with heat pipes. *Renew. Energy* **2019**, *138*, 90–105. [CrossRef]
15. Kim, J.-H.; Park, S.-H.; Kang, J.-G.; Kim, J.-T. Experimental performance of heating system with building-integrated PVT (BIPVT) collector. *Energy Procedia* **2014**, *48*, 1374–1384. [CrossRef]
16. Abdullah, A.L.; Misha, S.; Tamaldin, N.; Rosli, M.A.M.; Sachit, F.A. Photovoltaic thermal/solar (PVT) collector (PVT) system based on fluid absorber design: A review. *J. Adv. Res. Fluid Mech. Therm. Sci.* **2018**, *48*, 196–208.
17. Singh, H.P.; Jain, A.; Singh, A.; Arora, S. Influence of absorber plate shape factor and mass flow rate on the performance of the PVT system. *Appl. Therm. Eng.* **2019**, *156*, 692–701. [CrossRef]
18. Hu, J.; Sun, X.; Xu, J.; Li, Z. Numerical analysis of mechanical ventilation solar air collector with internal baffles. *Energy Build.* **2013**, *62*, 230–238. [CrossRef]
19. Kim, S.-M.; Kim, J.-H.; Kim, J.-T. Experimental study on the thermal and electrical characteristics of an air-based photovoltaic thermal collector. *Energies* **2019**, *12*, 2661. [CrossRef]
20. International Electrotechnical Commission. *Crystalline Silicon Terrestrial Photovoltaic (PV) Module—Design Qualification and Type Approval*; IEC: London, UK, 2016.

Article

Factors Influencing Fertility Intentions of Newlyweds in South Korea: Focus on Demographics, Socioeconomics, Housing Situation, Residential Satisfaction, and Housing Expectation

Seran Jeon, Myounghoon Lee and Seiyong Kim *

Department of Architecture, Korea University, Anam-dong, Seongbuk-gu, Seoul 02841, Korea; seran217@hanmail.net (S.J.); 83myul@hanmail.net (M.L.)
* Correspondence: kksy@korea.ac.kr; Tel.: +82-2-3290-3914

Abstract: Since 2001, South Korea has experienced sustained lowest-low fertility. This phenomenon has persisted despite the implementation of several social policies aimed at increasing fertility rates. The purpose of this study was to quantitatively analyze the demographics, socioeconomics, housing situation, residential environment, and housing expectation of newlyweds in terms of their fertility intentions in South Korea (within 5 years of marriage) in order to help the development of more effective housing policies. We extracted the factors on the basis of fertility theories and previous related studies and identified differential characteristics of the impact on fertility intentions for the first and for additional child(ren). The results show that fertility intention was higher in non-metropolitan and rental households. There was also a significant relationship between the anticipated period of a home purchase and fertility intention. In particular, for one-child families, the second child fertility intention was significantly affected. In conclusion, we quantitatively confirmed various factors that significantly impact the fertility plans of newlyweds. We suggest that the government implements housing policies on the basis of economic stability, the number of children, and the residential environment of newlywed couples.

Keywords: newlyweds; fertility intention; demographics; socioeconomics; housing situation; residential satisfaction; housing expectation; housing policy

Citation: Jeon, S.; Lee, M.; Kim, S. Factors Influencing Fertility Intentions of Newlyweds in South Korea: Focus on Demographics, Socioeconomics, Housing Situation, Residential Satisfaction, and Housing Expectation. *Sustainability* **2021**, *13*, 1534. https://doi.org/10.3390/su13031534

Academic Editor: Colin A. Jones

Received: 13 January 2021
Accepted: 27 January 2021
Published: 1 February 2021

1. Introduction

In 2001, South Korea recorded a total fertility rate of 1.30, becoming the lowest-low fertility society, with this social phenomenon persisting to this day—the total fertility rate was 0.98 in 2018, 0.92 in 2019, and is expected to drop to 0.86 by 2021 [1]. In response, the government implemented several policies to address these low fertility issues, such as the "Third Master Plan of Low Fertility Aged Society" and "Housing Special-Provision Policy". However, most of these policies focused on multi-child families with three or more children, while fertility support for households with less than three children has been excluded.

According to the 2015 Newlyweds Panel Analysis of Housing Conditions by the Ministry of Land, Infrastructure and Transport, 16.2% of newlyweds currently do not have children, with more than 74.9% of households delaying their fertility plans due to difficulties in their careers, the burden of parenting, and economic circumstances [2]. Moreover, the average number of children of newlyweds is 1.16. Although the birth of a first child after marriage is most common for newlyweds, considering that newlyweds (defined as within five years of marriage) are the population group having children, it is necessary to implement targeted residential environment and housing policies to increase the fertility rate of newlyweds [3].

Recently, the residential environment has experienced rapid changes due to alterations in the social environment. The residential patterns of housing type, housing expenses, housing tenure type, and residential period have had a significant influence on marriage and fertility [4,5]. Moreover, socio-cultural problems such as insufficient childcare support, parenting expenses, and marriage delays have been suggested as the cause of lowest-low fertility. Among these causes, housing issues are a critical factor directly contributing to the low fertility problem in South Korea.

Previous studies analyzed social, economic, and residential behavioral impacts on the fertility intention of married women to prepare measures counteracting low birthrates. Chun examined the cause of low fertility and the current state of housing policies related to fertility support and emphasized a necessity of the compatible residential setting for work and childcare, housing provision to support childcare and housework, social interest, and a shifted perception in the direction of residential policy for revitalizing the fertility rate [5]. In addition, Chun proposed that residential policies are the basis for supporting childbirth by empirically identifying the effects of housing policies on fertility support and the influences of residential charges on the birth rate [5]. Jeong critically reviewed the contents and problems of counter plans against the low fertility and highlighted the importance of political promotion to create an advantageous setting for childbirth via the reduction of childcare responsibility, the expansion of public childcare services, and the initiation of parental leaves [6]. Seo suggested age, parental value, the burden of costs, and career responsibility as the principal factors influencing the fertility intention of married women, i.e., the number of planned children [7]. Moreover, Seo underlined the significance of selective approaches across the number of born children, in contrast to a comprehensive approach, in order to deal with the issue of low fertility, as the beneficial capacity and child value were found to vary when women with a child plan to have additional children [7]. Jeong noticed the need for policies to alleviate the burden of delivering and parenting a child and the necessity of plans to promote the family support system to aid parenting, as suggested by the fertility age for the first child, educational levels, health status, and marriage satisfaction as the major influencing factors for the second childbirth [8]. Kang highlighted the necessity to create an overall environment that can reduce education and child-rearing costs, rather than temporary support, on the basis of a survey revealing that factors for the intention of the subsequent fertility were age, academic background, income, family make-up, and the number and gender of children [9].

These studies have crucial implications not only from an academic perspective but also from a political perspective, for what kind of and how much housing will affect marriage or fertility. However, housing itself has an impact, i.e., via residential patterns, rather than influencing marriage and fertility. Park et al. analyzed the relationship between housing and fertility [10]. Describing housing stability and parenting affinity as positive influencing factors on the fertility intention, Park states that facilities related to childbirth and parenting should meet the aspect of life convenience and increase the overall birth rate because these factors directly impact the fertility rate. Lee highlighted that the impact of housing on marriage and fertility differs depending on house ownership, housing type, residential period, housing purchase cost, and housing size, and that, thereby, differentiated housing policies should be approached by the types of housing [11]. Mulder and Billari investigated the association between fertility rates and homeownership regimes in Western countries [12]. They analyzed four types of homeownership regimes in terms of the owner-occupied housing and mortgage accessibility, and argued that countries with homeownership regimes with a high share of owners and low access to mortgages have the lowest fertility rates [12]. Masanja et al. studied factors driving the decrease in fertility rates in urban and rural areas in Tanzania [13]. The authors suggested the need for appropriate government policies and programs to match social changes that affect the fertility rate, such as small family sizes, improving the level of children's education, and a change in the social status of women. Vignoli et al. investigated the relationship between the fertility rate and the economic situation [14]. They found that the psychological stability of residents in the current residential environment is an important factor in fertility intentions. In addition, they argued that social and economic policies should be supported to increase the fertility rate because employment security and economic conditions are related to a sense of stability in the residential environment. Sági and Lentner studied Hungarian pro-birth policies and reported a policy gap in housing subsidies [15]. The authors evaluated family policy interventions such as housing support, tax allowances, and other child-raising benefits and concluded that an optimal mix of family policy incentives could maintain sustainable level of birth rate levels, but not necessarily increase them.

Previous studies exploring the factors influencing fertility intention and childbirth planning have focused on demographic factors such as wife's age, income, education level, and family composition [8,9]; socioeconomic factors such as income and childcare expenses [5–7,13]; and housing factors such as housing type, size of housing, housing satisfaction, and housing costs [4,10–12,14,15]. In addition, the studies explored parental leave, employment stability, childcare-friendly environment, social awareness trends, and economic activities of women [5,6,10,13]. Table 1 summarizes the outcomes of prior studies on fertility and its effects on demographics, socioeconomics, and housing. Most of these previous studies analyzed influencing factors on the fertility intention on the basis of ordinary families, although the main subjects of pregnancy planning and childbirth.

Table 1. Previous studies on fertility and the influences of demographics, socioeconomics, and housing.

Major Study Focus	Authors	Influences
Demographics	Jeong [8]	Educational levels and period, healthy status, marriage satisfaction
	Kang [9]	Age, educational background, income, family composition, number of children, the gender of children
Socioeconomics	Chun [5]	Income, values of marriage and family, women's economic circle, parenting expenses, housing expenses
	Jeong [6]	Reduction of childcare responsibility, expansion of public childcare services, activation of parental leave
	Seo [7]	Women's age, income, child value, parental value, charge of expenses, the approval rating of policy, beneficial capability
	Masanja et al. [13]	Women's empowerment, social transformation, differentials in education
Housing	Lee [4]	Housing tenure, housing type, extent of parental support for housing purchase, housing size, residential period
	Park et al. [10]	Satisfaction in the community environment, residential satisfaction, stability, life convenience, childcare-friendly environment
	Lee [11]	House ownership, housing type, residential period, housing purchase cost, housing size
	Mulder and Billari [12]	Housing purchase cost, housing ownership
	Vignoli et al. [14]	The psychological stability in the current residential environment
	Sági and Lentner [15]	Housing prices, regulatory gap in the housing market, family policy incentive

To overcome the limitations of previous studies, we first based our study on the influences commonly emphasized in fertility theories and previous studies. Available variables were extracted; classified into the characteristics of demographic, economic, and residential environments; and used for analysis. Second, factors affecting the fertility intention and plans of newlyweds, the main subjects of counter plans against low fertility, not ordinary households, were considered on the basis of national survey data. Third, to deal with low fertility issues, it is more important to selectively approach the number of born children rather than to comprehensively approach the fertility intention. On the basis of these assumptions, we identified whether there are differentiated characteristics between the influences on the period of the initial childbirth planning and the affecting factors on the plans for the first and second child.

The remainder of this study is structured as follows. Section 2 presents the research method, the selected model for empirical analysis, selected variables and the data analysis, and a descriptive analysis of the variables. In Section 3, on the basis of the influences revealed in prior fertility theories and studies, we extract and analyze the characteristics of demographics, socioeconomics, housing situation, residential environment, and housing expectation as derived from the Newlyweds Panel Analysis of Housing Conditions dataset. In Section 4, the significance of this study and future research directions are highlighted.

2. Materials and Methods

Publicly available microdata from the 2015 Newlyweds Panel Analysis of Housing Conditions released by the Ministry of Land, Infrastructure and Transport (https://mdis.kostat.go.kr) was used in this study [1,2]. In total, data on 2702 first-married couples within 5 years of marriage were selected for the analysis, whose marriages were reported from 1 January 2010 to 31 December 2014. Factors, such as the characteristics of demographics, socioeconomics, housing situation, residential satisfaction, and housing expectation were extracted on the basis of influences revealed in prior fertility theories and studies [3–6,14].

The demographic characteristics were categorized by the age of the wife, the duration of the marriage, and the residential region (metropolitan/non-metropolitan). The Republic of Korea is divided into 5 districts as follows: Seoul area (Seoul Metropolitan City), Gyeongin area (Incheon, Gyeonggi-do, Gangwon-do), Chungcheong area (Daejeon, Sejong Special Self-Governing City, Chungcheongnam-do, Chungcheongbuk-do), Jeolla area (Gwangju, Jeollabuk-do, Jeollanam-do, Jeju-do), and Gyeongsang area (Daegu, Ulsan, Busan, Gyeongsangbuk-do, Gyeongsangnam-do). The metropolitan region includes Seoul city, Incheon city, and Gyeonggi province. The wife's age value was generated by converting the date of birth. Economic characteristics were defined as income, mortgage (monthly expenses), and dual-income status. Pre-tax gross annual salary statements were used for income, and mortgages were applied to the analysis on the basis of mortgage statement as an average monthly expense.

The residential attributes were defined as housing ownership, rental status, and apartments or non-apartments. Residential satisfaction consisted of the satisfaction of the housing setting and environment of residential area. The housing setting satisfaction was based on the house location, housing condition, and management cost. The satisfaction of the environment of residential area was surveyed on satisfaction with local safety, local markets, transportation, neighboring nature, and child-friendly environment. The value of residential environment was calculated by averaging the satisfaction of housing setting and environment of residential area. The newlyweds' expected years of house purchase was classified as follows: less than 1 year, 1 to 3 years, 3 to 5 years, 5 to 10 years, more than 10 years, impossible, and unknown. Further residential circumstances were applied for the analysis using statements of the anticipated period for housing purchase.

We set up model 1, model 2, and model 3 to analyze factors affecting the fertility intention of newlyweds. Model 1 analyzed whether newlyweds had the intention to plan their fertility in particular situations, regardless of whether they have a child. Via model 2, the factors influencing the fertility intention for the first child were identified, and model 3 was applied to analyze the factors influencing additional childbirth, on the basis of first-married couples with one child only.

We used a binomial logistic regression for our statistical analysis. Logistic regression analysis was developed to alleviate the challenge of calculation, which is a disadvantage of the Probit model, with logistic regression being a model of selected probability assuming that the probabilistic utility is an independent distribution with a Weibull distribution. The purpose and procedure of the analysis are similar to linear regression analysis but differ in that ostensible-typed variables are applied as dependent variables. Logistic regression analysis utilizes odds, a ratio between the probability of occurrence and the probability of non-occurrence. The corresponding formula is expressed as $\text{Odds} = \text{p}/(1-\text{p})$. The concept of odds cannot be used for the general regression analysis with values between 0 and 1.

The concept of odds has 2 problems: the first is that it does not have negative (-) values, and the second is that the relationship among probabilities reveals an asymmetry around 1. As a method to solve these problems, natural logs are assigned to the values of odds, which is called logit. On the condition of a given explanatory variable, if the S-shape of a logistic function with a maximum value of 1 and a minimum value of 0, which represents the probabilities that a particular choice occurs or not, is converted to logit, it appears linearly. The formula is expressed as follows:

$$\text{Odds} = \tfrac{p}{1-p} = \exp[\alpha + B_1X_1 + B_2X_2 + \cdots + B_pX_p],$$
$$\ln\left(\tfrac{p}{1-p}\right) = \alpha + B_1X_1 + B_2X_2 + \cdots + B_pX_p$$

The concept of the odds ratio is used in the interpretation of the logistic regression analysis. The odds ratio refers to a change as X_i increases by a unit, given that the explanatory variable is constant. The formula is expressed as follows:

$$\text{Odds Ratio} = \frac{\text{Exp}(a + B_1X_1 + \cdots + B_i(X_i + 1) + \cdots + B_pX_p)}{\text{Exp}(a + B_1X_1 + \cdots + B_iX_i + \cdots + B_pX_p)} = \text{Exp}(B_i)$$

If the odds ratio is less than 1, the explanatory variable X_i has a negative (-) impact on the dependent variables, and if the odds ratio is larger than 1, X_i has a positive (+) influence. All analyses were performed using IBM SPSS Statistics ver. 22.0 (IBM Corp., Armonk, NY, USA).

3. Results

3.1. Descriptive Analysis

The descriptive analysis results of the 2015 Newlyweds Panel Analysis of Housing Conditions are shown in Table 2. The average age of the wives of the newlyweds was 32.24 years. The average annual income for households was 4810.5 million won (approximately EUR 38,262) for households, and the average monthly mortgage was 28.5 million won (approximately EUR 226). Note that the average basic rate of exchange in 2015 was used for the currency conversion (Korean Statistical Information Service, https://kosis.kr).

According to the characteristics of newlyweds, the proportion of households living in a rental relationship (70.6%) during the study period was higher than that of households owning their home (29.4%), and the proportion of households living in an apartment (61.7%) was higher than that of households living in a non-apartment (38.3%). Although this can be interpreted as newlyweds tending to prefer to live in an apartment, it is regarded as due to the surveys being conducted in urban areas in which many of the participants resided under the identified characteristics of newlyweds [2].

3.2. Factors Influencing Fertility Intention of Newlyweds

Whether newlyweds have fertility intentions in particular situations, regardless of already having children, was analyzed using model 1 (Table 3). The results show that during the study period, the fertility plan was correlated with the age of the wife and marriage duration. Non-metropolitan residents had a 1.369 times higher fertility plan than residents in metropolitan areas. This can be interpreted as being a result of economic factors, such as high housing costs, child support expenses, education expenses, and income instability, despite the relatively large proportion of young people in their 20s and 30s living in metropolitan areas [10,16–18].

Table 2. Descriptive analysis.

Variable Type		Variable Name		Frequency	Proportion (%)	Min	Max	Average	Standard Deviation
Dependent variables		Presence of children	Yes	2114	21.8	-	-	-	-
			No	588	72.2	-	-	-	-
		Presence of a fertility plan	Yes	1889	69.9	-	-	-	-
			No	813	30.1	-	-	-	-
Independent variables	Demographics	Age of wife (years)		-	-	19	49	32.24	4.03
		Marriage period	1st to 2nd year	1367	50.6	-	-	-	-
			3rd to 5th year	1335	49.4	-	-	-	-
		Residential region	Metropolitan	1519	56.2	-	-	-	-
			Non-metropolitan	1183	43.8	-	-	-	-
	Socioeconomics	Annual income (pre-tax/million won)		-	-	0	28,000	4810.5	2064.9
		Mortgage (monthly/million won)		-	-	0	650	28.5	46.8
		Dual-income status	Single	1169	43.3	-	-	-	-
			Dual	1533	56.7	-	-	-	-
	Housing situation	Ownership status	Owned	795	29.4	-	-	-	-
			Rent	1907	70.6	-	-	-	-
		Residential type	Apartment	1666	61.7	-	-	-	-
			Non-apartment	1036	38.3	-	-	-	-
	Residential satisfaction	Residential environment (average of housing setting and environment of residential area)		-	-	1	4	2.94	0.53
		Housing setting		-	-	1	4	2.93	0.57
		Environment of residential area		-	-	1	4	2.91	0.56
	Housing expectation	Anticipated years of housing purchase		-	-	0	11.7	8.6	50.15

Table 3. Factors influencing fertility intention of newlyweds (model 1).

Independent Variable		Demographics			Socioeconomics			Housing Situation		Residential Satisfaction			Housing Expectation
		Age of Wife	Marriage Period	Residential Region	Annual Income	Mortgage	Dual-Income Status	Ownership Status	Residential Type	Residential Environment	Housing Setting	Environment of Residential Area	Anticipated Years of Housing Purchase
1 to 2 years of marriage	Beta	0.046	0.484	0.329	-2.12×10^{-4}	-3.21×10^{-4}	−0.010	0.519	−0.071	0.146	−0.128	−0.155	0.001
	SE	0.014	0.073	0.120	2.88×10^{-5}	130×10^{-5}	0.110	0.141	0.120	0.124	0.119	0.116	4.23×10^{-4}
	Wald	11.195	44.057	7.479	0.544	6.090	0.008	13.504	0.350	1.400	1.161	1.783	6.072
	p-value	**0.001** [b]	**<0.001** [c]	**0.006** [b]	0.461	**0.014** [a]	0.927	**<0.001** [c]	0.554	0.237	0.281	0.182	**0.014** [a]
	Odds	1.047	1.622	1.389	0.999	0.999	0.990	1.680	0.932	1.158	0.880	0.856	1.001
	VIF	1.094	1.096	1.198	1.149	1.097	1.089	1.067	1.193	1.661	1.679	1.446	1.195
3 to 5 years of marriage	Beta	0.042	0.358	0.219	-4.69×10^{-4}	-2.23×10^{-4}	0.047	0.149	−0.082	0.101	0.043	−0.264	0.001
	SE	0.017	0.080	0.135	2.91×10^{-5}	1.35×10^{-5}	0.129	0.164	0.142	0.142	0.149	0.143	0.001
	Wald	6.225	19.985	2.643	2.585	2.693	0.134	6.493	0.334	0.459	0.089	3.853	0.456
	p-value	0.013	**<0.001** [c]	0.104	0.108	0.101	0.714	**0.011** [a]	0.564	0.498	0.765	0.050	0.499
	Odds	1.043	1.430	1.245	0.999	0.999	1.048	1.521	0.921	1.106	1.044	0.168	1.001
	VIF	1.034	1.074	1.232	1.125	1.099	1.037	1.033	1.128	1.187	1.297	1.211	1.184
Both groups	Beta	0.050	0.430	0.314	-3.60×10^{-4}	-2.71×10^{-4}	0.017	0.478	−0.036	0.174	−0.017	−0.195	0.001
	SE	0.012	0.037	0.099	2.27×10^{-5}	1.01×10^{-5}	0.093	0.018	0.117	0.106	0.101	0.097	3.97×10^{-4}
	Wald	18.053	137.753	10.059	2.521	7.103	0.035	16.845	0.130	2.704	0.027	4.036	5.946
	p-value	**<0.001 c**	**<0.001** [c]	**0.002** [b]	0.112	**0.008** [b]	0.852	**<0.001** [c]	0.718	0.100	0.869	**0.045** [a]	**0.015** [a]
	Odds	1.052	1.538	1.369	0.999	0.999	1.017	1.614	0.964	1.190	0.984	0.823	1.001
	VIF	1.052	1.130	1.200	1.142	1.097	1.039	1.030	1.139	1.325	1.426	1.281	1.171

Bold values denote statistical significance at the a: $p < 0.05$, b: $p < 0.01$, and c: $p < 0.001$ levels. SE indicates standard error ($\sigma/\sqrt{n}$). VIF indicates variance inflation factors, and all variables were within acceptable range (VIF < 10). The *p*-value of Hosmer–Lemeshow goodness-of-fit test was $p = 0.22$ (good fit). The value of Durbin–Watson independence of errors test was 2.01 (no autocorrelation). In Box–Tidwell linearity test, no interaction terms were significant ($p > 0.05$).

In terms of housing characteristics, the fertility plans of newlyweds living in rental households were 1.614 times higher than those of newlyweds living in their own homes. More specifically, newlyweds living in rental households had fertility intentions 1.680 times (1–2 years of marriage) and 1.521 times (3–5 years of marriage) higher than those who owned their home. Although this is inconsistent with the general perception that renting would more negatively impact fertility than owning a home, these results are consistent with the data (24.4% of owners, 43.9% of renters) of the 2015 Newlyweds Panel Analysis of Housing Conditions, which examined whether newlyweds planned on having a child according to housing tenure type [2,16].

Our analysis shows that households with a short anticipatory period for home purchases and/or those that are satisfied with their residential area are more likely to have children. This can be explained by the fact that housing stability for raising children is closely related to fertility rates.

3.3. Factors Influencing Fertility Intention of Newlyweds for the First Child

Model 2 shows the factors affecting the fertility intention of the first child (Table 4). Demographic characteristics have shown that the age of the wife and the marriage duration have a significant effect on the birth of their first child. This is in line with results of previous studies, which have shown that ordinary households, including newlyweds, have lower birthrates and newlyweds are older on average when they get married [2,18].

As shown in Table 4, the fertility plans of non-metropolitan residents were 1.483 times higher than of those living in metropolitan areas. Regardless of marriage duration (1–2 and 3–5 years of marriage), plans for the first child depended on the housing characteristics, with rental residents having 1.529 times higher birth plans than homeowners. Non-apartment residents had 1.623 times higher birth plans than apartment residents.

3.4. Factors Influencing Fertility Intention of Newlyweds for Additional Children

The factors affecting the fertility plans for additional children of newlyweds with a child were analyzed using model 3 (Table 5). Although there were differences in the fertility plans for an additional child in terms of the level of income of the newlyweds, specific patterns were unclear. In the past, an increase in the level of income was generally recognized to increase the fertility rate due to the younger age of the couple getting married and of the wife at childbirth. However, the phenomenon of giving up on an additional child with an increase in income observed today is considered to be caused by a higher desire for an increased quality of life than for additional children [14]. A more in-depth analysis is required between these economic variables and family planning.

According to the results of the 2015 Newlyweds Panel Analysis of Housing Conditions, newlyweds without children showed a large difference regarding income, with 6.0% of single-income households and 31.5% of dual-income households having no children [2]. This difference indicates difficulties associated with the newlyweds' working life, responsibilities of raising children, and economic causes. As an increasing number of women enter the social circle, new subsidy fertility policies are required to promote childbirth and reduce the burden of parenting. We also found that a higher level of satisfaction with the environment of the residential area, such as safety, childcare facilities, and the living and transportation infrastructure, have a positive impact on having additional children. A healthy work–life balance between work and childcare is important for the overall fertility rate, as shown in the analysis of the relationship between family characteristics, residential regional environment, and the rate of additional child fertility [16].

Table 4. Factors influencing fertility intention of newlyweds for the first child (model 2).

Independent Variable		Demographics			Socioeconomics			Housing Situation		Residential Satisfaction			Housing Expectation
		Age of Wife	Marriage Period	Residential Region	Annual Income	Mortgage	Dual-Income Status	Ownership Status	Residential Type	Residential Environment	Housing Setting	Environment of Residential Area	Anticipated Years of Housing Purchase
1 to 2 years of marriage	Beta	−0.005	1.005	0.380	-2.69×10^{-4}	−2.79e−4	−0.066	0.331	0.453	0.118	−0.168	0.062	-197×10^{-4}
	SE	0.015	0.134	0.133	3.46×10^{-5}	1.13e−5	0.122	0.161	0.131	0.140	0.131	0.131	3.88×10^{-4}
	Wald	0.120	56.556	8.224	60.455	6.035	0.288	4.210	11.945	0.711	1.638	0.222	0.259
	p-value	0.729	**<0.001** [c]	**0.004** [b]	**<0.001** [c]	**0.014** [a]	0.591	**0.040** [a]	**0.001** [b]	0.399	0.201	0.637	0.611
	Odds	0.995	2.731	1.463	0.999	1.000	0.937	1.392	1.574	1.125	0.845	1.064	0.999
	VIF	1.198	1.109	1.503	1.338	1.311	1.071	1.368	1.234	1.382	1.368	1.504	1.198
3 to 5 years of marriage	Beta	−0.076	0.200	0.609	-1.00×10^{-4}	-4.29×10^{-4}	−0.100	0.879	0.654	−0.049	−0.005	0.226	-278×10^{-4}
	SE	0.027	0.137	0.225	4.11×10^{-5}	3.05×10^{-5}	0.219	0.321	0.229	0.257	0.243	0.239	0.002
	Wald	7.717	2.136	7.344	5.980	1.980	0.207	7.493	8.161	0.036	0.001	0.894	0.025
	p-value	**0.005** [b]	0.144	**0.007** [b]	**0.014** [a]	0.159	0.649	**0.006** [b]	**0.004** [b]	0.849	0.982	0.344	0.873
	Odds	0.927	1.222	1.838	0.999	1.000	0.905	2.407	1.923	0.952	0.995	1.253	0.999
	VIF	1.689	1.508	1.513	1.429	1.501	1.468	1.456	1.614	2.557	3.124	3.847	1.500
Both groups	Beta	0.026	0.804	0.394	-2.03×10^{-4}	2.72×10^{-4}	−0.090	0.425	0.484	0.064	−0.132	0.101	-3.27×10^{-4}
	SE	0.013	0.052	0.112	2.52×10^{-5}	1.11×10^{-5}	0.105	0.139	0.112	0.120	0.113	0.113	3.63×10^{-5}
	Wald	3.840	240.484	12.463	65.151	5.958	0.746	9.403	18.649	0.286	1.367	0.799	0.814
	p-value	**0.049** [a]	**<0.001** [c]	**<0.001** [c]	**<0.001** [c]	**0.0015** [a]	0.388	**0.002** [b]	**<0.001** [c]	0.592	0.242	0.371	0.367
	Odds	0.975	2.233	1.483	0.999	1.000	0.914	1.529	1.623	1.067	0.876	1.106	0.999
	VIF	1.192	1.237	1.383	1.269	1.237	1.080	1.133	1.264	1.244	1.503	1.439	1.477

Bold values denote statistical significance at the a: $p < 0.05$, b: $p < 0.01$, and c: $p < 0.001$ levels. SE indicates standard error ($\sigma/\sqrt{n}$). VIF indicates variance inflation factors, and all variables were within acceptable range (VIF < 10). The *p*-value of Hosmer–Lemeshow goodness-of-fit test was $p = 0.36$ (good fit). The value of Durbin–Watson independence of errors test was 2.07 (no autocorrelation). In Box–Tidwell linearity test, no interaction terms were significant ($p > 0.05$).

Table 5. Factors influencing the fertility intention of newlyweds for additional children (model 3).

	Independent Variable	Beta	SE	Wald	*p*-Value	Odds Ratio	VIF
Demographics	Age of wife	−0.006	0.014	0.169	0.681	0.994	1.095
	Marriage period	0.810	0.046	306.131	**<0.001** [c]	2.247	1.130
	Residential region	−0.262	0.118	4.902	**0.027** [a]	0.769	1.185
Socioeconomics	Annual income	-1.69×10^{-4}	3.64×10^{-5}	21.650	**<0.001** [c]	0.999	1.161
	Mortgage	1.27×10^{-5}	-9.75×10^{-5}	1.698	0.193	1.000	1.117
	Dual-income status	−0.263	0.110	5.743	**0.017** [a]	0.768	1.057
Housing situation	Ownership status	0.056	0.146	0.148	0.701	1.058	1.033
	Residential type	0.067	0.121	0.306	0.580	1.070	1.140
Residential satisfaction	Residential environment	0.182	0.124	2.151	0.142	1.199	1.461
	Housing setting	0.122	0.121	1.018	0.313	1.130	1.500
	Environment of residential areas	−0.273	0.111	6.037	**0.014** [a]	0.761	1.299
Housing expectation	The anticipated years of housing purchase	−0.001	0.001	2.109	0.146	0.999	1.175

Bold values denote statistical significance at the a: $p < 0.05$ and c: $p < 0.001$ levels. SE indicates standard error ($\sigma/\sqrt{n}$). VIF indicates variance inflation factors, and all variables were within acceptable range (VIF < 10). The *p*-value of Hosmer–Lemeshow goodness-of-fit test was 0.51 (good fit). The value of Durbin–Watson independence of errors test was 1.84 (no autocorrelation). In Box–Tidwell linearity test, no interaction terms were significant ($p > 0.05$).

4. Discussion and Conclusions

This study was conducted to identify the factors affecting the fertility rate of newlyweds in South Korea. On the basis of our quantitative research, we suggest that the following policies should be considered to increase the fertility rate of newlyweds:

First, housing policies promoting economic stability of their first home should be ensured for newlywed couples who plan their first child. For newlywed couples who plan additional children, customized policies are necessary to improve the residential environment, which is directly related to fertility intention. Newlywed households planning to start a family should be targeted by a housing policy that provides economic stability for their initial settlement, and newlyweds who already have children should be targeted by customized policies to improve the residential environment in the area that they live in.

Second, flexible mortgage plans must be implemented for newlyweds in their first 1–2 years of marriage because they generally have few economic assets. It is necessary to lower the interest rate and raise the mortgage limits for long-term rental deposits or home purchases. In particular, increasing the mortgage limit and extending the payment term length is crucial to reduce the burden on metropolitan households [19]. If the government were to loosen the marital income standards for mortgages, more double-income newlywed couples could benefit from mortgages. In order to improve the fertility rate of apartment residents, it is necessary to take measures to ease the interest rate on loans from rent subsidy programs for housing purchases. Furthermore, regarding the period of housing, to induce multiple births through housing stability, as in some Organization for Economic Co-operation and Development (OECD) countries, an institutional mechanism is required to suppress excessive increases in monthly rent.

Third, appropriate residential and economic policies based on income status should be supported. For example, the results showed that the fertility intention of newlyweds in rental households was higher than that in own households, which can be explained by the complex influence of income, the women's educational background, and the residential environment in which children can be raised [20]. Our data show that the income of newlyweds in self-owned households was higher than that of newlyweds in rental households. The higher the income, the higher the women's educational levels and the higher their willingness to engage in social activities, resulting in a lower fertility intention of newlyweds who own their own households. On the contrary, a higher proportion of newlyweds in rental households had a lower income. Since the women's educational level and frequency of social activities were found to be lower than those in self-owned households, the fertility intention was relatively higher [21]. Therefore, personalized support policies based on the income status are needed, such as extended support of childcare expenses for low-income households and a safe environment and daycare facilities for high-income households.

Forth, it is necessary to establish a maternity-friendly urban residential environment since fertility plans of newlyweds are influenced by residential satisfaction. In a previous study, the parenting-friendly urban residential environment was defined as "a convenient and safe environment for individual households to decide whether to have children" [22]. Recently, local governments have made efforts to establish locally differentiated policies by developing "maternity environment indicators" [4].

Comparing South Korea and Japan, the time when the birth rate declined below the replacement fertility level ($\approx$2.1) and the time when the low birth rate policy commenced occurred roughly 10 years earlier in Japan. Still, there are many similarities to Korea, such as housing policy trends. Japan's population policy to respond to low birthrates began with the Angel Plan in 1996, but the active policy began after the enactment of the Basic Act on Countermeasures for Small Self-Socialization in 2003. In Japan, the children's allowances are universally paid, adequate housing is provided on the basis of income status, and childcare facilities are installed near the place of residence to create a better environment for raising children. Moreover, the government supports the costs of constructing or remodeling childcare facilities or playgrounds through the multi-dwelling unit certification system [23].

On 15 December 2020, Korea announced the "4th Basic Plan for Low Fertility and Aged Society". The government plans to pay the infant allowance and provide a lump sum of 2 million won (approximately EUR 1486, in terms of average basic rate of exchange in 2020) for childbirth to solve the low birthrate problem [24]. However, it is difficult to rebound the fertility rate simply by reducing the burden of childcare because fundamental solutions related to the labor problem are missing. Although the government has set up an additional support policy for parents with children under 12 months of age that pays up to 3 million won (approximately EUR 2229) per month (100% of normal wages) for a parental leave of 3 months, most of parents are unable to take parental leave due to their working environment. Additionally, the low fertility rate issue is the result of a complex combination of various factors, such as unstable employment, the burden of education costs, and gender discrimination experienced by young people. Labor market gaps, unstable employment problems, and low wages are factors that hamper marriage and childbirth. The gender-discriminatory structure in which women are burdened with parenting and housework also contributes to the low birth rate [13]. It is important not only to raise the fertility rate but to prepare a new vision and adaptative policy for society.

This study has the following limitations. First, since we focused on married couples, the factors affecting the marriage and fertility of unmarried couples were not analyzed. Therefore, follow-up studies are needed on the fundamental factors of unmarried people. Second, the low birthrate problem is a complex result of various factors in addition to demographic, social, and residential environment factors [23]. Therefore, follow-up studies are warranted on socio-structural causes and solutions on various aspects, such as labor, economy, and politics.

We quantitatively confirmed various factors such as demographics, socioeconomics, housing situation, residential environment, and housing expectation that significantly impact the fertility intentions of newlyweds. Housing policies, such as lower interest rates, higher loan limits, extended repayment periods, and eased loan qualifications, could help newlyweds settle in the early stages and increase their fertility intentions. Childcare support policies should be improved, such as the expansion of reliable childcare facilities and the introduction of an equal parental leave system that enables joint parenting of couples. Since the low birthrate problem is the result of complex factors such as population, society, economy, and housing, an integrated policy should be implemented that considers various aspects of the issue, not just one-time support for subsidies. The results of this study are expected to substantially contribute to raising the fertility rate of newlyweds in South Korea by meeting the needs of families that have children and reinforcing housing policies in the future.

Author Contributions: Conceptualization, S.J., M.L., and S.K.; methodology, S.J. and M.L.; validation, S.J. and M.L.; formal analysis, S.J. and M.L.; investigation, S.J. and M.L.; data curation, S.J. and M.L.; original draft preparation, S.J. and M.L.; review and editing, S.J., M.L., and S.K.; supervision, S.K. All authors have read and agreed to the published version of the manuscript.

Funding: This research received no external funding.

Institutional Review Board Statement: Not applicable.

Informed Consent Statement: Not applicable.

Data Availability Statement: Data are publicly and freely available from the Newlyweds Panel Analysis of Housing Conditions released by the Ministry of Land, Infrastructure and Transport (https://mdis.kostat.go.kr).

Acknowledgments: This article was prepared on the basis of the award-winning work of the 2016 Residential Environment Statistical Research Contest using the microdata of the Korea Research Institute for Human Settlement [25].

Conflicts of Interest: The authors declare no conflict of interest.

References

1. The Korean Statistical Information Service (KOSIS). *The Population Trend Research*; KOSIS: Daejeon, Korea, 2019.
2. The Ministry of Land, Infrastructure and Transport. *Second Year Newlyweds Panel Analysis Report*; The Ministry of Land, Infrastructure and Transport: Sejong City, Korea, 2015.
3. Shin, Y. An Analysis on Determinants of Newlywed's Hosing Tenure Choice. Master's Thesis, Konkuk University, Seoul, Korea, 2016.
4. Lee, S. *A Study on the Effects of the Residential Environment one Fertility and Its Political Counterplan*; Korea Institute for Health and Social Affairs: Yeongi-gun, Korea, 2013; Volume 183, pp. 1–4.
5. Chun, H. *The Political Plan of Housing Support for Dealing with Low Fertility Issues: Special Directions of Urban Policy and Housing Policy in the Low Fertility Times*; The Korea Research Institute of Human Settlements: Yeongi-gun, Korea, 2012.
6. Jeong, S. The counterplan of low fertility, what is the problem? *Korea J. Popul. Stud.* **2015**, *38*. [CrossRef]
7. Seo, J. Factors Influencing married women's childbearing willingness based on number of children-ever-born. *Korea J. Popul. Stud.* **2015**, *38*, 3–6.
8. Jeong, E. *A Study on the Factors Related to the Fertility and Pregnancy for the Second Child*; Korea Institute for Health and Social Affairs: Yeongi-gun, Korea, 2013; Volume 137, pp. 5–8.
9. Kang, H. A Study on the Influencing Factors on the Subsequent Fertility of Employed Mothers. Ph.D. Thesis, Jung-an Theological Seminary, Seoul, Korea, 2014.
10. Park, T.; Choi, H.; Han, Y. *Effect Analysis of Neighborhood Environment on Childbearing-Women's Residential Satisfaction and Childbirth Intention*; The Korea Research Institute of Human Settlements: Yeongi-gun, Korea, 2016; Volume 89, pp. 111–130.
11. Lee, S. *Paradigm Shifts in Family Policy on Changes in Marriage and Fertility Behavior 2016-44*; Korea Institute for Health and Social: Sejong, Korea, 2016.
12. Mulder, C.H.; Billari, F.C. Homeownership regimes and low fertility. *Hous. Stud.* **2010**, *25*, 527–541. [CrossRef]
13. Masanja, G.F.; Lwankomezi, E.; Emmanuel, C. The effects of declining fertility on household socioeconomic conditions in Tanzania: A comparative study of urban versus rural areas of Kwimba District, Mwanza Region. *Int. J. Popul. Res.* **2016**, 4716432. [CrossRef]
14. Vignoli, D.; Guetto, R.; Bazzani, G.; Pirani, E.; Minello, A. A reflection on economic uncertainty and fertility in Europe. *Genus* **2020**, *76*, 1–27. [CrossRef] [PubMed]
15. Sági, J.; Lentner, C. Certain aspects of family policy incentives for childbearing—A hungarian study with an international outlook. *Sustainability* **2018**, *10*, 3976. [CrossRef]
16. Lee, S. *The National Fertility and the Welfare Patters of Family Health*; Korea Institute for Health and Social Affairs: Yeongi-gun, Korea, 2015; Volume 8, pp. 99–111.
17. Jeong, S. Review of theoretical approach to low fertility. *Korea J. Popul. Stud.* **2009**, *49*, 167–176.
18. Lee, S.; Choi, H. *Analysis on Association between Housing and Fertility*; Korea Institute for Health and Social Affairs: Yeongi-gun, Korea, 2012; Volume 47, pp. 73–89.
19. Byeon, D. Predictors of Fertility-Intention Among the Married Man and Women: Utilization of the Theory of Planned Behavior. Ph.D. Thesis, Konkuk University, Seoul, Korea, 2015.
20. Chun, H.; Kim, Y.; Jeong, H.; Kim, H.; Ha, S.; Kim, J.; Youn, Y.; Oh, M.; Kim, T. *Housing and Urban Policy Strategies Responding to Low Fertility*; The Korea Research Institute of Human Settlements: Sejong City, Korea, 2012.
21. Noh, S. Predictors of Childbearing Intention and Behavior in Married Women: Longitudinal Analysis of Korean Longitudinal Survey of Women and Families. Ph.D. Thesis, Junganng University, Seoul, Korea, 2015.
22. Moon, S. *Measures to Build a Childbirth-Friendly Community Environment*; Busan Women's Family Development Institute: Busan, Korea, 2016; pp. 9–29.
23. Lee, S.L.; Lee, J.H. *Association between Housing and Fertility among Newly Married Couples*; Annual Report 2017; Korea Institute for Health and Social Affairs: Yeongi-gun, Korea, 2017; pp. 192–193.
24. *The 4th Basic Plan for Low Fertility and Aged Society*; Presidential Committee on Ageing Society and Population Policy: Seoul, Korea, 2020.
25. Jeon, S.; Lee, M. *Effect Analysis of Residential Environment on Newly Married Couples' Childbirth Intention, 2016 Residential Environment Sta-tistical Research Contest using the Microdata*; Korea Research Institute for Human Settlement: Yeongi-gun, Korea, 2016.

Article

Building Envelope Thermal Defects in Existing and Under-Construction Housing in the UAE; Infrared Thermography Diagnosis and Qualitative Impacts Analysis

Kheira Anissa Tabet Aoul *, Rahma Hagi, Rahma Abdelghani, Monaya Syam and Boshra Akhozheya

Architectural Engineering Department, College of Engineering, United Arab Emirates University, Al Ain 15551, United Arab Emirates; r-hassan@uaeu.ac.ae (R.H.); 201350278@uaeu.ac.ae (R.A.); 201350087@uaeu.ac.ae (M.S.); 201450126@uaeu.ac.ae (B.A.)

* Correspondence: kheira.anissa@uaeu.ac.ae; Tel.: +971-3-713-5184 (ext. 5190)

Abstract: The built environment accounts for the highest share of energy use and carbon emissions, particularly in emerging economies, caused by population growth and fast urbanization. This phenomenon is further exacerbated under extreme climatic conditions such as those of the United Arab Emirates, the context of this study, where the highest energy share is consumed in buildings, mostly used in the residential sector for cooling purposes. Despite efforts to curb energy consumption through building energy efficiency measures in new construction, substantial existing building stock and construction quality are left out. Construction defects, particularly in the building envelope, are recognized to affect its thermal integrity. This paper aims, first, to detect through thermography field investigation audit construction defects bearing thermal impacts in existing and under-construction residential buildings. Then, through a qualitative analysis, we identify the resulting energy, cost, and health impacts of the identified defects. Results indicate that lack or discontinuity of insulation, thermal bridging through building elements, blockwork defects, and design change discrepancies are the recurrent building and construction defects. The qualitative review analysis indicates substantial energy loss due to lack of insulation, thermal bridging with cost and health implications, while beneficial mitigation measures include consideration of building envelope retrofitting, skilled workmanship, and the call for quality management procedures during construction.

Keywords: building envelope thermal defects; construction defects; thermography; qualitative analysis; new construction; existing building; construction quality; thermal bridging; impacts; energy; housing; UAE

Citation: Tabet Aoul, K.A.; Hagi, R.; Abdelghani, R.; Syam, M.; Akhozheya, B. Building Envelope Thermal Defects in Existing and Under-Construction Housing in the UAE; Infrared Thermography Diagnosis and Qualitative Impacts Analysis. *Sustainability* **2021**, *13*, 2230. https://doi.org/10.3390/su13042230

Academic Editors: Ali Bahadori-Jahromi and Jun-Tae Kim

Received: 9 December 2020
Accepted: 25 January 2021
Published: 19 February 2021

1. Introduction

The quest for energy efficiency in the built environment has driven substantial research and technology advances to curb the ever-increasing energy demand needed to service human needs. Developing countries and emerging market economies, in particular through population and economic growth, drive the demand for fossil fuel [1], which certainly brings more prosperity but, at the same time, aggravates the resulting impacts of climate change [2]. The building industry accounts for the largest percentage of the total energy use and carbon emissions globally [3], a fact often further exacerbated by extreme climatic conditions and late or inadequate implementation of building energy efficiency measures. Nevertheless, the worldwide agenda of an energy-efficient built environment has been embraced globally, yet with different contextual challenges and variable accomplishments.

The United Arab Emirates (UAE), the context of this study and similar to many fast growing economies, has witnessed high population and economic growth and an extremely fast urbanization in the last four decades, a result of its immense oil revenues. The UAE population stands at 9.8 million in 2020 from a mere 1 million in 1980 [4]. The total population of Dubai, for example, has grown by 1000% over the last 40 years alone [5].

Most of the population (over 85%) lives in urban settings [6]. The result has been an unprecedented urban growth with, however, minimal consideration for non-renewable resources and energy-related repercussions until 2010 [7]. The construction sector in the UAE stands as a leading economic sector that has garnered global interest regarding the quality and energy impacts of the resulting booming construction. In terms of energy usage, the built environment in the UAE accounts for 70% of energy consumption, mainly used in cooling, compared to the global average of 40% [8]. For instance, Abu Dhabi, the largest emirate, contributes up to 22% of active projects and 38% of the total value of projects [9]. The Abu Dhabi region occupies the first position in terms of the number of constructed buildings in the UAE, where around 80% of electricity consumption is attributed to buildings alone [10], a fact that has positioned the UAE as one of the world's largest energy consumers per capita, with a demand trend expected to intensify [11].

Amid the country's overall growth, the main component of the urban fabric in the UAE is the residential sector, which leads the way in energy usage and carbon emissions. The residential market sector is mainly in the form of extensive government-sponsored housing programs and large privately developed rental developments [10]. Within the residential sector a significant amount includes existing and newly constructed detached or semi-detached houses, amounting to about 65% of the urban fabric, according to the National Statistics Center [10], widely recognized as the most demanding type of buildings in terms of cooling, especially under the local extreme hot climate [12]. The UAE's desert climate, characterized by extreme high summer temperatures, with a maximum annual temperature average of 45 °C in August, high solar radiation, and high humidity on the coastal zones, imposes serious challenges to both designers and owners alike. The design and construction challenges reside in adapting the intended building to the extreme hot climate of the UAE, while building owners have the challenging task to alleviate the high running energy cost. This climatic condition, along with building design, construction type, quality and materials, plays a critical role in the overall building energy performance. Unlike the climatically adapted vernacular houses, contemporary internationally styled houses in the UAE have disregarded the climatic construction methods and cultural context and relied on active cooling and ventilation systems for thermal comfort. The construction methods and materials used in housing were not controlled until rather recently, despite the harshness of climate and its impact on a building's cooling needs and increasing electrical demands [13].

The central government acknowledged the importance of targeting building energy use as key to reducing both the country's energy consumption and carbon emission and introduced several energy conservation measures and control procedures that apply to new construction. In fact, it was not until 2010 that the Urban Planning Council in the Emirate of Abu Dhabi established Estidama, the local sustainability framework [14], which aims at achieving sustainability and energy conservation in buildings through the provision of guidelines for newly constructed buildings. Similarly, the Green Regulations and Specifications in the Emirate of Dubai (2011) was established as a first step toward implementing green building strategies. It came into effect in 2011 as mandatory for governmental buildings while remaining voluntary for private ones. The focus in both resides in the specification of minimum U-values for walls and roofs. For example, in Dubai, Green Building Regulations limit U-values for the roof and walls to a maximum of 0.3 W/m^2 K and 0.57 W/m^2 K, respectively; whereas in Abu Dhabi, the Estidama PEARL code prescribes (at its lowest rating) maxima for roof and wall U-values of 0.14 W/m^2 K and 0.32 W/m^2 K, respectively [15]. Alongside an active regulatory implementation of the required measures in new constructions, there has been a wide government call to address the existing building stock in a collective target to reduce energy demand by 30% by 2030. Recently, in 2017, the Ministry of Energy introduced the political feasibility of policy options for the country's Energy Transition. It announced a new UAE Energy Strategy 2050 that outlines a number of energy targets for 2050, including: energy efficiency targeting a 40% improvement relative to the current annual growth in electricity. Additionally, it targets increased implementation of energy efficiency (EE) standards with monitored building performance and audits to achieve greater EE technology adoption and demand site management [16].

The momentum to address building energy efficiency in UAE targeted first and foremost new constructions with a primary focus on wall and roof insulation, leaving out the status and performance of the relatively new but extensive existing building stock and construction quality and processes. Further, rapid growth puts building and construction quality under pressure but contextual research has so far focused on issues affecting delays and cost [17–19]. Under the UAE climate, the building envelope highly contributes to the total heat gain in a building [20]. This factor is important as the building envelope can contribute 50% or more of the embodied energy distribution in major building elements in residential buildings and between 50 to 60% of the total heat gain [21].

Construction defects in housing, specifically the ones occurring in the building envelope, are recognized to contribute to the energy performance gap of the building [22,23] The generally accepted definition of construction defects refers to a deficiency in the construction process, from either design, materials and systems, or workmanship, that leads to some form of failure (financial, safety, performance, or other). The construction defects are extensively addressed in the literature from identification of type, classification and standardization [24,25]; source of the defect (design, workmanship) and origin of defect (change, error, omission, damage) [26,27]; type of defects including missing, misaligned or incorrect installation in different building elements [24,25,28] or, of more relevance, the relationship between quality defects and thermal performance of buildings [22].

However, in the UAE context, research within housing defects, especially those occurring in the building fabric, are extremely limited. The nature, type, and origin of defects have not been explored, either in existing buildings to identify and prioritize mitigation strategies or in the new construction to ensure code and standards deliver as per target. Hence, the aim of this paper is to first identify construction defects that may affect the thermal integrity of the building's envelope in both existing and under-construction residential buildings via thermal imagery diagnosis. Second, through a qualitative analysis we explore the energy, economic, and health impacts of the identified defects. The intent is to open up ways of addressing them either in terms of optimal retrofitting measures for existing buildings or construction quality processes during construction. Identifying defects in existing buildings caused by building age, deterioration, and workmanship defects may guide or prioritize actions and type of retrofitting measures, yielding optimum benefits. Defects identified and categorized in housing during construction can assist in establishing adequate and early mitigation strategies during the construction process.

2. Materials and Methods

This study comprises a field investigation and a qualitative review analysis. First, the field investigation through Infra-Red Thermography (IRT) auditing in both existing residential buildings and under-construction housing units is presented. Thus, thermal imaging rationale, validity, and procedure are presented first. Then the various impacts of the identified construction defects are assessed through a qualitative review analysis.

2.1. Thermal Imaging and Building Envelope

Infra-Red Thermal (IRT) imaging is a non-destructive technique utilized to perform qualitative and quantitative tests on the building envelope to identify various defects that contribute to energy loss [29]. IRT is used to obtain empirical data for building envelope performance in heat loss/gain, air leakage, and moisture problems through thermal images that indicate external and internal surface temperature differences. Kylili et al. [30] recognized IRT as one of the most appropriate nondestructive imaging techniques, a widely used tool to quickly identify potential building defects among researchers and practitioners [30]. Since IRT is used to obtain empirical data for the actual thermal bridging, it has become particularly useful at the post-construction stage in assessing energy efficiency where the designed U-values are compared to the measured values [31]. Further, in already built structures walkthrough internal thermography assessments were found to substantially detect more defects than other IRT approaches [32]. It can also be a useful tool for retrofitting, thus enabling grant-providing authorities to assess gains in thermal performance of the retrofitted building envelope. [33]. As such, Hopper et al. [34] investigated the possibilities and limitations of using IRT as qualitative test method for assessing the installation of retrofitted external wall insulation to pre-1919 dwellings with solid exterior walls [34].

IRT has also been used in under-construction buildings to detect defects. The application of thermography at various phases of the construction process has mainly emerged from field tests on housing projects in Wales, UK [35]. The scope for four types of "in-construction" tests was identified in: (a) making early stage checks on the installation of insulation, (b) identifying air leakage through the building envelope, (c) assessing insulation continuity and the severity of thermal bridges, and (d) investigating the performance of building services [35]. Similarly, Littlewood [36] developed the in-construction testing protocol using thermography to inspect the dwelling envelope under construction, aiming to identify when defects occur that detrimentally affect operational energy use for heating.

2.2. Case Studies

All case studies considered in this study were single-family, detached, or semi-detached residential buildings in the city of Al Ain located in the Emirate of Abu Dhabi. Al Ain city is located inland and is characterized by an extremely hot and arid climate. The case studies included existing buildings and construction sites at different stages of the construction process. Two main types of residential buildings in Al Ain dominate the landscape. The first one is in the form of detached private villas, mostly for nationals, whereas the private rental market is dominated by housing developments of relatively similar layouts. This research investigated both housing typologies for construction defects spread across different sites within the city (Figure 1).

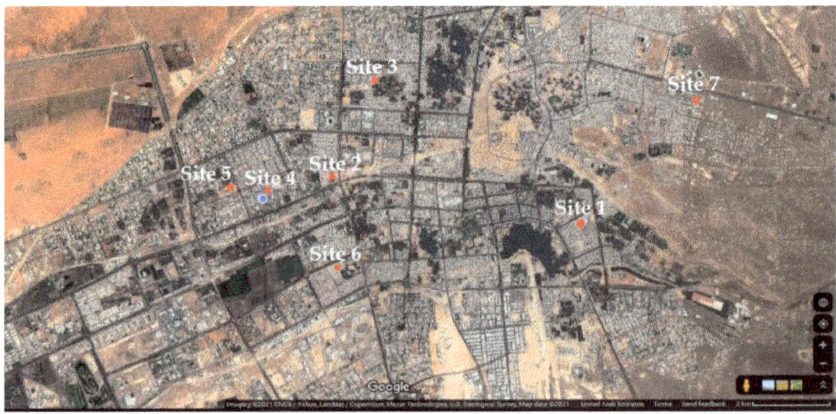

Figure 1. Site locations of the investigated residential units in Al Ain city (UAE).

The dominant construction method in the housing sector, present in six of the seven test sites, is concrete blockwork infill with a reinforced concrete post and beam structural system. This is also the dominant construction method across the Mediterranean and Gulf region. Site 7 is an exception, as it uses precast concrete structure, which is a rare occurrence in the UAE housing sector. The existing building sites (Site 1 and Site 2) were built over 10 years apart (Figure 2) and were selected to identify the impact of time, workmanship, and construction quality on the building envelope thermal efficiency. These housing units were built before the implementation of the Estidama Building Regulation Code [14], therefore, the exterior walls and roofs were likely to lack any kind of thermal insulation, a shortcoming which is duly expected to highly affect the house energy consumption. In the considered two sites, the U-value of the exterior walls and roof were 2.319 and 2.849 W/m^2 K, respectively, per submitted values of building permit. On the other hand, the exterior envelope of the residential units under construction (Figure 3) adhered to building Estidama energy code [15], where the U-value of the exterior walls and roof were 0.29 and 0.14 W/m^2 K, respectively. Separate thermal thermography investigations conducted in these buildings are summarized in Table 1.

(**a**) Site 1

(**b**) Site 2

Figure 2. Existing residential buildings in (**a**) Site 1 completed in 1999, (**b**) Site 2 completed in 2010.

(**a**) Site 6

(**b**) Site 7

Figure 3. Under-construction residential units in (**a**) Site 6 and (**b**) Site 7.

Table 1. Construction status and residential units' characteristics in each investigated site.

Building Status		General Information	Building Phase	N. of Tested Units	Construction Method
Existing Buildings (pre-energy efficiency code)	Site 1	Attached and semi-attached units/residential compound/196 units	Existing building competed in 1999	5 units	Non-insulated concrete blockwork infill with reinforced concrete post and beam structural system
	Site 2	Semi-attached units/residential compound/30 units	Existing building completed in 2011	2 units	
Under-construction buildings (post-energy efficiency code)	Site 3	Single-family house/two levels	Under construction—blockwork stage	1 unit	Insulated concrete blockwork infill with reinforced concrete post and beam structural
	Site 4	Single-family house/Three levels	Under construction—finishing stage	1 unit	
	Site 5	Single-family house/two levels	Under construction—blockwork stage	1 unit	
	Site 6	Semi-attached units/residential compound/80 units	Under construction—blockwork stage	6 units	
	Site 7	Detached unit/residential compound/300 units	Handover stage	3 units	Insulated precast concrete panel and frame

2.3. Testing Procedure

The Infra-Red Thermography (IRT) investigation requires adherence to specific environmental conditions. The temperature gradient between the interior and the exterior of the investigated building should be at least 10 °C [37]. In this study, the passive approach was used as an analysis scheme for the investigation of thermal patterns on the building envelope to detect temperature differences under natural conditions [30]. Similarly, in Bauer et al.'s study [38], a thermographic survey using the passive approach was used to detect defects on ceramic tiles and cracks in the mortar. The most significant outcome of the study was that the identification of the surface temperature abnormalities depended on various conditions such as orientation of the building or direction of solar radiation. The results specified the correct inspection time after the defect type on the surface and correct direction of heat flow were defined.

The thermal investigation included an internal and external survey of selected elevations of the buildings with close-up thermograms for specific analysis, using a thermal imaging camera following the infrared methodology of the qualitative detection of thermal irregularities in the building envelope standard BS EN 13187:1999 [39]. For the existing building thermal survey, the passive analysis scheme was used; the target is typically an exposed structural element and measurement was performed from the interior of the building. Residents were asked to keep the air conditioning on for four hours prior to the test time to keep the indoor temperature constant. In the thermal investigation of buildings under construction, the external energy source is the Sun, where buildings are illuminated by periodically changing solar radiation and exposed in an atmospheric temperature. The north elevation was selected for investigation for each building to avoid direct solar radiation. The physics of periodic heating/cooling by solar radiation was used by Mathur et al. [40] to illuminate a concrete slab by a periodic changing of solar radiation and bathing in an atmospheric temperature [40]. For analyzing and measuring thermal images in this research, the qualitative analysis was based on the evaluation of differences in measured radiation by inspection of color patterns within a thermal image [41].

3. Results of Field Investigation: Building Defects Detection through Thermal Imaging

The thermal imaging audit revealed several recurrent issues in the form of thermal bridging within the exterior envelope in most of the investigated buildings, as shown in Table 2. Among the issues encountered, the dominant one found in most buildings was the non-insulated structural frames, followed by blockworks defects, then non-compliant design changes. These issues relate to workmanship errors, as well as to design and construction non-compliant decisions. The thermal investigation of Site 7 showed no significant areas of concern regarding thermal bridging, given the precast construction model and the minimal opportunities for workmanship errors or design changes. This is in line with the acknowledgment that prefabrication has been generally considered a more sustainable method in the building sector, supported by evidence of the demonstrated energy-saving potential of prefabrication based on life cycle analysis and thermal performance evaluation [42].

Table 2. Classification of identified construction defects in the investigated sites.

		Construction Defects			
	Site	**Construction Quality**	**Non-Insulated Structural Elements**	**Blockwork Defects**	**Design Change/ Discrepancies**
Existing buildings	Site 1		■		
	Site 2	■	■		
Buildings under construction	Site 3	■	■	■	■
	Site 4		■	■	
	Site 5		■	■	
	Site 6		■		■
	Site 7				

3.1. Construction Quality in Existing Buildings

The audit of selected units from the existing housing in Site 1 and Site 2, which were built 10 years apart, highlighted a number of common thermal behaviors and some variations. First, as expected, it identified the absence of insulation in the exposed building envelope, such as in the roof and exterior walls in both sample units (Figures 4 and 5), which critically calls for retrofitting as remediation. In the thermogram of the sliding door opening in the Site 1 unit, the line analysis on the concrete block wall around the opening showed the average temperature of the concrete block wall was 29 °C, while the line analysis on the concrete block wall around the sliding door opening in the Site 2 unit was 32.1 °C. The Site 2 unit showed more thermal anomalies between the sliding door frame and the exterior wall, compared to the Site 1 unit. This was likely caused by a construction defect related to poor sealing (Figure 6).

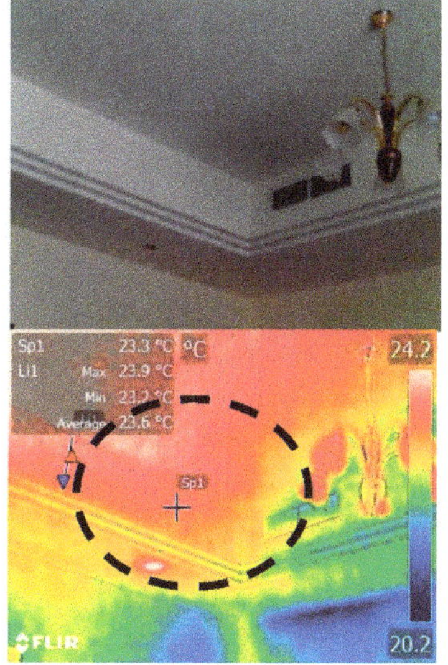

(a) Roof junction Site 1 SSW—202.5

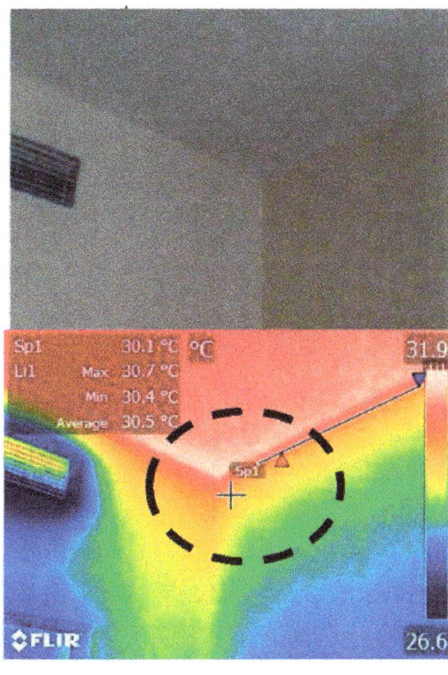

(b) Roof junction Site 2 ESE—112.5

Figure 4. Thermogram of roof junction showing lack of roof insulation in existing buildings of Site 1 and Site 2.

These results highlighted the strong relationship between workmanship, construction quality, and the thermal behavior of the building's envelope. Unexpectedly, the results showed more thermal anomalies in the newer units of Site 2 completed in 2010 than the older ones (Site 1, built in 1999), thus indicating that the thermal behavior of the two buildings was more a function of workmanship and construction quality than simply building age. This further reinforces the need to diagnose defects during construction, which, although costly, are preventable if detected and remediated early enough.

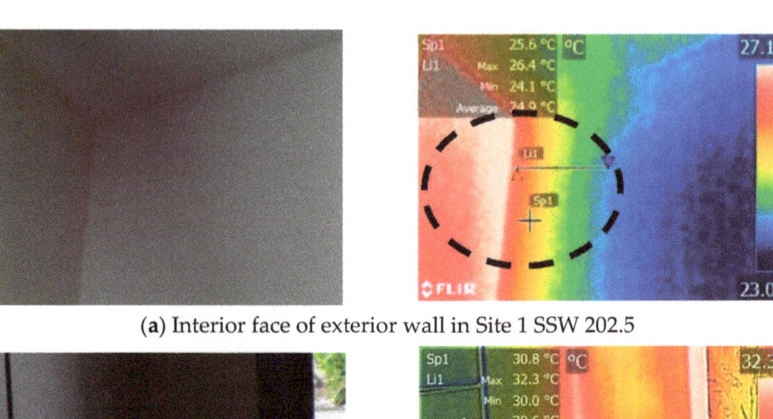

(**a**) Interior face of exterior wall in Site 1 SSW 202.5

(**b**) Interior face of exterior wall in Site 2 ESE 112.5

Figure 5. Thermogram showing higher surface temperature of the non-insulated exterior wall in existing buildings of Site 1 and Site 2.

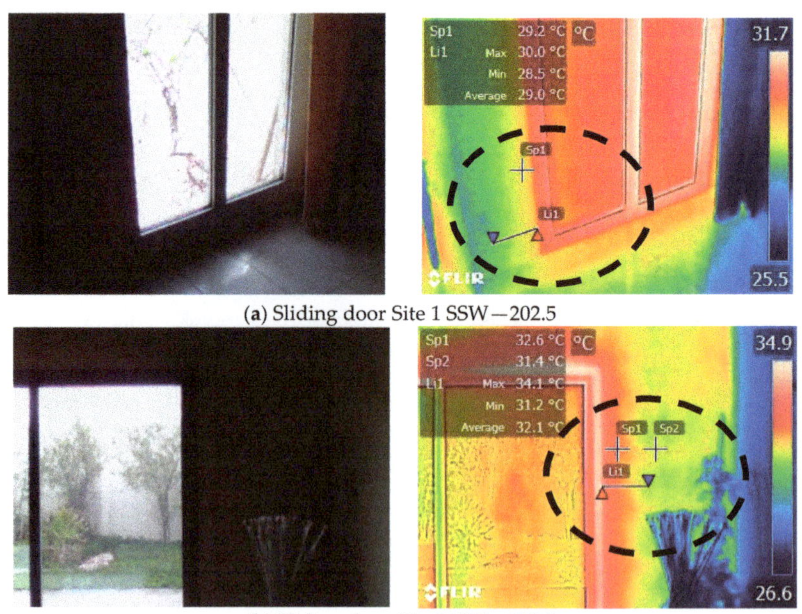

(**a**) Sliding door Site 1 SSW—202.5

(**b**) Sliding door Site 2 ESE—112.5

Figure 6. Thermogram of sliding door in Site 1 and Site 2.

3.2. Uninsulated Structural Elements

The thermal imaging investigation in under-construction sites indicated primarily the thermal bridging between the building components where structure composition changes. The exterior walls of the case study consisted of Autoclaved Aerated Concrete (AAC) blocks with a U-value of 0.29 W/m^2 K that met the required energy efficiency code, and a non-insulated reinforced concrete frame (2.398 W/m^2 K). In Figure 7, the thermogram showed the interior face of an exterior wall in Site 4 where both the concrete blocks and reinforced concrete frame indicated different thermal behaviors against absorbed solar energy, which led to a form of thermal bridging. Another example of an exterior wall can be seen in Figure 8, with a thermal variation between the concrete blocks and reinforced concrete frame in the range of 2 °C. The cavity-insulated concrete blocks limited the amount of stored solar energy, while the high thermal mass of the reinforced concrete frame structure stored and transferred more heat. The impact of such defects was explored by Friess et al. [13], who evaluated through simulation the role of thermal bridges between concrete blocks and the reinforced concrete frame of residential villas in Dubai. The results suggest the importance of adding insulation to the entire opaque envelope.

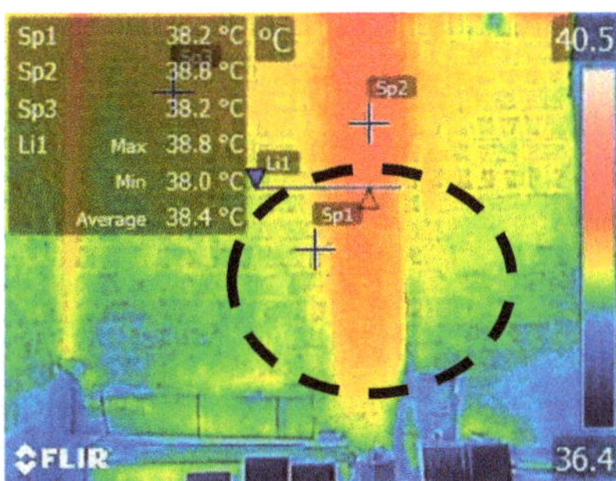

Figure 7. Thermogram of interior face of exterior wall in Site 4 (North façade).

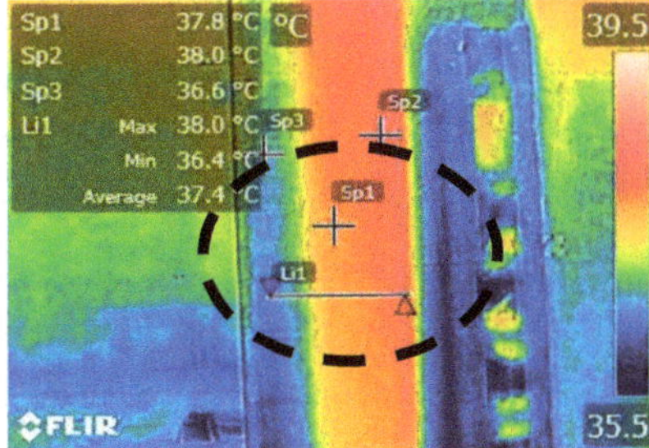

Figure 8. Thermal pattern of the exterior wall in Site 4 (North façade).

3.3. Blockwork Defects

In the under-construction sites, the Autoclaved Aerated Concrete (AAC) construction blocks can offer significant energy savings, thermal bridging control, and air-tightness of buildings due to their thermal mass and integrated insulation [43]. However, the thermal integrity of a wall can be compromised by poor workmanship, which would lead to thermal bridging. The thermal investigation exhibited different types of anomalies in the otherwise insulated blockwork. These were observed in three sites (Sites 3, 4, and 5) and the thermal defects were due to oversized service outlets and mortar joints. Figure 9 shows the thermograph of an exterior wall made of AAC blocks with provision of electricity outlets. The temperature of the spot check of the electricity outlet area was 38.2 °C, while the temperature of spot check of AAC block was 37.6 °C, with a temperature difference of 0.6 °C. In Figure 10, the thermogram shows a temperature variation of the interior face of the AAC exterior wall due to different mortar joint sizes. Both cases acted as thermal bridges that increased transmission loads and reduced wall thermal resistance (R-value). Al-Sanea and Zedan [44] reported experimental results of the mortar joints effect on R-value, indicating that for a typical wall with insulation thickness of 75 mm, mortar joints with Hmj = 10 mm (corresponding to 4.8% thermal bridge area) increase peak, daily, and yearly cooling and heating transmission loads by 62%, while the wall R-value decreases by 38%, compared to a similar wall with no mortar joints (Hmj = 0) [44].

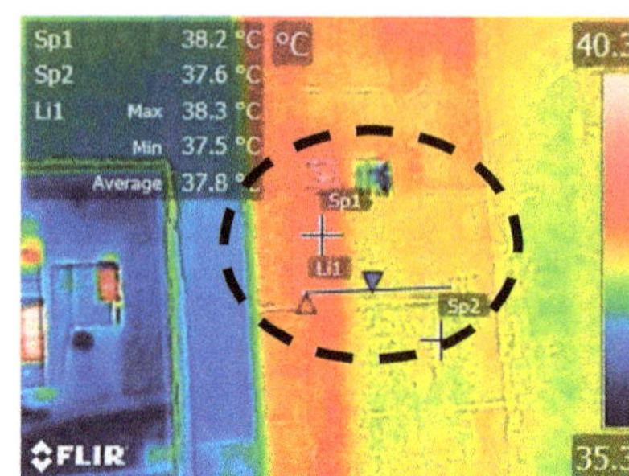

Figure 9. Thermogram showing oversized provisions for electrical services (exterior wall; Site 3).

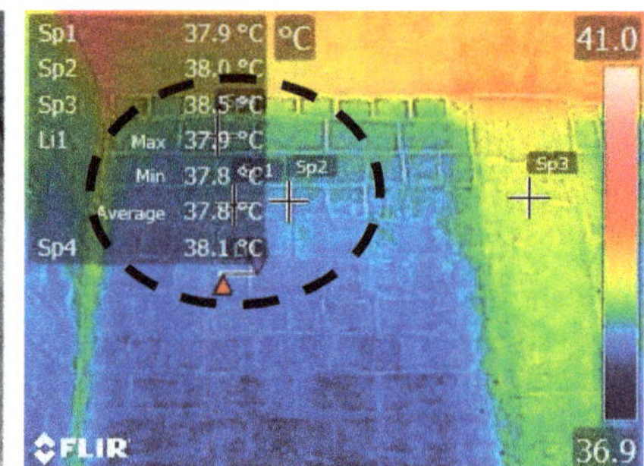

Figure 10. Thermogram showing oversized mortar joints in the blockwork (interior face of exterior wall; Site 5).

3.4. Design Discrepencies

Another defect identified during the testing of under-construction units was the result of design change with non-compliant construction details. This was not as frequent but still a recurring defect. In this investigation, the design change was caused by a client's choice, where exterior windows became fake windows kept for aesthetic reasons, but with non-insulated concrete fill (Figure 11). Thermographs showed discrepancies of the thermal behavior of the wall as a result of the non-insulted concrete fill and heat gain through convection of hot trapped air. The average temperature on the concrete block wall was found to be 34.6 °C in Site 6, and the difference between the concrete wall and the fake window area was 5.5 °C. In the thermograph of Site 3, there was a temperature variation in the range of 4.8 °C within the same wall due to the fake window and the non-insulated fill (Figure 12). Design changes in the construction phase with implementation defects may well be a main unaccounted-for source of heat gain in otherwise insulated buildings. Although they are more common than exceptional, their consequences have been largely ignored in simulations despite their possible significant reduction of the global thermal resistance of the envelope.

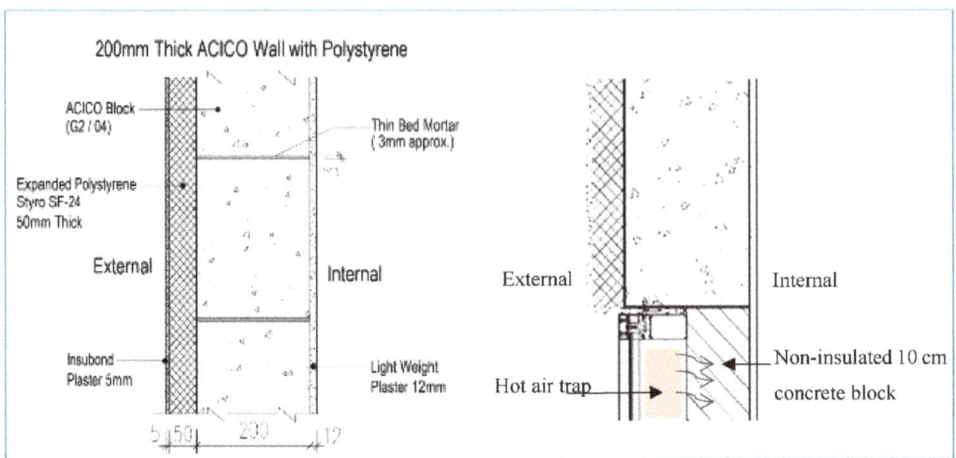

Figure 11. Construction details, (**left**) Typical wall construction (Site 6); (**right**) Heat gain through convection as a result of non-compliant design changes.

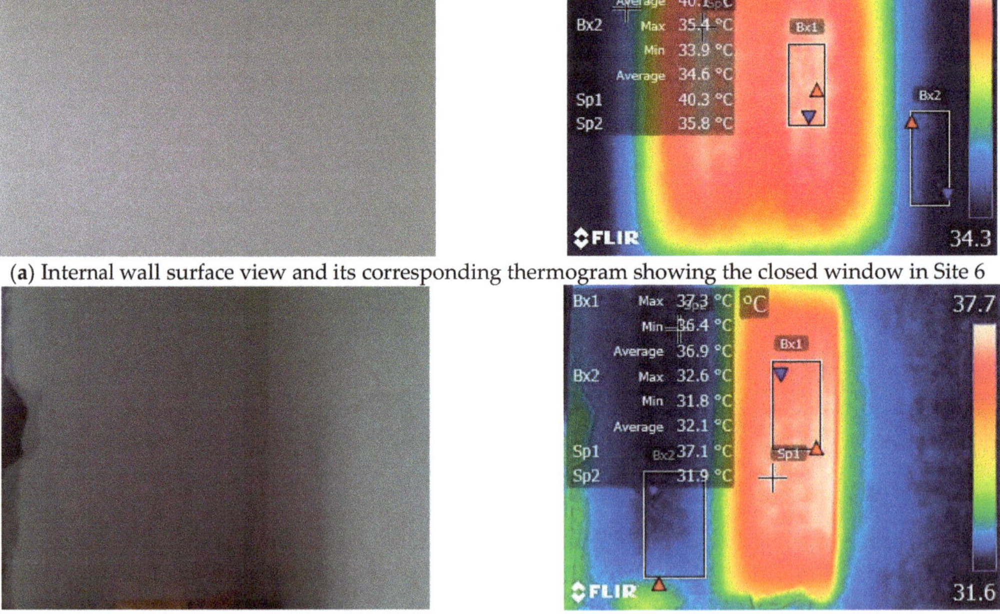

(**a**) Internal wall surface view and its corresponding thermogram showing the closed window in Site 6

(**b**) Internal wall surface view and its corresponding thermogram showing the closed window in Site 3

Figure 12. Thermograph of interior face of fake exterior window with non-insulated material fill (Site 6).

Finally, the building and construction defects identified in this field investigation may be best categorized as per their type, location, origin, possible source of error, and thermal impacts, as indicated in Table 3. Most of these defects are linked to workmanship, likely a lack of skilled workers, and insufficient site controls. The current construction quality control procedure does not include verification of construction thermal defects, relying solely on specifications and building energy performance simulations, whereas field studies and post-construction status highlighted major energy impacts, largely due to construction defects. The impacts of these defects are explored next through a qualitative review analysis.

Table 3. Summary synthesis of identified building defects.

Defect Type	Location	Origin	Source of Error	Impact
Missing insulation	Exterior Walls Roof	Lack of code (existing building prior to energy code implementation in 2010)		Heat gain
Insulation discontinuity		Design change Workmanship	Using non-insulated blocks	Thermal bridging
Structural–wall junction	Exterior Walls	Non-insulated structural elements (column, beam)		Thermal bridging
Window–wall junction	Exterior Walls	Workmanship	Poor sealing	Thermal bridging
Blockwork defects	Exterior Walls	Workmanship	Oversize electric services Oversize mortar joints between blocks	Thermal bridging

4. Qualitative Impacts Analysis of Identified Building and Construction Defects

The construction defects identified in the field investigation conducted in the UAE are not unique, but recurrent in many other countries. A large body of literature shows that the same construction defects, led by limited energy standards, thermal leaks and bridging, and design discrepancies, induce multiple impacts. Hence, the qualitative review analysis carried out next aimed to unveil the relationships between the defects and their impacts, leading to higher energy consumptions, poor building systems performance, occupants' discomfort, and additional issues. Table 4 summarizes the literature findings grouped under the three major construction defects identified through the thermal imaging field investigation and their respective impacts, which are discussed in detail next.

Table 4. Summary of the literature on impacts of building and construction defects.

Construction Defects	Impacts	References
Lack of insulation and energy standards	Energy	[13,20,45–51]
	Cost	[52–54]
	Comfort	[13]
Thermal bridging	Energy through walls	[55–61]
	Energy through windows and openings	[47,62–64]
	Energy through building junctions	[29,65–72]
	Comfort and health	[56,72–79]
Design discrepancies	Workmanship quality	[24,28,58,80–92]

4.1. Non-Insulated Building Envelope

The audited existing housing built prior to the implementation of the energy building code in UAE all had non-insulated walls and roof. Lack of insulation is a well-reported issue in the literature as dramatically reducing the thermal performance of buildings [22,45,87,93–95]. In the UAE as well as in numerous other contexts, this is mainly due to the lack or late implementation of regulations that enforce the installation of thermal insulation [46]. The lack of insulation clearly results in high heat loss or gain, leading to high energy consumption [13,20,45,50], high running cost [52–54], increase in CO_2 emissions [20,53] as well as negatively affecting building users' comfort [96]. The impact on energy was found to be substantial, ranging from almost 25% [13] to 78% [48]. Even in a less extreme climate, such as the Mediterranean region, the lack of insulation significantly contributed in raising the energy consumption where the opaque external elements of the building without insulation contributed approximately 50% of the total cooling load [52]. Several studies proved that the impact of the missing insulation in the building walls and roof is not limited to high energy usage, but consequently increases the annual energy cost [52,54] as well as limits thermal comfort to very short periods [96].

The absence of building insulation is a scenario that calls for multiform actions: at the forefront, the establishment of rigorous standards and codes that will govern new construction, a practice that has been embraced in the UAE through Estidama in the Emirate of Abu Dhabi, with other emirates following the same path [97]. On the other hand, addressing the low performance of the existing building stock calls for retrofitting measures [48–51]. Researchers have stressed the importance of establishing high energy standards and strict regulations for upgrading the existing design to acceptable thermal insulation requirements. The benefits of using thermal insulation in buildings have been extensively addressed [20,49]. A study in the UAE estimated, through energy modeling of an existing villa, a 37.2% reduction in the total annual cooling load when thermal insulation refurbishment is upgraded to the Estidama 1 Pearl requirements [46]. Similarly, in another context, Altun et al. [54] estimated that adding insulation to meet current standards improves the building's annual heating energy up to 75%, 70% for life-cycle cost, and 73% for life-cycle greenhouse gas emissions, with a payback period under two years for the greenhouse gas and under seven years for the initial investment cost [54].

A number of retrofitting measures to mitigate the negative impacts of the lack of insulation on the energy consumption, cost, and users' comfort have been commonly addressed. In their extensive research, Brannigan and Booth [53] addressed the selection criteria for the most suitable mitigation strategy, taking into consideration cost savings, minimal disruption during installation process, and resulting aesthetics of the property. The findings indicate that the challenge resides in the type and the thickness of the insulating layer [96]. Material types and the thickness of the thermal insulation of exterior walls exhibited a material-based variable optimal economic thickness [98]. Different thermal insulation materials have been tested to examine the most effective ones, with a high agreement on the efficiency of extruded polystyrene in reducing heat loss and saving energy [13,48,52]. However, the increased insulation may lead to higher levels of indoor pollutants and condensation issues calling for ventilation upgrades [51].

Heat loss or gain is not limited to the missing thermal insulation in the building envelope, but as revealed in the thermography diagnosis and supported by the reviewed literature, thermal bridging in the structural elements as well as at slab–wall junctions, window–wall junctions and balcony joints is a leading cause of the thermal energy performance gap.

4.2. *Thermal Bridging*

A thermal bridge is a part of the building envelope where, relative to the surrounding area, the thermal transmittance is considerably greater at the local level [75]. Thermal bridges are primarily caused by geometric or structural effects or by the discontinuation of thermal insulation in whole or in part. Thermal bridges are a major cause of poor energy performance, increasing heating and cooling loads, poor durability, surface damages, poor indoor air quality, and hygiene problems such as mold growth, surface condensation, and the staining of surfaces, resulting in different impacts.

4.2.1. Impacts on Energy

The various construction defects identified in the field study lead primarily to thermal bridging, a result of heat loss or gain that impacts energy usage. Numerical and experimental studies have assessed the negative thermal impacts of different types of thermal bridges. The qualitative review presented next analyzed thermal bridges according to the type of defect and uncovered a pattern of negative impacts within building connections, windows, walls, and other miscellaneous openings.

The overall effect of thermal bridges in building junctions increased the thermal losses through the building envelope by about 9% in an experimental building setup in which infrared thermography assessment was used [29]. In a similar UAE climatic context, a numerical study for a typical villa in Kuwait showed that due to uninsulated columns and beams the thermal resistance of buildings can be reduced by 48% [66]. In a milder Mediterranean climate, a similar conclusion was reached, where the correction of thermal bridges studied on pillars, balconies, and slab–wall junctions resulted in an overall annual energy saving of 8.5%. The savings were also reflected in the primary energy needs for heating, where 25% and 17.5% reductions were achieved for terraced houses and semi-detached houses, respectively [71]. Similarly, the analysis of a building's thermal behavior showed that the design and construction choice of balconies and foundations had the biggest impact on the value of linear thermal bridges. It was shown that correct design solutions can improve the value of linear thermal bridges up to 13 times [67]. Moreover, a study done on Greek single-family houses showed that the impact of thermal bridging on the overall annual heating load was estimated at 13%. The investigated retrofit solutions applied to balconies and window edges achieved a decrease of the annual heating energy requirement from 4 to 10% [69].

A similar pattern of results was reached under extreme climate conditions, cold in this instance, where the results of the simulation of a high-rise residential building in British Columbia displayed an increase in the annual heating energy load by 37.4–42.2% from different thermal bridge connections, such as: intermediate floor junctions, intermediate wall/window junction, balcony junctions, balcony sliding door junction, partition wall junction, roof junction, and basement wall junction [99]. The previous results also aligned with the results found by Jedidi and Benjeddou [65] in which thermal bridges caused additional heat losses exceeding 40% of the total heat losses through the envelope. Increasing the insulation on the corners showed a positive effect on thermal bridging [65].

In studies exploring the impact of added insulation on building thermal performance, the thickness, material type, and insulation location exhibited different results. An increase of 50% and 100% in the thickness of the insulation considerably decreased the U-value, linear heat loss coefficient ψ, and surface temperature factor fRsi [65]. Karabulut et al.'s [70] study indicated that the heat transfer rate did not decrease in the heat bridge region with internal insulation, while the most suitable insulation model was the outer insulation. The previous studies were supported by Erdem and Pinar [56], who studied the impact of internal aerogel retrofitting on the thermal bridges of residential buildings; their findings confirmed that the simple internal retrofit by insulation is not the optimal solution to reduce the heat losses from residential buildings if appropriate attention is not paid to non-insulated building elements [56]. In this regard, another dominant type of thermal bridging caused by construction defects lies with windows and openings. Marincioni et al. [64] found that the junctions, particularly window jambs and lintels, show the highest transmission heat transfer coefficient. Uninsulated corners and side reveals account for 40–52% of the transmission heat transfer coefficient at junctions for thin walls, while they account for 53–69% for thick walls. When the reveals are insulated, the transmission heat transfer coefficient is lowered by more than two-thirds [64]. Similarly, Ibrahim et al. [47] highlighted that the percentage of the windows' thermal bridge energy load of the total house load is approximately 4–8% in the case of exterior walls having no interior insulation, and nearly 2–5% in the case of exterior walls having a 5-cm thick interior insulation. Furthermore, when using more exterior insulation, the relative effect of the thermal bridge was higher. As a mitigation strategy, applying a layer of 1 and 2 cm of coating on these thermal bridges lowered the energy consumption by about 36% and 50%, respectively, for the exterior walls with no internal insulation, and by about 24% and 33% in the situation of exterior walls having interior insulation [47]. An increase of the air-tightness level in external windows can also contribute to the reduction of the annual heating energy consumption per unit area of the building. Based on simulations and tests done by Liu et al. [62], the heating energy consumption was reduced by approximately 49%. On another note, the incorrect positioning of the window frame toward the external side or the internal side of the wall was found to increase the linear thermal transmittances by 70–75%. Depending on the jamb's configuration, the shift from the internal position to the intermediate position can lower the linear thermal transmittance by 31–50%, considering the wall with external insulation. Thermal losses can be further reduced by 52–75% by the shift from the internal to the external position [63].

Walls as diagnosed in the field investigation of this study can be a receptacle of construction defects that can negatively impact the energy loads within a building. Besides poor workmanship, these thermal bridges can occur through walls and affect the thermal gains or losses depending on the construction type of the wall. A study done in the UK measured the thermal properties of 25 new dwellings and concluded that, in terms of construction type, the worst performing construction form is partial fill masonry, which can be particularly susceptible to thermal bypassing and wind washing if it is not assembled correctly or if the levels of workmanship are poor [58]. Supporting the previous study, an analysis conducted by Zedan et al. [55] showed that mortar joints in masonry construction increased the cooling and heating loads almost linearly with the thermal bridge-to-wall-area ratio in a typical two-story villa in Riyadh. The same study concluded that the effects of thermal bridges resulting from mortar joints in walls can be eliminated by using tongue and groove insulated building blocks [55]. In a similar study about commonly used masonry veneer constructions, Finch et al. [100] revealed that masonry ties occupy 0.04% of the overall wall area, and the rules within some energy codes would permit such thermal bridge consequences to be ignored. The findings described in the paper demonstrated that this code compliance generalization results in exaggerated R-values reaching up to 30%. Nevertheless, the correct selection of tie material and tie design, such as carbon fiber or basalt, can reduce the effect of thermal bridging to less than 10% [100].

Comparing the results of traditional wall construction methods with those of the advanced ones, a study on the effect of thermal bridging in ventilated facades presented that thermal bridges in metal cladding systems can lower the effectiveness of the thermal insulation if no specific attention is given during the design and the construction process [57]. The use of a chemical anchor in the ventilated façade can decrease thermal bridge heat flows up to 10% in comparison to steel anchors. In a comparable type of construction, Theodosiou et al. [60] studied the magnitude of problems with the thermal bridge in double skin facades (DSFs). The overall influence on DSFs exceeds 25% of the total heat flow of the envelope, causing a substantial underestimation of the actual heat flow through the envelope of the building [60]. An alternative study on the effectiveness of precast sandwich panel wall construction in four different climatic boundary conditions showed that air circulation around the diagonal tie connectors between the panels can lead to an increase of room heating energy demand by about 7.0–10.3% in all studied climatic regions. Therefore, in precast sandwich panels installation of additional insulation around diagonal tie connectors (DTCs) is required to prevent natural air convection occurring in cavities between two insulation sections [61].

Looking at the structural material used in buildings, Real et al. [59] studied the effect of normal weight concrete (NWC) on thermal bridging and the contribution of lightweight aggregates (LWAs) in concrete to the reduction of the thermal bridging effect in buildings. The results indicated that, depending on the type of LWA, the contribution of thermal bridging in an apartment in Lisbon, Portugal, was 11–19% lower in the case with structural lightweight aggregate concrete elements than in the case with NWC elements. This finding emphasized the importance of selecting a high performing aggregate type in the concrete used in structures to reduce overall thermal bridging in walls [59].

4.2.2. Impacts on Health and User Comfort

The impact of thermal bridges is not limited to energy and resulting operational costs; thermal bridges can have an impact on the health and user comfort. Thermal bridging resulting in lower thermal performance and surface condensation lead to hygiene problems such as mold growth and the staining of surfaces, which affects the user's comfort levels [75].

Lower thermal performance caused by thermal bridging had a negative impact on building users, as exemplified through a survey conducted in Turkey exploring users' opinions about air quality conditions in terms of how they feel in winter and summer. Of all the users, 37% complained about the low heating level in winter, while 41% complained about overheating in summer. An investigation of the thermal performance of skeleton structural frame, wall openings, roof, and ground floor was ensued. It was shown that the thermal effectiveness of the wall was diminished by the thermal bridging in reinforced concrete structural elements, contributing to the highest heat loss through the wall [72].

Another potential consequence of thermal bridging impacting human comfort is condensation, which can arise if the internal insulation is improperly installed [73]. Some air-conditioned buildings in warm climates are susceptible to condensation risk [101]. The UAE has both arid and hot, humid climatic zones in which the peak average temperature and humidity can reach 45 °C and 85%, respectively [77]. Infiltration of warm, moist outside air through cracks and holes in the enclosure causes condensation and moisture to develop on materials that are cooler or have been cooled by air-conditioning systems [102]. Moisture issues are sources for indoor air pollutants in the UAE, which highly affects the integrity of indoor air quality (IAQ) of a building and severely impacts human health. This is a critical concern in the UAE where individuals spend most of their time at home due to extreme temperatures throughout the seasons [103]. In hot and humid climates, the overall thermal transmittance performance analysis uncovered envelope failure on surface condensation standards at the present operative conditions, which has also led to mold growth [75]. Similar studies in different climatic conditions have also investigated the effect of thermal bridging on the rise of condensation and have agreed upon the criticality of thermal bridges in terms of surface condensation and mold growth [56,74,75]. An alternative mitigation strategy to counteract the effect of mold formation in walls due to thermal bridging is increased thermal insulation, which raises the indoor temperatures and lowers average relative humidity levels and thus the potential for condensation and mold growth [76]. Moreover, insulating glazed surfaces, such as employing double-glazing, can reduce the likelihood of surface condensation occurring due to minimization of the temperature differential [79].

4.3. Construction Quality and Workmanship

Design changes are recurrent and an expected occurrence in building projects [104]. Design changes are a potential source of building defects, as also identified in the audited cases. They typically originate from any change in dimension or location of building elements, change orders, or workmanship quality [80]. Many design defects can be detected during the project's construction phase, enabling corrective measures or rework [82]. According to Han et al. [81], design errors and inconsistencies leading to rework and/or design changes are considered the primary contributor to timeline delays and budget overruns in construction projects and are non-value-adding efforts. Numerous studies tried to predict the frequency of occurrence of design discrepancies [27,83–85]. According to Love and Li [83], 54% of the defect's costs are attributed to design changes. In addition, Fayek et al. [84] reported that design changes and engineering reviews constitute 55% of the defects and 62% of the retrofit expenses. Similarly, Love and Li [83] and Josephson et al. [85] concluded that 55% of the defects recorded originated from errors related to poor workmanship during the construction stage.

Similar results were reached by Georgiou [86]. He analyzed the quality defects observed during the construction and post-handover stages of 100 domestic building projects in Australia and concluded that workmanship and incomplete work were the most frequent defects, accounting for 40% and 20% of the occurrences, respectively. The term workmanship suggests the same defect nature as inappropriate installation or incorrect installation, as referred to in other studies [24,87], which is in line with and corroborates the main findings. According to Forcada et al. [26], omission and workmanship constitute 64% of the sources of defects discovered at the post-handover stage in Spain's housing sector. Although widely accepted as a significant source of defect, Aïssani et al. [88] argued that workmanship suffers from limitations such as the difficulty to evaluate its impact on energy predictions due to variations from one construction site to another. In practice, unlike other industries, the building industry cannot specify performance with zero allowances due to workmanship discrepancies. In practice, many technical reports and scientific papers have discussed workmanship anomalies. Most of them are aware that the observed deviation in building performance is due to faulty workmanship. It is therefore essential to consider the sources of uncertainty [89].

In terms of the impact of design discrepancies, Zero Carbon Hub [90,91] emphasized a lack of focus and understanding of the implications of designers' design decisions on the building energy performance during the design stage. This lack of awareness of the design team is likely to impact various aspects of building energy performance [105]. Palmer et al. [106] investigated a building project of 76 UK homes. They concluded that the lack of literacy of the design team toward energy-related aspects added to an uncoordinated approach of the different design disciplines and resulted in non-intended thermal bridges and constructability issues, which increased the air permeability of the building envelope. Additionally, authors such as Oreszczyn et al. and Palmer et al. [92,106] suggested that design defects are related to the quality and accuracy of the information embedded in construction drawings and details, resulting in incorrect interpretation and unnecessary amendments by the team working on-site. If not addressed with the right approach, these misinterpretations can lead to faulty construction details, affecting, in turn, the intended building energy performance. These changes in the building's originally designed energy performance are rarely assessed as part of the process. For instance, Johnston et al. [58] measured the thermal properties of 25 new dwellings in the UK and concluded that the whole fabric U-value was 1.6 greater than predicted in the design stage, caused by discontinuity of the insulation panels due to poor workmanship management. Supporting that, the analysis of quality failures identified in Northern China cases found that the 10 most common quality failures had poor workmanship sources such as "Incorrect size of the new window frame and door frame," or "Untreated wall around the new windows." These failures can cause unaccounted losses. The direct costs range from 5 to 20% of the contract value [107].

Regarding improving design and management to achieve the predicted building energy performance, the inclusion of an energy professional in the project team should be considered. This stakeholder would be appointed to monitor the project progress to ensure ongoing compliance with the relevant energy performance targets during the design and construction, handover and close-out stages [91]. There seems to be scarce information in terms of quantifying the variations of workmanship errors and their implications and which type of defect has a more significant impact in the building energy use, concerning both the actual contribution to heat loss and in respect to the frequency of occurrence in construction projects [22].

5. Conclusions

The aim of this paper was first, to diagnose construction defects that may affect the thermal integrity of the building's envelope in both existing and under-construction housing projects in Al Ain in the United Arab Emirates. Second, we intended to identify the prevailing and subsequent impacts of the identified building defects through a qualitative review analysis.

The thermal behavior and construction quality of residential building envelopes were assessed through an infrared thermography audit performed on several existing and under-construction residential units under the extreme hot climate of Al Ain (UAE). The analysis highlighted three major areas of defects: lack or discontinuity in building envelope insulation, thermal bridging, and discrepancies due to non-compliant design changes. Unexpectedly, in existing buildings the results exhibited more thermal anomalies in newer units than in older ones, indicating that building thermal performance was more a function of workmanship and construction quality than mere building age. The most frequent recurring defect in units under construction was thermal bridging between the non-insulated reinforced concrete structural frames and insulated concrete block, which occurred in all cases examined except in the housing using precast panels, a defect that will compromise the targeted thermal performance of the exterior walls. Another diagnosed defect was the oversized service outlets and mortar joints, which were recurring defects that, in turn, affected the thermal integrity of walls. Finally, non-compliant design changes in the construction phase with improper or non-compliant implementation resulted in clear thermal bridging. In the under-construction housing units, the analysis indicated poor workmanship, absence of coordination during design changes, and incorrect construction details as the root causes of the defects. Hence, these may well be related to the general lack of understanding of thermal insulation, scarcity of skilled labor, and insufficient or inadequate site coordination, supervision, and construction quality control during construction.

Next, a qualitative review analysis identified the impacts of these defects. The existing literature provided evidence on the occurrence frequency of the identified defects in different contexts. The lack of insulation in the building envelope was found to be a common issue in all pre-code era housing, an expected outcome in context due to the late implementation of strict building energy efficiency regulations in the UAE that mandated, from 2010, thermal insulation. This issue dramatically increased heat transfer and energy consumption while it also affected energy cost and user comfort. Besides the overall envelope insulation, thermal bridging through walls, roof, and wall openings as well as at joints and connections was a recurrent issue with numerous measurable negative impacts. Thermal bridges were analyzed per type of defect, highlighting in all cases the negative impact on energy loss, increased energy cost, and substantial impact on the health and comfort of users. Research indicated that the presence of lower thermal performance and surface condensation led to hygiene problems such as mold growth and the staining of surfaces, which, in turn, affected the users' comfort levels. On the other hand, poor workmanship management and design discrepancies were found to be critical sources of defects. Predicting and evaluating the impact of the construction workmanship is difficult; therefore, allowances should be considered in the overall predicted building energy performance balance.

These results and their multiple impacts on energy, cost, health, and comfort call for engaging an energy professional to ensure compliance with the targets during all the construction stages as well as an effective reconsideration of the quality control process during construction to ensure the designed building meets its intended standards, expected thermal performance, and users' well-being.

Author Contributions: Conceptualization, K.A.T.A.; methodology, K.A.T.A. and R.H.; formal analysis, K.A.T.A., R.H., R.A., M.S., B.A.; investigation, R.H., R.A., M.S., B.A., resources, All; writing—original draft preparation, All; writing—review and editing, K.A.T.A.; visualization, R.H., R.A.; supervision, project administration; funding acquisition, K.A.T.A. All authors have read and agreed to the published version of the manuscript.

Funding: The authors gratefully acknowledge financial support from the United Arab Emirates University through the Emirates Centre for Energy and Environment Research funded research project N. 31R102 and the SURE Plus undergraduate research funded projects Grant N. G00002503 and G00002866.

Conflicts of Interest: The authors declare no conflict of interest.

References

1. Berardi, U. A Cross-Country Comparison of the Building Energy Consumptions and Their Trends. *Resour. Conserv. Recycl.* **2017**, *123*. [CrossRef]
2. Bidwai, P. The Emerging Economies and Climate Change: A Case Study of the BASIC Grouping. In Shifting Power—Critical Perspectives on Emerging Economies. 2014. Available online: https://www.tni.org/en/search/field_publication_series/shifting-power-critical-perspectives-on-emerging-economies-44/language/en (accessed on 30 October 2020).
3. Allouhi, A.; El Fouih, Y.; Kousksou, T.; Jamil, A.; Zeraouli, Y.; Mourad, Y. Energy Consumption and Efficiency in Buildings: Current Status and Future Trends. *J. Clean. Prod.* **2015**, *109*. [CrossRef]
4. World Population Review. United Arab Emirates Population. Available online: https://worldpopulationreview.com/countries/united-arab-emirates-population. (accessed on 14 November 2020).
5. Elessawy, F.M. The Boom: Population and Urban Growth of Dubai City. *Horizons Hum. Soc. Sci.* **2017**, *2*. [CrossRef]
6. The World Bank. Urban Population (% of Total Population). United Arab Emirates. Available online: https://data.worldbank.org/indicator/SP.URB.TOTL.IN.ZS?locations=AE. (accessed on 14 November 2020).
7. Shahbaz, M.; Sbia, R.; Hamdi, H.; Ozturk, I. Economic Growth, Electricity Consumption, Urbanization and Environmental Degradation Relationship in United Arab Emirates. *Ecol. Indic.* **2014**, *45*. [CrossRef]
8. World Energy Council. World Energy Trilemma Index | 2020. Available online: https://www.worldenergy.org/publications/entry/world-energy-trilemma-index-2020 (accessed on 1 December 2020).
9. A.C. Overview. UAE Construction Analytic Reports & Forecasts. Available online: https://www.bncnetwork.net/Access-UAE-Construction-Projects-News-Analytic-Reports-and-Forecasts (accessed on 1 December 2020).
10. Abu Dhabi Government Statistics Center. Statistical Yearbook Energy and Water. Available online: https://www.scad.gov.ae/ReleaseDocuments/StatisticalYearbookEnergyandWater_2019_Annual_Yearly_en.pdf (accessed on 30 October 2020).
11. Energy Information Administration. Country Analysis Executive Summary: United Arab Emirates. Available online: https://www.eia.gov/international/content/analysis/countries_long/United_Arab_Emirates/uae_2020.pdf (accessed on 30 October 2020).
12. St. Clair, P. *Low-Energy Design in the United Arab Emirates—Building Principles*; Royal Australian Institute of Architects: Melbourne, Australia, 2009.
13. Friess, W.A.; Rakhshan, K.; Hendawi, T.A.; Tajerzadeh, S. Wall Insulation Measures for Residential Villas in Dubai: A Case Study in Energy Efficiency. *Energy Build.* **2012**, *44*. [CrossRef]
14. Department of Municipalities and Transport. Pearl Building Rating System. Available online: https://www.dmt.gov.ae/en/Urban-Planning/Pearl-Building-Rating-System (accessed on 30 October 2020).
15. Department of Urban Planning and Municipalities. Estidama Insulation Products and Systems for PEARL Villa Rating. Available online: http://estidama.upc.gov.ae/estidama-villa-products-database/insulationproducts (accessed on 14 November 2020).
16. Efird, B.; Griffiths, S.; Mollet, P.; Al-Mubarak, I.; Sgouridis, S.; Tsai, I. *The Political Feasibility of Policy Options for the UAE's Energy Transition*; King Abdullah Petroleum Studies and Research Center: Riyadh, Saudi Arabia, 2017.
17. Motaleb, O.; Kishk, M. An Investigation into Causes and Effects of Construction Delays in UAE. Association of Researchers in Construction Management, ARCOM 2010. In Proceedings of the 26th Annual Conference, Leeds, UK, 24 January 2010.
18. AL Mousli, M.H.; El-Sayegh, S.M. Assessment of the Design–Construction Interface Problems in the UAE. *Archit. Eng. Des. Manag.* **2016**, *12*. [CrossRef]
19. Johnson, R.M.; Babu, R.I.I. Time and Cost Overruns in the UAE Construction Industry: A Critical Analysis. *Int. J. Constr. Manag.* **2020**, *20*. [CrossRef]
20. Tabet Aoul, K.A.; Hagi, R.; Abdelghani, R.; Akhozheya, B.; Karaouzene, R.; Syam, M. The Existing Residential Building Stock in UAE: Energy Efficiency and Retrofitting Opportunities. *Proc. Annu. Int. Conf. Archit. Civil. Eng.* **2018**. [CrossRef]
21. Mwasha, A.; Williams, R.G.; Iwaro, J. Modeling the Performance of Residential Building Envelope: The Role of Sustainable Energy Performance Indicators. *Energy Build.* **2011**, *43*. [CrossRef]
22. Alencastro, J.; Fuertes, A.; de Wilde, P. The Relationship between Quality Defects and the Thermal Performance of Buildings. *Renew. Sustain. Energy Rev.* **2018**. [CrossRef]
23. Gupta, R.; Kotopouleas, A. Magnitude and Extent of Building Fabric Thermal Performance Gap in UK Low Energy Housing. *Appl. Energy* **2018**, *222*. [CrossRef]
24. Macarulla, M.; Forcada, N.; Casals, M.; Gangolells, M.; Fuertes, A.; Roca, X. Standardizing Housing Defects: Classification, Validation, and Benefits. *J. Constr. Eng. Manag.* **2013**. [CrossRef]
25. Forcada, N.; Macarulla, M.; Gangolells, M.; Casals, M. Handover Defects: Comparison of Construction and Post-Handover Housing Defects. *Build. Res. Inf.* **2016**, *44*, 279–288. [CrossRef]
26. Forcada, N.; Macarulla, M.; Gangolells, M.; Casals, M.; Fuertes, A.; Roca, X. Posthandover Housing Defects: Sources and Origins. *J. Perform. Constr. Facil.* **2013**. [CrossRef]
27. Josephson, P.E.; Hammarlund, Y. Causes and Costs of Defects in Construction a Study of Seven Building Projects. *Autom. Constr.* **1999**. [CrossRef]

28. Forcada, N.; Macarulla, M.; Love, P.E.D. Assessment of Residential Defects at Post-Handover. *J. Constr. Eng. Manag.* **2013**, *139*. [CrossRef]
29. Bianchi, F.; Pisello, A.L.; Baldinelli, G.; Asdrubali, F. Infrared Thermography Assessment of Thermal Bridges in Building Envelope: Experimental Validation in a Test Room Setup. *Sustainability* **2014**, *6*. [CrossRef]
30. Kylili, A.; Fokaides, P.A.; Christou, P.; Kalogirou, S.A. Infrared Thermography (IRT) Applications for Building Diagnostics: A Review. *Appl. Energy* **2014**. [CrossRef]
31. Dall'O', G.; Sarto, L.; Panza, A. Infrared Screening of Residential Buildings for Energy Audit Purposes: Results of a Field Test. *Energies* **2013**, *6*. [CrossRef]
32. Fox, M.; Goodhew, S.; De Wilde, P. Building Defect Detection: External versus Internal Thermography. *Build. Environ.* **2016**, *105*. [CrossRef]
33. O'Grady, M.; Lechowska, A.A.; Harte, A.M. Infrared Thermography Technique as an In-Situ Method of Assessing Heat Loss through Thermal Bridging. *Energy Build.* **2017**, *135*. [CrossRef]
34. Hopper, J.; Littlewood, J.R.; Taylor, T.; Counsell, J.A.M.; Thomas, A.M.; Karani, G.; Geens, A.; Evans, N.I. Assessing Retrofitted External Wall Insulation Using Infrared Thermography. *Struct. Surv.* **2012**, *30*. [CrossRef]
35. Taylor, T.; Counsell, J.; Gill, S. Energy Efficiency Is More than Skin Deep: Improving Construction Quality Control in New-Build Housing Using Thermography. *Energy Build.* **2013**, *66*. [CrossRef]
36. Littlewood, R.J. Chapter 4—Assessing and Monitoring the Thermal Performance of Dwellings. In *Architectural Technology: Research in Practice*; Emmitt, S., Ed.; Wiley-Blackwell: Oxford, UK, 2013.
37. FLIR Systems. *Thermal Imaging Guidebook for Building and Renewable Energy Applications*; FLIR: Wilsonville, OR, USA, 2011.
38. Bauer, E.; Pavón, E.; Barreira, E.; De Castro, E.K. Analysis of Building Facade Defects Using Infrared Thermography: Laboratory Studies. *J. Build. Eng.* **2016**, *6*. [CrossRef]
39. British Standard BS EN. *Thermal Performance of Buildings. Qualitative Detection of Thermal Irregularities in Building Envelopes. Infrared Method*; BSI: Brussels, Belgium, 1999.
40. Mathur, S.S.; Umesh, G.; Seth, A.K.; Sharma, R.P.; Kaushik, S.C. Periodic Heating/Cooling by Solar Radiation. In *Sun: Mankind's Future Source of Energy*; Elsevier: Amsterdam, The Netherlands, 1978.
41. Kirimtat, A.; Krejcar, O. A Review of Infrared Thermography for the Investigation of Building Envelopes: Advances and Prospects. *Energy Build.* **2018**. [CrossRef]
42. Yu, S.; Liu, Y.; Wang, D.; Bahaj, A.B.S.; Wu, Y.; Liu, J. Review of Thermal and Environmental Performance of Prefabricated Buildings: Implications to Emission Reductions in China. *Renew. Sustain. Energy Rev.* **2020**. [CrossRef]
43. Radhi, H.; Eltrapolsi, A.; Sharples, S. Will Energy Regulations in the Gulf States Make Buildings More Comfortable—A Scoping Study of Residential Buildings. *Appl. Energy* **2009**, *86*. [CrossRef]
44. Al-Sanea, S.A.; Zedan, M.F. Effect of Thermal Bridges on Transmission Loads and Thermal Resistance of Building Walls under Dynamic Conditions. *Appl. Energy* **2012**, *98*. [CrossRef]
45. Bojić, M.; Djordjević, S.; Stefanović, A.; Miletić, M.; Cvetković, D. Decreasing Energy Consumption in Thermally Non-Insulated Old House via Refurbishment. *Energy Build.* **2012**, *54*. [CrossRef]
46. AlNaqbi, A.; AlAwadhi, W.; Manneh, A.; Kazim, A.; Abu-Hijleh, B. Survey of the Existing Residential Buildings Stock in the UAE. *Int. J. Environ. Sci. Dev.* **2012**. [CrossRef]
47. Ibrahim, M.; Biwole, P.H.; Wurtz, E.; Achard, P. Limiting Windows Offset Thermal Bridge Losses Using a New Insulating Coating. *Appl. Energy* **2014**, *123*. [CrossRef]
48. Evangelisti, L.; Guattari, C.; Gori, P. Energy Retrofit Strategies for Residential Building Envelopes: An Italian Case Study of an Early-50s Building. *Sustainability* **2015**, *7*, 445. [CrossRef]
49. Akhozheya, B.; Syam, M.; Abdelghani, R.; Aoul, K.A.T. Retrofit Evaluation of a Residential Building in UAE: Energy Efficiency and Renewable Energy. In Proceedings of the 5th International Conference on Renewable Energy: Generation and Application, ICREGA 2018, Al Ain, United Arab Emirates, 26–28 February 2018.
50. Khoukhi, M.; Darsaleh, A.F.; Ali, S. Retrofitting an Existing Office Building in the UAE towards Achieving Low-Energy Building. *Sustainability* **2020**, *12*, 2573. [CrossRef]
51. Jahed, N.; Aktaş, Y.D.; Rickaby, P.; Altinöz, A.G.B. Policy Framework for Energy Retrofitting of Built Heritage: A Critical Comparison of UK and Turkey. *Atmosphere* **2020**, *11*, 674. [CrossRef]
52. Aktacir, M.A.; Büyükalaca, O.; Yilmaz, T. A Case Study for Influence of Building Thermal Insulation on Cooling Load and Air-Conditioning System in the Hot and Humid Regions. *Appl. Energy* **2010**, *87*. [CrossRef]
53. Brannigan, A.; Booth, C.A. Building Envelop Energy Efficient Retrofitting Options for Domestic Buildings in the UK. *WIT Trans. Ecol. Environ.* **2013**, *179*. [CrossRef]
54. Altun, M.; Akgul, C.M.; Akcamete, A. Effect of Envelope Insulation on Building Heating Energy Requirement, Cost and Carbon Footprint from a Life Cycle Perspective. *J. Fac. Eng. Archit. Gazi Univ.* **2020**, *35*. [CrossRef]
55. Zedan, M.F.; Al-Sanea, S.; Al-Mujahid, A.; Al-Suhaibani, Z. Effect of Thermal Bridges in Insulated Walls on Air-Conditioning Loads Using Whole Building Energy Analysis. *Sustainability* **2016**, *8*, 560. [CrossRef]
56. Erdem, C.; Pinar, M.C. The Impact of Internal Aerogel Retrofitting on the Thermal Bridges of Residential Buildings: An Experimental and Statistical Research. *Energy Build.* **2016**. [CrossRef]

57. Theodosiou, T.; Tsikaloudaki, K.; Bikas, D. Analysis of the Thermal Bridging Effect on Ventilated Facades. *Procedia Environ. Sci.* **2017**. [CrossRef]
58. Johnston, D.; Miles-Shenton, D.; Farmer, D. Quantifying the Domestic Building Fabric "Performance Gap". *Build. Serv. Eng. Res. Technol.* **2015**, *36*. [CrossRef]
59. Real, S.; Gomes, M.G.; Moret Rodrigues, A.; Bogas, J.A. Contribution of Structural Lightweight Aggregate Concrete to the Reduction of Thermal Bridging Effect in Buildings. *Constr. Build. Mater.* **2016**. [CrossRef]
60. Theodosiou, T.; Tsikaloudaki, K.; Tsoka, S.; Chastas, P. Thermal Bridging Problems on Advanced Cladding Systems and Smart Building Facades. *J. Clean. Prod.* **2019**. [CrossRef]
61. Kiil, M.; Käärid, M.S.; Klõšeiko, P.; Võsa, K.V.; Simson, R.; Sarevet, H.; Thalfeldt, M.; Kurnitski, J. PCSP's Diagonal Tie Connectors Thermal Bridges Impact on Energy Performance and Operational Cost: Case Study of a High-Rise Residential Building in Estonia. In *E3S Web of Conferences*; EDP Sciences: London, UK, 2020.
62. Liu, Y.; Li, B.; Cao, X. The Research on the Influence of Building Air Tightness to Energy Consumption of Residential Building in a Hot Summer and Cold Winter Zone in China. In *IOP Conference Series: Materials Science and Engineering*; IOP Publishing: Bristol, UK, 2019; Volume 609.
63. Cappelletti, F.; Gasparella, A.; Romagnoni, P.; Baggio, P. Analysis of the Influence of Installation Thermal Bridges on Windows Performance: The Case of Clay Block Walls. *Energy Build.* **2011**. [CrossRef]
64. Marincioni, V.; Altamirano-Medina, H.; May, N.; Sanders, C. Estimating the Impact of Reveals on the Transmission Heat Transfer Coefficient of Internally Insulated Solid Wall Dwellings. *Energy Build.* **2016**. [CrossRef]
65. Jedidi, M.; Benjeddou, O. Effect of Thermal Bridges on the Heat Balance of Buildings. *IJSRCE* **2018**, *2*, 41–49.
66. Al-Sayed Omar, E. *Impact of Columns and Beams on the Thermal Resistance of the Building Envelope*; Energy Systems Laboratory: College Station, TX, USA, 2002; Available online: http://esl.tamu.edu (accessed on 30 October 2020).
67. Levinskyte, A.; Banionis, K.; Geleziunas, V. The Influence of Thermal Bridges for Buildings Energy Consumption of "A" Energy Efficiency Class. *J. Sustain. Archit. Civ. Eng.* **2016**, *15*, 47–58. [CrossRef]
68. *Dynamic Effect of Thermal Bridges on the Energy Performance of Residential Buildings*; Concordia University: Montreal, QC, Canada, 2015.
69. Kotti, S.; Teli, D.; James, P.A.B. Quantifying Thermal Bridge Effects and Assessing Retrofit Solutions in a Greek Residential Building. *Procedia Environ. Sci.* **2017**. [CrossRef]
70. Karabulut, K.; Buyruk, E.; Fertelli, A. Numerical Investigation of Heat Transfer for Thermal Bridges Taking into Consideration Location of Thermal Insulation with Different Geometries. *Stroj. Časopis Teor. Praksu Stroj.* **2009**, *51*, 431–439.
71. Evola, G.; Margani, G.; Marletta, L. Energy and Cost Evaluation of Thermal Bridge Correction in Mediterranean Climate. *Energy Build.* **2011**. [CrossRef]
72. Kaymaz, E.; Sezer, F.S. Thermal Performance Assessment of Existing Buildings: A Case Study from Turkey. *Int. J. Humanit. Soc. Sci.* **2015**, *6*, 43–54.
73. Collins, M.; Dempsey, S. Residential Energy Efficiency Retrofits: Potential Unintended Consequences. *J. Environ. Plan. Manag.* **2019**. [CrossRef]
74. Ilomets, S.; Kalamees, T. Evaluation of the Criticality of Thermal Bridges. *J. Build. Pathol. Rehabil.* **2016**. [CrossRef]
75. Miasik, P.; Lichołai, L. The Influence of a Thermal Bridge in the Corner of the Walls on the Possibility of Water Vapour Condensation. In *E3S Web of Conferences*; EDP Sciences: London, UK, 2018.
76. Willand, N.; Ridley, I.; Maller, C. Towards Explaining the Health Impacts of Residential Energy Efficiency Interventions—A Realist Review. Part 1: Pathways. *Soc. Sci. Med.* **2015**. [CrossRef] [PubMed]
77. Al-Alili, A.; Al Qubaisi, A. Efficient Residential Buildings in Hot and Humid Regions: The Case of Abu Dhabi, UAE. *Int. J. Therm. Environ. Eng.* **2018**, *17*. [CrossRef]
78. Ali, M.; Oladokun, M.O.; Osman, S.B.; Samsuddin, N.; Hamzah, H.A. Moisture Condensation on Building Envelopes in Differential Ventilated Spaces in the Tropics: Quantitative Assessment of Influencing Factors. In *MATEC Web of Conferences*; EDP Sciences: London, UK, 2016; Volume 66.
79. WHO Regional Office for Europe. *WHO Guidelines for Indoor Air Quality: Dampness and Mould—WHO Regional Office for Europe, World Health Organization—Google Books*; Heseltine, E., Rosen, J.J., Eds.; World Health Organisation: Geneva, Switzerland, 2009.
80. Choudhry, R.M.; Gabriel, H.F.; Khan, M.K.; Azhar, S. Causes of Discrepancies between Design and Construction in the Pakistan Construction Industry. *J. Constr. Dev. Ctries* **2017**, *22*. [CrossRef]
81. Han, S.; Love, P.; Peña-Mora, F. A System Dynamics Model for Assessing the Impacts of Design Errors in Construction Projects. *Math. Comput. Model* **2013**. [CrossRef]
82. Oyewobi, L.O.; Oke, A.A.; Ganiyu, B.O.; Shittu, A.A.; Isa, R.B.; Nwokobia, L. The Effect of Project Types on the Occurrence of Rework in Expanding Economy. *J. Civ. Eng. Constr. Technol.* **2011**, *2*, 119–124.
83. Love, P.E.D.; Li, H. Quantifying the Causes and Costs of Rework in Construction. *Constr. Manag. Econ.* **2000**, *18*. [CrossRef]
84. Fayek, A.R.; Dissanayake, M.; Campero, O. Developing a Standard Methodology for Measuring and Classifying Construction Field Rework. *Can. J. Civ. Eng.* **2004**. [CrossRef]
85. Josephson, P.-E.; Larsson, B.; Li, H. Illustrative Benchmarking Rework and Rework Costs in Swedish Construction Industry. *J. Manag. Eng.* **2002**. [CrossRef]
86. Georgiou, J. Verification of a Building Defect Classification System for Housing. *Struct. Surv.* **2010**. [CrossRef]

87. Forcada, N.; MacArulla, M.; Gangolells, M.; Casals, M. Assessment of Construction Defects in Residential Buildings in Spain. *Build. Res. Inf.* **2014**, *42*. [CrossRef]
88. Aïssani, A.; Chateauneuf, A.; Fontaine, J.P.; Audebert, P. Quantification of Workmanship Insulation Defects and Their Impact on the Thermal Performance of Building Facades. *Appl. Energy* **2016**, *165*. [CrossRef]
89. de Wilde, P.; Tian, W.; Augenbroe, G. Longitudinal Prediction of the Operational Energy Use of Buildings. *Build. Environ.* **2011**, *46*. [CrossRef]
90. Zero Carbon Hub. *Closing the Gap between Design and As-Built Performance, End of Term Report*. 2014. Available online: https://www.zerocarbonhub.org/sites/default/files/resources/reports/Closing_the_Gap_Bewteen_Design_and_As-Built_Performance_Interim_Report.pdf (accessed on 30 October 2020).
91. Zero Carbon Hub. Closing the Gap between Design and As-Built Performance—Evidence Review Report. 2014. Available online: https://www.zerocarbonhub.org/sites/default/files/resources/reports/Closing_the_Gap_Between_Design_and_As-Built_Performance-Evidence_Review_Report_0.pdf (accessed on 30 October 2020).
92. Oreszczyn, T.; Mumovic, D.; Davies, M.; Ridley, I.; Bell, M.; Smith, M.; Miles-Shenton, D. *'Condensation Risk—Impact of Improvements to Part L and Robust Details on Part C', Final Report*; Office of the Deputy Prime Minister: London, UK, 2005.
93. The "Warm Houses" Program: Insulating Existing Buildings through Compulsory Retrofits. *Sustain. Energy Technol. Assess.* **2015**, *9*. [CrossRef]
94. Lucchi, E. Applications of the Infrared Thermography in the Energy Audit of Buildings: A Review. *Renew. Sustain. Energy Rev.* **2018**. [CrossRef]
95. Underhill, L.J.; Milando, C.W.; Levy, J.I.; Dols, W.S.; Lee, S.K.; Fabian, M.P. Simulation of Indoor and Outdoor Air Quality and Health Impacts Following Installation of Energy-Efficient Retrofits in a Multifamily Housing Unit. *Build. Environ.* **2020**. [CrossRef]
96. Al-Homoud, M.S. The Effectiveness of Thermal Insulation in Different Types of Buildings in Hot Climates. *J. Build. Phys.* **2004**, *27*. [CrossRef]
97. Sustainability and Green Building Rating Systems: LEED, BREEAM, GSAS and Estidama Critical Analysis. *J. Build. Eng.* **2017**. [CrossRef]
98. Zhang, L.; Liu, Z.; Hou, C.; Hou, J.; Wei, D.; Hou, Y. Optimization Analysis of Thermal Insulation Layer Attributes of Building Envelope Exterior Wall Based on DeST and Life Cycle Economic Evaluation. *Case Stud. Therm. Eng.* **2019**. [CrossRef]
99. Baba, F.; Ge, H. Dynamic Effect of Balcony Thermal Bridges on the Energy Performance of a High-Rise Residential Building in Canada. *Energy Build.* **2016**, *116*. [CrossRef]
100. Finch, G.; Wilson, M.; Higgins, J. *Thermal Bridging Of Masonry Veneer Claddings & Energy Code Compliance*; ASTM International: Philadelphia, PA, USA, 2013.
101. Özdeniz, M.B.; Hançer, P. Suitable Roof Constructions for Warm Climates—Gazimăusa Case. *Energy Build.* **2005**. [CrossRef]
102. Seppänen, O.; Kurnitski, J. Moisture Control and Ventilation. In *WHO Guidelines for Indoor Air Quality: Dampness and Mould*; WHO: Geneva, Switzerland, 2009.
103. Bani Mfarrej, M.F.; Qafisheh, N.A.; Bahloul, M.M. Investigation of Indoor Air Quality inside Houses From UAE. *Air Soil Water Res.* **2020**, *13*. [CrossRef]
104. Mohamad, M.I.; Nekooie, M.A.; Al-Harthy, A.B.S. Design Changes in Residential Reinforced Concrete Buildings: The Causes, Sources, Impacts and Preventive Measures. *J. Constr. Dev. Ctries* **2012**, *17*, 23–44.
105. van Dronkelaar, C.; Dowson, M.; Spataru, C.; Mumovic, D. A Review of the Regulatory Energy Performance Gap and Its Underlying Causes in Non-Domestic Buildings. *Front. Mech. Eng.* **2016**, *1*. [CrossRef]
106. Palmer, J.; Godoy-Shimizu, D.; Tillson, A.; Mawditt, I. *Building Performance Evaluation Programme: Findings from Domestic Projects—Making Reality Match Design*; Innovate: London, UK, 2016.
107. Qi, Y.; Qian, Q.K.; Meijer, F.M.; Visscher, H.J. Identification of Quality Failures in Building Energy Renovation Projects in Northern China. *Sustainability* **2019**, *11*, 4203. [CrossRef]

Article

Conceptual Parametric Relationship for Occupants' Domestic Environmental Experience

Sajal Chowdhury [1,*], Masa Noguchi [1] and Hemanta Doloi [2]

[1] ZEMCH EXD Lab, Faculty of Architecture, Building and Planning, The University of Melbourne, Melbourne, VIC 3010, Australia; masa.noguchi@unimelb.edu.au
[2] Smart Villages Lab, Faculty of Architecture, Building and Planning, The University of Melbourne, Melbourne, VIC 3010, Australia; hdoloi@unimelb.edu.au
* Correspondence: sajal.chowdhury@unimelb.edu.au or sajalc@student.unimelb.edu.au; Tel.: +61-40-154-3154

Abstract: Today's architectural design approaches do not adequately address the relationship between users' spatial, environmental and psychological experiences. Domestic environmental experience generally indicates users' cognitive perceptions and physical responses within dwelling spaces. Therefore, without a clear perception of occupants' experiences, it is difficult to identify proper architectural solutions for a domestic environment. To understand notions of these domestic experiences, the current study explores the theoretical relationship between spatial and environmental design factors within domestic settings which led to the concept of "Environmental Experience Design (EXD)". Extensive data exploration was conducted using a combination of thirty keywords through different databases (e.g., Scopus, ScienceDirect, PubMed, Google Scholar, Mendeley and Research Gate) to categorise the relevant literature regarding thematic study areas such as human perception and phenomenology, environmental design and psychology, residential environment and design, health-wellbeing and user experiences. This study has identified theoretical associations between spatial and environmental design factors of different domestic spaces that can stimulate occupants' satisfaction and comfort by reviewing eighty-seven studies from the literature. However, occupants' contextual situations significantly impact domestic spaces, where spatial and environmental design attributes may be connected to diverse sociocultural factors. The scope of explanation about user context is limited, to some extent, in environmental design theories. Thus, combining occupants' contexts with spatial and environmental design factors will be a future research direction used to explore the notion of "*Domestic Environmental Experience Design*"

Keywords: domestic environment; spatial factors; environmental factors; occupants' experiences; theoretical relationship

Citation: Chowdhury, S.; Noguchi, M.; Doloi, H. Conceptual Parametric Relationship for Occupants' Domestic Environmental Experience. *Sustainability* **2021**, *13*, 2982. https://doi.org/10.3390/su13052982

Academic Editor: András Reith

Received: 14 February 2021
Accepted: 4 March 2021
Published: 9 March 2021

Publisher's Note: MDPI stays neutral with regard to jurisdictional claims in published maps and institutional affiliations.

1. Introduction

Generally, people spend more than 60% of their time in domestic indoor environments [1]. In the domestic setting, occupants' living experiences and household activities are diverse in every step of their daily lives and there are numerous preferences related to occupants' spatial needs and demands [2–4]. These preferences are connected to multiple aspects of domestic settings and are perceived through occupants' living experiences. Every component of the domestic environment has a negative or positive impact on occupants' psychological responses. Several studies also identified that these factors stimulate humans' mediative capacities in their living environments [5]. Therefore, it is necessary to explore occupants' perceptions and living experiences in domestic settings to enhance their mental wellbeing [6]. Domestic environmental experience generally indicates users' experiences of cognitive perceptions and physical responses in domestic settings [6]. The environmental design concept may be enriched by integrating rigorous perceptions and systematic data associated with occupants' experiences in different spaces of the built environment [7].

The literature has identified that the domestic environment serves various purposes and has meaningfulness related to occupants' spatial and environmental aspects. A domestic setting generally has different spatial zone distributions. Each area has specific characteristics related to its spatial and environmental factors that affect occupants' physical and psychological wellbeing in living environments [6]. Thus, every domestic component has a significant spatial and environmental relationship that may enhance occupants' emotions [6,8,9]. These spatial and environmental factors may vary from one space to another, associated with occupants' multidimensional preferences (e.g., needs and demands) [6]. Consequently, the valuation of these factors is critical in designing domestic environments according to different living spaces. However, without a clear perception of occupants' spatial and environmental experiences of different areas in a domestic setting, it is difficult to identify overall environmental design solutions to enhance their mental wellbeing.

This study has been conducted using a comprehensive literature review based on occupants' domestic environments and their household experiences. Various aspects of occupants' spatial and environmental experiences in domestic settings have been elaborated in this study. This study's primary research question is "*What is the theoretical parametric relationship for occupants' domestic environmental experiences that enhance their wellbeing?*" This study's main objective is to explore the theoretical parametric relationship between occupants' spatial and environmental design factors through the household experience of different areas in domestic settings that may stimulate occupants' quality of life. To develop a conceptual framework or associations to identify a correlation among various spatial and environmental design factors in a domestic setting, emphasising space-wise household experiences is the primary focus of this study.

2. Literature Selection Criteria and Research Methods

A literature review is a pedagogical study linked to a particular theme or research question [10,11]. Several studies found significant impacts of domestic environments on human perceptions [1,6,8,12]. This study explores theoretical relationships of different domestic spaces between spatial and environmental design factors through occupants' household experiences based on the literature, which may stimulate living quality. The study was conducted using a comprehensive background of 30 keywords based on occupants' domestic environments and their household experiences. The keywords encompassed domestic environment and occupants' experiences as well as a diversity of psychological and behavioral aspects related to residential settings (Figure 1). "Human Perception and Phenomenology, Environmental Design and Psychology, Residential Environment and Design, Health and Wellbeing and User Experience" are the main thematic study areas that were considered to explore occupants' spatial and environmental experiences in domestic settings in this study.

The following 30 keywords associated with the thematic study domains have been used: domestic environment, housing, dwelling, home, occupant experiences, occupant spatial experiences, occupant environmental experiences, household activities, high rise residential apartment, dwelling environment, apartment building, residential function, household functions and activities, space use behavior, occupant psychology, occupant behavior, residential comfort, residential satisfaction, indoor environment, environmental quality, mental wellbeing, physical environment, post-occupancy evaluation, human perception and phenomenology, ecological design, human emotions, feelings and moods, consumer behavior, user experience, and product user experience.

This study is limited to scholarly research articles published between 1970 and 2020. An extensive data search was conducted using a combination of different keywords related to domestic environmental experiences through Scopus, Science-Direct and PubMed databases to categorise relevant studies based on the research theme, titles, abstracts, keywords and findings that fell into the thematic study areas. Supplemental cross-searches were also conducted through Google Scholar, Mendeley, Research Gate and other academic search engines. All the literature referring to the domestic environment and occupants' psychological or behavioral experiences in residential buildings in their title, abstract or keywords were categorised for screening. The collected studies were separated according to the following five criteria: (a) focus on the domestic environment and occupant perception, (b) occupant experiences in different spaces of a domestic environment, (c) environmental design and occupant's psychology in a domestic setting, (d) human factors in built environments, and (e) peer-review. After the final screening, duplicate and non-relevant studies were omitted from the selection and the significant relevant references list was formed according to this study's scope and limitation.

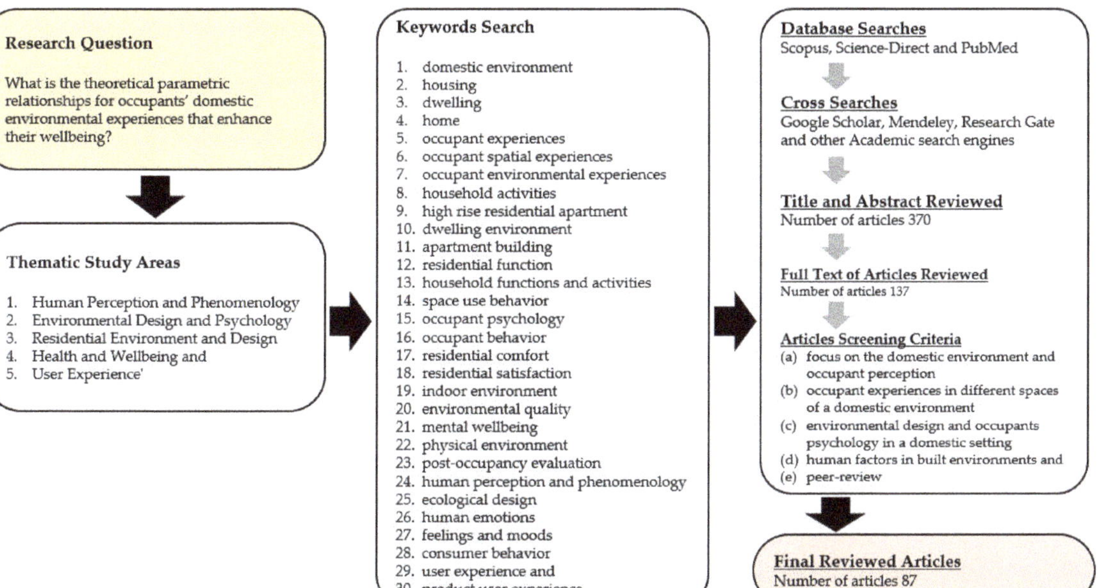

Figure 1. Literature selection strategies and research methods.

The following sections first analyse the background to spatial and environmental design factors regarding occupants' experiences. It then analyses these according to the different spaces of a domestic environment based on the literature. The review provides a systematic and comprehensive assessment of the domestic environment and occupants' experiences regarding spatial and environmental design aspects to enrich the state-of-the-art of existing knowledge and explore the potential for future investigations into "*Domestic Environmental Experience Design*".

3. Theoretical Background

This literature review mainly explores the theoretical relationships between spatial and environmental design factors in the domestic living environment, addressing occupants' wellbeing. In this study, the component considering perceived spatial factors focuses on users' spatial experiences linked to user preferences (needs and demands) in their living environments. The other component, which concerns the environmental design factors, deals with architectural design elements primarily related to indoor environmental physical design aspects. It also encompasses not only indoor environmental qualities but also the psychological aspect of occupants' feelings. In the following sections, space-related user preferences (spatial factors) and environmental design components (environmental factors) were elaborated through the literature to explore the theoretical parametric relationship in occupants' experiences in their domestic living environments.

3.1. Spatial Factors for Occupants' Experiences

McClure et al. mentioned seven domains: "products, interiors, structures, landscapes, cities, regions and earth" in the built environment. Every field has its own identity and these identities closely correlate to each other. The domestic environment design reflects interior and structure domains as part of the total built environment [13]. Additionally, indoor architecture's history suggests that human social, psychological and physical perceptions attach numerous qualities of living to environments, where the core concern is human scale and performance [14]. According to Blossom, human nature has two characteristics, functional and behavior; thus, designers need to realise how the built environmental design element affects individual perception, considering psychological aspects, for the different zones, as Hall mentioned [14].

Philosophically, space reflects human perceptions of physical existence. According to the earliest theoretical perspective of space, perception reflects the human-centric dimension. German philosopher August Schmarsow addressed the spatial interaction of beings with the world and quoted space as a place for physical and mental projection [2]. Further, Edward T. Hall, founder of the study of anthropological space named "Proxemics," in the book "The Hidden Dimension", mentioned the relationship between spatial setting and human beings [2]. According to environmental psychology, human beings always interact and perceive their immediate environments by sensory dimensions such as smell, vision, touch, hearing, haptic and kinesthesia [14]. When experiencing a space or place, sensory organs play a fundamental role for a human being. In that sense, spatial design attributes impact human perceptions in numerous ways within the total built environments [2,5].

Again, the literature review shows that place-attachment theory is a vital perception and the core concept of environmental psychology which affects people and places [15]. Human beings create a robust understanding of place attachment within their immediate environment, which supports physical and psychological wellbeing [16]. Place attachment is the personal interaction with the environment and the central concept of human emotional responses [16]. According to Stedman, an individual's satisfaction is another element of place attachment, which defines the value of meeting basic human needs in a living environment [15]. Stokols describes "Home" as a dwelling place where individuals can fulfill their psychological, physical and social needs to keep themselves connected [17]. Seemingly, Hayward emphasised psychological concepts, with significance given to privacy, identity, socialisation, continuity and personalisation, as a home attachment [18]. Dovey highlighted the phenomenon of "Home" considering three themes: order, identification and dialectic processes, in the article entitled "Home and Homelessness" [19].

While most home environmental studies begin with people who already live in a home and deal with satisfaction, Rapoport identified a previous question about how they reach this point. Rapoport observed that the consequence of the environment was addressed improperly. In reality, the near environment's real effect on a human being is habitat selection according to their needs and preferences [20]. In the meantime, Pennartz described the home atmosphere as focusing on communication, accessibility, relaxation and individual experiences [21]. Mallett addressed the notion of "Home" in an article entitled "Understanding Home: A Critical Review of the Literature" considering people's relationship with spaces and objects. Home is a dwelling interaction space between people, places and things associated with a comfortable feeling, intimacy, security, relaxation and persecution [22]. The importance of human needs and spatial hierarchy, such as territoriality, physical and psychological comfort, privacy and function, provide the designer with an understanding of human nature and satisfaction [13].

In a nutshell, "Home" as a domestic setting indicates varied meaningfulness, functions, purposes and aims. Human needs include identity, control, security, privacy, order, variety, sociability, aesthetics and choice, integrated into environmental psychology and interpreted by human experiences (Figure 2) [5,23]. However, it is critical to understand the assessing and prioritising of these human physical and psychological needs [2]. In short, the domestic experience is a medium that may connect occupants' needs and demands with their living environments.

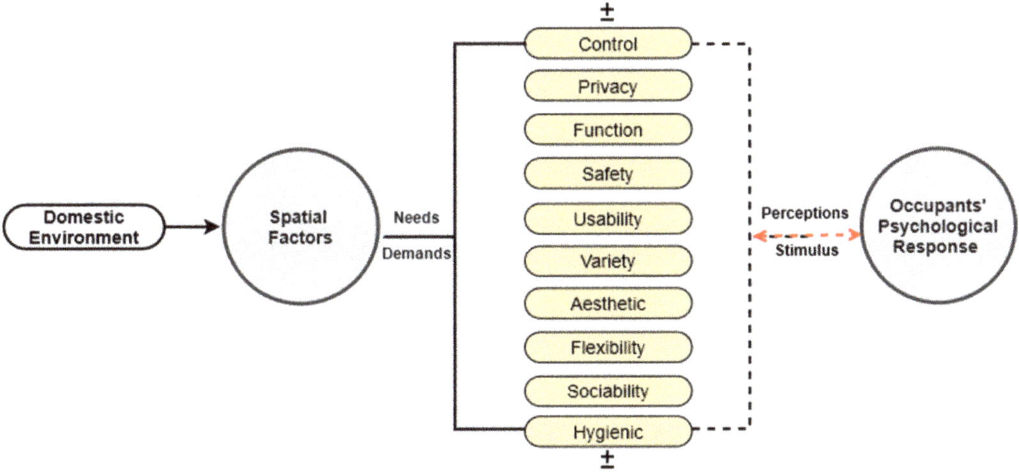

Figure 2. Relationship between spatial factors and occupants' response. Here, the sign (±) indicates adding/decreasing the number of factors according to user needs and demands. (Illustration: Author, based on literature)

3.2. Environmental Factors for Occupants' Experiences

The domestic setting has indoor and outdoor environmental characteristics that reflect occupants' spatial needs and demands in their living environments. Several studies indicate that each design element directly or indirectly impacts the occupants' overall physical and psychological wellbeing [16]. Several studies focusing on indoor environmental conditions such as noise, lighting, material, air, odors and color; conclude that environmental psychology bridges design and human response at the indoor environment scale.

Kaplan's "Attention Restoration Theory (ART)" proposed a framework that differentiates between stress and attentional components of human experiences in their environments. Emphasising the critical role of natural environments, this integration contributes towards human-environment interaction. According to Kaplan, experiencing a natural environment reduces human stress [24]. In addition, Ulrich's "Stress Reduction Theory (SRT)", focusing on the role of nature in wellbeing, indicates an evolutionary perspective that suggests that natural experiences have an immediate benefit on human mental wellbeing. Ulrich emphasises affective and aesthetic human responses to the natural environment [25]. Both Kaplan and Ulrich identify natural settings or environments as stimulating components for human wellbeing in numerous ways.

In the book entitled "A Home for the soul: a guide for dwelling with spirit and imagination", Lawlor describes the interaction of human emotions and feelings with architectural design components based on human spiritual perceptions in their dwelling environments [26]. Lawlor also discussed interconnectedness and perception of human cognition and architectural design elements such as earth, fire, air, space and water, as five spiritual elements, through human experiences in their dwelling environments. Consequently, Lawlor reveals how the eight fundamental building elements of architecture can be related to different aspects of human thinking and feeling. The author mainly explores everyday household traditions and symbolism, rather than exploring dwellers' situational or practical experiences with restrictions or limitations.

Furthermore, Evans argues that every element of the built environment directly or indirectly affects occupants' mental health [8]. According to Evans, most research on housing focused mainly on physical health. Nonetheless, different house types (e.g., highrise) and housing quality impact occupants' mental wellness. Evans identified that the natural environment affects occupants' psychological perceptions in highrise residential environments [8]. Poor quality of housing and indoor environments increases the negative impact of psychological stress and illness. Still, this is not enough to draw a clear methodological perception and conclusion.

Ergan et al. examined that occupants' emotional reactions to color, light, noise, air quality and crowding are distinctive and momentary in the living environment [27]. Levels of illumination, pollution and daylight exposure affect occupants' psychological wellbeing in numerous ways [8,28]. Several studies and randomised experiments are prioritised to evaluate the physical environment's potential role in occupants' mental wellbeing. As Evan stated, some methodological problems may create conflicting prejudice, triggering the undervaluation of housing–wellbeing associations. The author noted that ambient environmental interactions with architectural components positively impact human physical, biological and psychological aspects [28]. For instance, noise affects users' privacy, while smell impacts human memory [28]. Meanwhile, indoor lighting variation also affects and triggers human moods, feelings and psychological growth [28].

Moreover, other studies also describe that features of different domestic spaces, such as the bedroom, kitchen, dining, living, toilet, game room, guest room, guest bath, study, media room, entrance, utility room, backyard and garage, stimulate occupants' daily household activities and interact with emotional states [9,29]. In contrast, crowded enclosed spaces with no ventilation increase personal psychological stress, while space adjustability decreases mental stress [9]. Other services and utility facilities also impact the occupants' mental satisfaction in the residential environment [2]. Seemingly, spatial ergonomics also affect usability and occupants' emotional perceptions, such as relaxation and pleasantness within domestic environments [2].

According to Amérigo and Aragones' interpretation, domestic satisfaction is essential for the quality of life and indoor environment elements can stimulate human feelings [30]. In their research, a theoretical approach was applied to explore a general view of a person's satisfaction within a residential environment and a conceptual framework of individual interaction in a residential setting was presented. They developed an empirical model of residential satisfaction; a question arises regarding the trustworthy dimensions for residential pleasure or happiness, which might direct future environmental research on domestic settings. However, the concept of residential quality integrating different human factors and occupants' comfortable domestic environments indicates a direction for future research [30]. Considering the literature reviewed, the environmental experience is a multilayered phenomenon requiring different kinds of sensory involvement and interaction. The domestic environment influences occupants' wellbeing and needs to accommodate occupants' daily household activities. As discussed above, the synthesis of previous research studies leads to major categories of human experiences in various architectural design spaces. The literature gap still exists in the understanding of how spatial and environmental design features impact occupants' experiences in different spaces of domestic settings according to their household activities.

In Figure 3, a conceptual relationship is derived from the literature describing the environmental factors that contribute to stimulating occupants' emotions in their domestic environments [6,23].

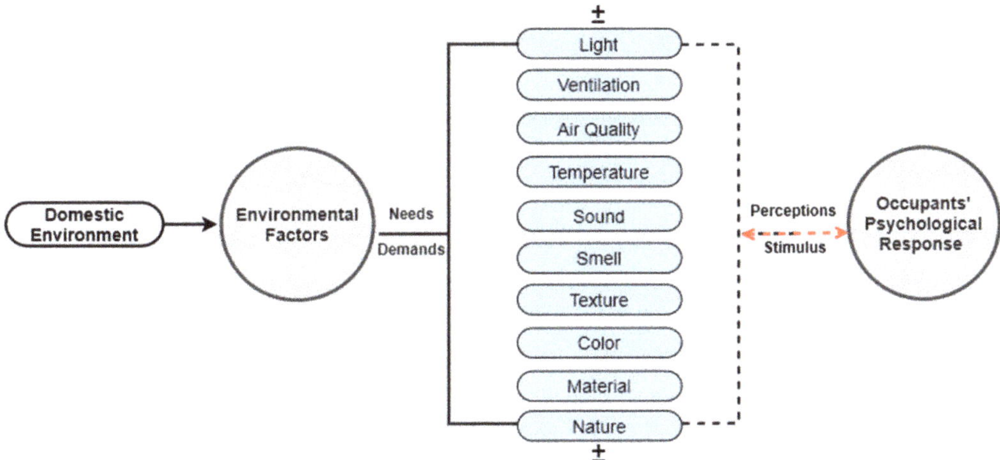

Figure 3. Relationship between environmental factors and occupants' response. Here, the sign (±) indicates adding/decreasing the number of factors according to user needs and demands. (Illustration: Author, based on the literature).

4. Domestic Spaces and Occupants' Experiences

4.1. Domestic Environment and Its Functional Aspect

Several psychological and phenomenological studies have been conducted to define the meaning of "Home" as a domestic environment [31]. According to Pallasmaa, the phenomenology of "Home" is not just an architectural effort. It has an aesthetic view, considering physical, psychological and sociocultural phenomenon. Pallasmaa believes that "Home" has multilayered characteristics, integrating memories, desires, intimacy, privacy, identity, function and even language [32]. Continuing this exploration, the domestic environment becomes an essential feature of "Self-identification" for peoples, where privacy, comfort and domesticity are the occupants' core achievements [31,33]. Moore and Caan described their views within psychological and sociological debates: domestic environments reflect numerous human behaviors and preferences because of different physical, psychological and social contextual human experiences. These experiences are essential to mediating tangible and intangible design aspects in the living environment [2,31]. Considering several studies, the term "*Domestic Environmental Experience*" was defined briefly as user experiences of cognitive perceptions and physical responses to their domestic built environment with a diversity of daily household activities [6]. In short, domestic environmental experience connects occupants' physical, psychological and social needs and demands, correlated with different factors of the built environment, such as spatial and environmental factors.

The above literature identified that the domestic environment has various purposes, meaningfulness and aims related to occupants' different spatial and environmental aspects. A domestic setting generally has mainly three types of spatial zones distribution. For example, private areas (e.g., bedroom, study, attached toilet and balcony), used independently by family members only; semi-private areas (e.g., family space, kitchen space, dining room, storage, utility and prayer room), commonly used by only family members; and public areas (e.g., foyer, living room, guest room, balcony and powder room), used by guests beyond family members [6]. It has also been identified that these indoor spatial arrangements can be connected with external additions, such as a balcony, garden and porch, which fall into private, semi-private or public spaces [6]. A domestic setting is interconnected with its different areas, for example, the kitchen has close connectivity with the dining space. The guest room, toilet and dining room have a positive relationship with the living room. Each area also has spatial characteristics related to environmental design factors that accelerate occupants' physical and mental wellbeing in living circumstances [6,9]. Thus, every indoor area of the domestic environment has a significant spatial and environmental relationship that may enhance occupants' psychological responses. These factors, in the domestic setting, are associated with occupants' needs and demands and vary from space to space; they are elaborated below according to the literature.

4.1.1. Entrance

The entrance is a transitional space from outside to inside or one room to another in a dwelling [34]. The front entrance door is the most noticeable demarcation between the public and private realm, where the inhabitants' culture or tradition has a strong impact [35]. The indoor privacy of a residence depends on how people enter it. Internal privacy is compromised if such a place has too many entrances [36]. Evans, Kalantari and Shepley identified that the chance of social contact is better when entering residential units that are adjacent or directly connected to significant pedestrian paths [8,37]. Graham et al. identified the invitation approach as the most frequently selected ambiance for an entryway, whereas other factors, such as sophistication, family, quiet and cosy, also impact human psychology [9]. Ochodo et al. mentioned that the materials (i.e., steel) used for the entry doors reduce occupants' stress and susceptibility compared to a wooden door, regarding safety issues. According to the study, inhabitants living in homes with wooden entrances experience anxiety and distress from attacks by thieves or robbers at night [38]. Oswald et al. also mentioned that entrance and accessibility impact occupants' behavior, significantly enhancing positive wellbeing and satisfaction for aged people and children in a family. From several studies, the researchers found that negative psychological symptoms increase among inhabitants because of poor accessibility, which is also noticeably connected to diverse characteristics of healthy aging [39]. The relationship between spatial and environmental design factors of the domestic entrance and occupants' response is illustrated in Figure 4.

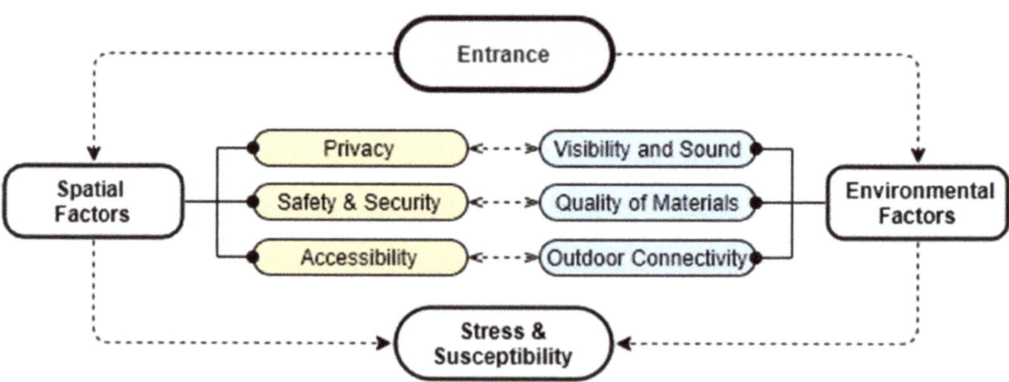

Figure 4. Relationship between spatial and environmental factors of entrance. (Illustration: Author, based on literature).

4.1.2. Living Room

In 1981, Alexander described the living room as the occupants' relaxation and entertainment space and traditionally, the largest area seen upon entering a dwelling [34]. The living room is primarily used to meet and share leisurely events with family members and others [12,40]. Hereafter, interactions and communication are the two main activities experienced in the living room, where comfort and relaxation are the occupants' primary preference [41]. Graham et al. also mapped relaxation, togetherness and comfort as the main factors in occupants' desired psychological ambiance in their living room [9].

Thus, indoor environmental quality and the spatial arrangement of furniture reflect the family's personality and preferences in the living room [42]. According to Saruwono et al., this space develops interaction opportunities between house owners and guests, where the room capacity, spatial organisation and furniture arrangement create different physiological responses among the users. Studies found that homeowners can control guests' communication boundaries through specific furniture layouts [42,43]. Comfort is vital for inhabitants' psychological and physical experience in their living room. Consequently, furniture choice and arrangement affect the occupants' diverse personal comfort levels and preferences and encourage social interaction [8]. As space is a primary point of emphasis in a domestic setting, the living room should also reflect safety for the users. Therefore, circulation by the flexible spatial arrangement of furniture is considered a priority [42].

In the living room, family members communicate with each other by doing various activities such as reading, watching television or just chatting [44]. Living room design is fundamental to supporting occupants' intimacy and comfort, with colors playing an essential role in stimulating mental wellbeing. Warm colors perform well in the living room. These colors evoke a sense of comfort and stimulate dialog [45]. As well as this, wall materials and lighting fixtures also have a strong correlation with occupants' overall satisfaction level in terms of spatial comfort issues [46].

Banaei et al. found that pleasure is an essential human factor for a living room, where room shape and size play a vital role in enhancing occupants' wellbeing [47]. According to the study, PAD correlation identified that daylight and nature-connectedness have a significant association with enhanced pleasure and arousal for the inhabitants. The author also found that the curved roof has a vital role in pleasure, affecting the inhabitant's emotional experiences [47]. Seemingly, studies also identified that using many curved lines may create stress. Flexible walls in living spaces enable individuals to create a friendly environment where people can perform activities and share their experiences with others according to their preferences [48].

Moreover, views of nature have a diverse effect on aesthetic value and the functional aspects. They provide cumulative influences and micro-restorative benefits to occupants' wellbeing in a shared or common space [49]. Built environmental design needs to be incorporated with nature and its components to improve this integrity [50]. In the housing context, it has also been proposed that sunlight and a view of nature, indoor potted plants and photos of plants or small landscapes enhance residents' sense of satisfaction and positive emotions [24,51,52]. The literature also identifies that highrise buildings with large windows may create discomfort, anxiety, stress and unhappiness among inhabitants [27].

Females in low and middle-income families, mainly involved in the indoor household and outdoor gardening, experience higher emotional wellbeing. However, in a real scenario, occupants with socioeconomic constraints often remain in compact living spaces and have limited ability to own indoor greenery [53]. By contrast, residents with higher incomes tend to reside in homes with more greenery [54]. Furthermore, several studies reported that room shape and size significantly impacted occupants' emotions in the smaller domestic setting during the quarantine period. Indoor gardening in a living space and maintenance can be one of the most effective enjoyable activities to mitigate social isolation's stressful and unpleasant impacts on emotional wellbeing in the COVID-19 situation [53].

Privacy is another essential human factor for male and female guests to maintain social aspects and safety for occupants and outcomes. Therefore, design interventions such as entrance door location, window size and position, room height, balconies and internal courtyard may help to achieve privacy in a living space [35]. Zanjani et al. elaborated on three essential factors, safety, aesthetics and memories, by evaluating participants' experiences to enhance relaxation feelings. Here, occupants' traditional values evoke security factors, the aesthetics value stimulates occupants' imagination and fosters individual personalization, and memories reflect familiar feelings. These factors may shape human experiences in a living environment [55]. The overall relationship is illustrated in Figure 5.

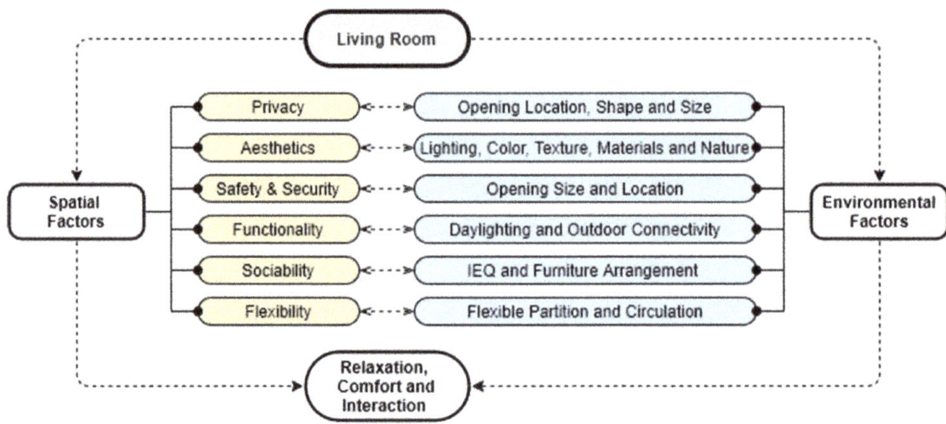

Figure 5. Relationship between spatial and environmental factors of living room. (Illustration: Author, based on the literature).

4.1.3. Dining Room

Habitually eating together as a family has a startling impact on occupants' health and wellbeing. Family bonds become more vigorous and children adjust better and are less likely to use drugs where the family members share meals. Therefore, the eating environment needs to be stimulated by developing a more pleasant dining area (e.g., outside view, good air and daylight) that is more accessible from the kitchen and living room [56].

The dining room is another prominent space where inhabitants gather in a domestic setting. It is a space to accommodate the activities of eating as well as everyday casual things like chatting [45]. The dining space is a crucial element of design in a domestic setting. In general, the dining space acts as a transitional space between the unit's private and public zone and indicates the center of activities in a dwelling setting. As the center of activities, adequate space is necessary for proper circulation to enhance occupant satisfaction [57]. Graham et al. emphasises family togetherness as the prominent psychological ambiance for occupants in a dining space. According to the study, other psychological ambiances, such as sophistication, entertainment and relaxation are closely interrelated in this particular space to the enhancement of occupants' positive emotions [9].

According to Hendrassukma, the dining room's indoor color can arouse occupants' eating habits and inspire conversation between family members and other guests. The author identified that warm colors such as red could stimulate appetite, whereas yellow can increase starving [45]. In another experiment, Ritterfeld and Cupchik identified that a decorative room has a complex and stimulating phenomenon, whereas a sophisticated room is perceived as logical, contemporary, and relaxed [58]. This perception may help develop a dining space in a domestic setting to promote occupants' positive emotions.

Madsen explained that this space is the family's social gathering place, where family members all take a seat, have dinner and make conversation [44]. Relaxing, sitting comfortably and enjoyment are the most critical factors that enhance occupants' positive psychology. Madsen also described that, in a dining space, displaying photos and quotes promotes homemaking attitudes that contribute to occupants' emotions and comfort: relaxing, reading, watching, drinking, eating, etc. [44]. Studies also identified that a short depth of space and high-density results in mental distress and social withdrawal. Another factor found in the literature was the ceiling height, where a room with a higher ceiling was observed as more spacious than a room with a lower roof, leading to a lower sense of stress and crowding [8]. The spatial and environmental design relationship of the dining room is demonstrated in Figure 6.

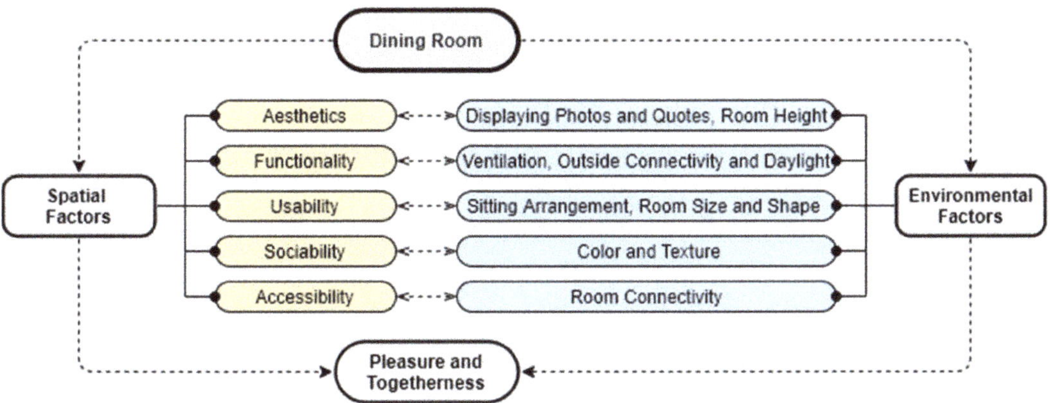

Figure 6. Relationship between spatial and environmental factors of dining room. (Illustration: Author, based on the literature).

4.1.4. Kitchen

Alexander, in 1981, elaborated on the cooking space, where occupants of the household prepare their food and defined it as a kitchen. Today, this place is more prominent and complicated in the domestic setting, where different household activities occur, such as food preparation, cooking, storing and garbage. The author identified that easy accessibility, convenient location, circulation and connection with other spaces within the domestic environment are the most significant factors to consider in a kitchen's functionality [34]. The kitchen is one of the homeliest places, where family members do activities together. It is the place where occupants spend most of their time cooking a meal or getting something. The kitchen is a space where occupants feel comfort, mostly because they enjoy cooking after coming home from outside work or activities [44]. Several studies identified that the kitchen promotes positive interaction and inspires healthy eating among family members. Food preparation can serve as a mediator of social activity and sufficient cooking space facilities improve positive collaboration between the occupants [56].

According to Altas and Özsoy, the kitchen's location in the domestic setting added value to occupants' satisfaction with their living environment [36]. Walters mentioned that an open kitchen (e.g., no wall or door between the kitchen and living-dining spaces) stimulates positive feelings and facilitates family activities and encourages family members to spend more time together [41]. Usually, a kitchen mainly consists of two functional spaces, such as the pantry and central food preparation zone. According to Graham et al., spatial organisation and abundance are the prominent factors in creating psychological ambiance for a pantry. In contrast, spatial organisation, family togetherness, productivity and richness are the most essential and frequent psychological aspects for the central kitchen zone [9].

Pleasant lighting is also essential for safety and creating an enjoyable environment in the kitchen, as well as a view from the windows, which is vital to the occupants' feeling of comfort. Studies also showed that suitable daylight levels in kitchen interiors connect occupants' moods and behavior during cooking activities [59]. There is a close connection between the kitchen and dining space, where color can influence the occupant's feelings. Preferably, light colors in the kitchen prevent risk during food preparation. Clean paint can also stimulate passion during food-processing and eating. Cool colors create a hygienic appearance and warm colors evoke a positive mood when applied to the kitchen walls or cabinet [45]. In 2011, Cho et al. emphasised the importance of a proper ventilation system in the kitchen [60]. Without adequate ventilation (opening windows or using extractor fans), pollution, dampness, mold and fungi growth can be seen, which may create structural defects and have respiratory effects on children [61,62]. This may also impact the behavior of the occupants [63]. Therefore, proper ergonomics, as well as enough maintenance and waste management facilities, can prevent this problematic scenario [41,64,65]. The overall relationship is illustrated in Figure 7.

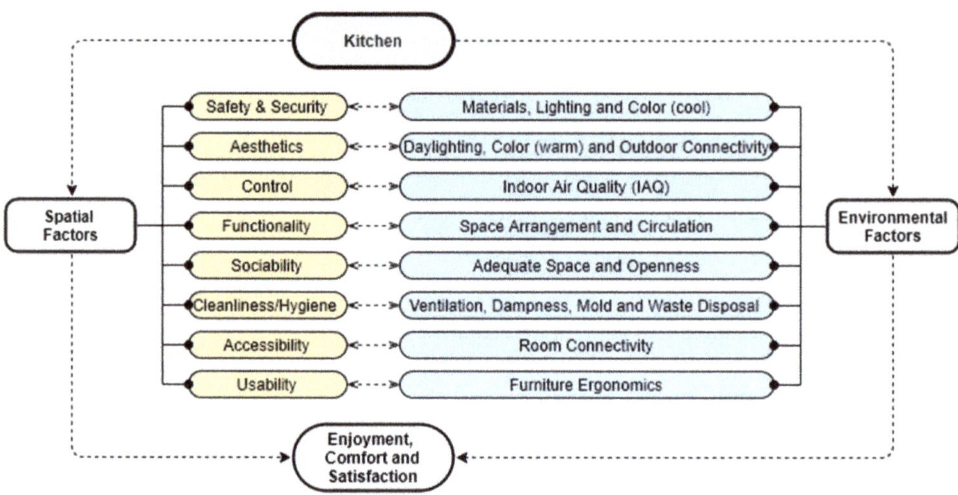

Figure 7. Relationship between spatial and environmental factors of domestic kitchen. (Illustration: Author, based on the literature).

4.1.5. Bedroom

In general, residents use the bedroom for sleeping, dressing activities, quiet retirement and socialising with dearest friends and family members [66]. However, there are various debates and complexities regarding the bedroom's location and function concerning other spaces and occupants' psychological aspects in a domestic setting. According to data from multiple studies, a domestic environment may have three types of bedrooms: the master bedroom, child bedroom and guest bedroom. There is a psychological interrelationship with each bedroom's physical environment, which has been described below based on literature.

Master Bedroom

People in high-density urban residential settings face social pressure in their living environments, so they prefer a particular space to remain alone when they return home [12]. A master bedroom's primary function is relaxation, rest and healthy sleep for adults or parents, where privacy is essential for confirming harmony in family life [12]. The definition of sleeping space changes over time and usually identifies a bedroom as one of the most private areas in a domestic environment, where freedom from excessive noise is desired [34]. It is a space where the occupant can take a step back from various in-house activities and enjoy privacy and noiselessness [48]. The bedroom also responds to human circadian rhythms [56]. According to Hendrassukma, the bedroom, for adults or parents, has the primary purpose of calming activities such as resting or sleeping, where the indoor environment strongly influences these activities [45].

The bedroom relates to occupants' feelings of comfort at home. Madsen explained that when a person becomes sick, the bedroom is the only homely spot in the domestic environment where the occupant feels comfort and relaxed. The author further described that suitable sounds and smells also make it a homelier place. If anyone changes the same bed to a different location, he/she may not feel the same comfort, with space security and belongingness influencing occupants' psychology [44]. This is because occupant habits such as lying, reading, relaxing, watching and other activities before going to sleep are closely correlated with existing bedroom scenarios. Consequently, the bedroom indicates an image of comfort, warmth and relaxation, with the bed as a prominent factor in that homely spot [44].

On the other hand, room shape and size influence the occupants psychologically in master bedrooms. A square room seems to be more crowded than a rectangle shape within an equal area. Not only that, but longitude direction may create vision rigidity and have a psychological impact [12]. According to Mridha, most developers generally prioritise making the master bedroom more comfortable and attractive than other domestic spaces within the apartment unit to attract local clients. In that case, the rooms' location, size and shape are essential to design factors in ensuring occupants' comfort in a master bedroom [57].

Indoor environmental elements, such as proper daylight, noise level, color and ventilation are essential for occupants' mental wellbeing in a master bedroom. According to Kennedy et al., noise at night in a bedroom is not desirable and another vital factor to consider for occupants' wellbeing. It is directly related to the quality of sound sleep. The author also recommended well-ventilated bedrooms for sleep health, whereas having a balcony creates extra facilities for the occupants' refreshment [67].

For daylight, window location is the most significant criterion for a bedroom. Not only the wall between the bedroom and balcony, but window louvers and height, and window glass type and opening system are correlated with ensuring natural ventilation, daylight and the external vision of a bedroom [41,57]. A room with outer vision seems more spacious than rooms without such an image—a window connected to the bedroom to guide people's sight [12]. Hendrassukma also mentioned that color significantly impacts occupants' psychology in personal spaces like a master bedroom [45].

Besides these, several studies emphasised the psychological comfort of familiar objects and pictures in bedrooms that may retell past events and positive memories. According to the literature, displaying memories and photographs is an integral part of visual relief and establishing personal spaces within the domestic setting. Bao also mentioned that the photos and mirrors on the wall could transfer vision and provide a sense of space expansion [12]. In personal spaces like bedrooms, occupants have different choices based on their sex, age and behavior. In general, women tend to beautify or decorate their private rooms more than men do in their domestic settings. Gosling elaborated that women have various choices, such as photos, lotions, jewelry, candles and others.

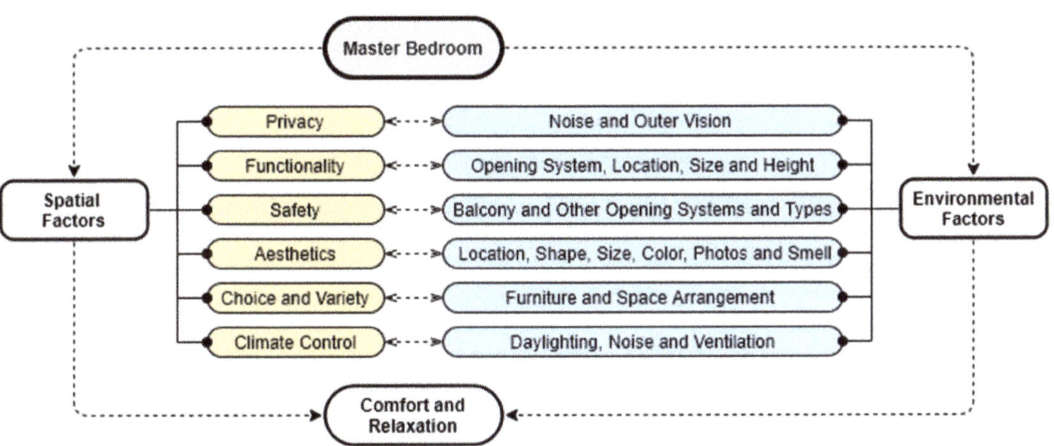

Figure 8. Relationship between spatial and environmental factors of master bedroom. (Illustration: Author, based on literature).

In contrast, men tend towards CDs, sports equipment and achievement-related items in personal spaces like bedrooms [68]. Finally, in 2015, Graham mapped occupants' desired psychological ambiance in their master bedrooms. According to the authors, romance, privacy, comfort, relaxation and love are the most prominent and frequent psychological ambiances of occupants for their master bedrooms [9]. The overall relationship between the spatial and environmental design factors of the master bedroom is illustrated in Figure 8.

Child Bedroom

According to Bao, self-cognition is an essential design element for the perfect children's space, where parents entering the room (e.g., bedroom) by knocking on the door first can promote children's positive psychological sense [12]. In that case, the visual contact formed by a sudden interruption will induce negative emotions due to privacy concerns [12]. Creating an exciting space encourages children to ask questions and learn from their living environment, such as children's bedrooms where space flexibility, a dramatic setting, natural elements and an adequate play area impact their mental wellbeing [69].

Here, the reason for space flexibility and functionality is to provide adaptability to contextual situations. Space can be utilised proportionately according to function by altering spatial aspects [69]. For example, a child's bedroom (space) can be used as a study room. This space can also be separated into different areas by using portable light dividers where other functions, such as painting, playing, reading and sleeping, can be applied. Graham et al. identified the most frequently selected ambiances, such as quiet, productivity, organisation, privacy and creativity or self-expression, for the study space, which may be recommended for the bedroom and study area for children in domestic settings [9]. Using appropriate materials and textures for children's spaces will enhance their imagination in living environments [70]. In this case, using natural materials and avoiding artificial fabrics will encourage originality and develop positive psychology [71].

Many studies found that children living in rigid spaces cannot express their creative capacities to discover new opportunities. Nevertheless, various shapes, colors and other indoor environmental factors, including furniture and visual detail, increase children's aptitude to learn, realise and cultivate inventiveness in their living spaces [69,71].

In the child's living space, color has a significant influence and a meaningful correlation with enhancing creative potential [71]. Enjoyable, colorful and stimulating pictures have been identified as the basis for human creativity and inspiration. The shape, size, layout and function of the spaces are important factors in social interactions and encourage skill enhancement in a child's room [72]. For example, using colored glass, different light spectrums, proper daylight, and play with water and natural elements may stimulate children's curiosity in their living environments [69]. The overall relationship between spatial and environmental design factors of a child's bedroom is illustrated in Figure 9.

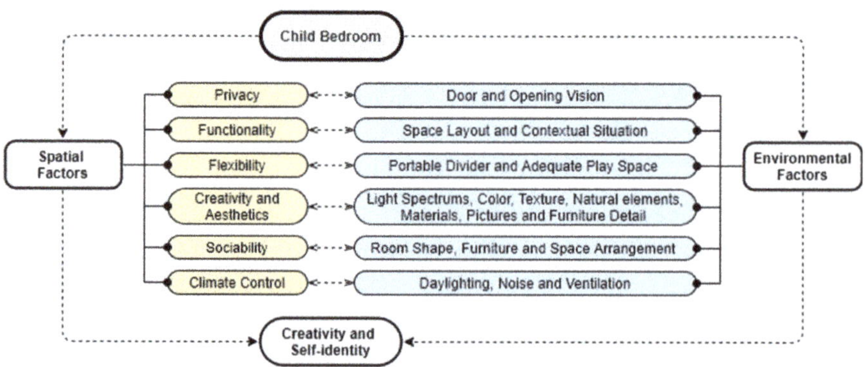

Figure 9. Relationship between spatial and environmental factors of child bedroom. (Illustration: Author, based on the literature).

Guest Bedroom

Besides the first two types of bedrooms (e.g., master bedroom and child bedroom), a guest room generally indicates a bedroom in a house for visitors or guests to stay and sleep in, attached to the domestic setting's public zone. Graham et al. also described the psychological ambiance of guest rooms. According to the authors, an inviting approach is the primary psychological ambiance of this particular space. Comfort, relaxation, attachment and convenience are other prominent and frequent psychological aspects of guest rooms [9]. According to the literature, a comfortable bed and functional furniture layout for closets and luggage are an essential part of a guest room to ensure guest comfort and privacy. Convenient power outlets, that can be reached without moving furniture around and easy access to the toilet are other significant issues in designing a guest bedroom. It is essential to have easy access to enough light in a guest bedroom during nighttime, such as bedside lighting to read and flexible space layout, especially for older guests [7,9,66]. The overall relationship between spatial and environmental design factors of the guest bedroom is illustrated in Figure 10.

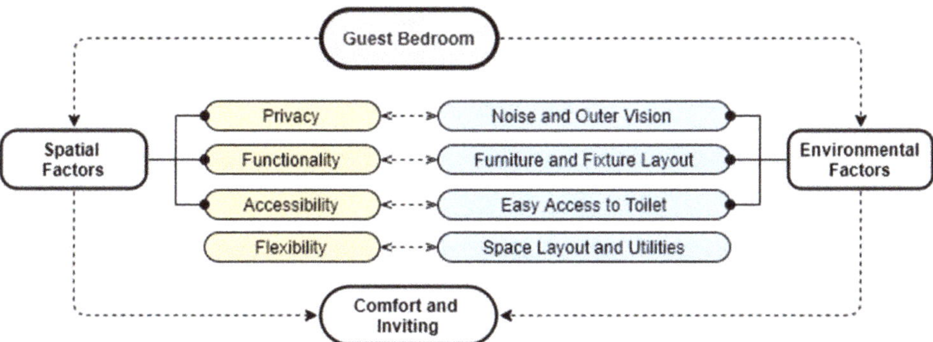

Figure 10. Relationship between spatial and environmental factors of guest bedroom. (Illustration: Author, based on the literature).

4.1.6. Bathing and Toilet

Bathing and toilet are the most private areas in a house. According to Alexander, this space is one of the most prominent functional rooms for its size and quality [34]. Amerigo et al. mentioned that an indoor toilet is the extreme design objective of residential excellence for specific cultures and socioeconomic levels [30]. In a high-density domestic setting, the quality, number and location of indoor toilets influence occupants' behavior, choice and freedom because of different situational contexts such as crowdedness, limited resources and usability [12].

Several studies found that increased privacy and cleanliness are the most vital human spatial factors for occupants' health and wellbeing in bathroom settings, where there is always a gender effect [73]. Suitable bathroom and toilet facilities are related to high levels of unhappiness and depression after changes in gender, marital status, age, engagement and migratory conditions [63]. Many studies found that occupant panic disorder increases due to a lack of proper toilet facilities in the living area [38]. According to Graham et al., privacy, relaxation and rejuvenation are the prominent psychological ambiances for bathing and toilet, whereas an inviting approach and quietness are also preferred by the inhabitants [9].

Bathrooms are essential for individual hygiene, mainly at the start and end of the day. Residents become restless if the toilet is inadequate and unhygienic. This situation creates anxiety, panic disorder and depression regarding the occupants' safety and personal privacy, especially in a crowded domestic environment [38]. According to UK GBC, avoiding moisture, pollutant and mold growth are essential issues when developing good indoor quality for bathing and toilet [56]. Cho et al. identified a strong psychological need for window ventilation in the bathroom to enhance the occupants' indoor quality [60]. Hendrassukma emphasised that color and lighting design issues are essential to improving the comfortable atmosphere in the bathroom. According to the author, the occupants need a relaxed, comfortable and safe atmosphere in the bathroom. Color preferences in the bathroom can improve the impression of calm, freshness and cleanliness. In that case, the author suggested a white color choice for a bathroom to enhance the impression of cleanliness and purity. The combination of white with other cool colors also makes the occupant feel relaxed and peaceful after bathing.

However, a bathroom should have sufficient indoor lighting to reduce the room's moisture and prevent risk caused by a darkened room [45]. Madsen further explained that a bathroom is a place where people should have a degree of privacy. The author also emphasised insulation, floor quality and heating systems to provide a high degree of comfort in the bathroom setting. Therefore, occupants may wear slippers for safety issues [44]. Mridha mentioned that bathroom size is a dominant predictor of occupants' satisfaction at present [57]. Several studies focus on accessibility and aesthetic beauty for bathrooms, which indicates the prominent need and demand among occupants for new built and modifications or renovations, where fixtures, functional layout, usability and greening for purification and relaxation need to be considered in detail in the design process [26,74]. The overall relationship is illustrated in Figure 11.

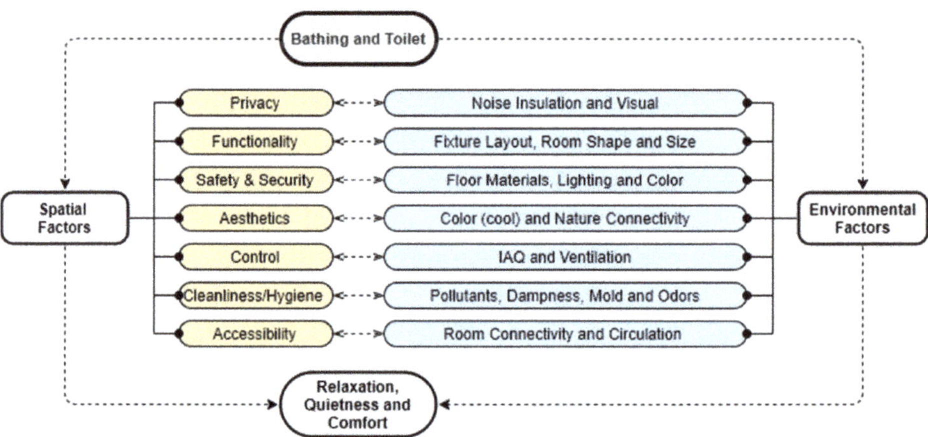

Figure 11. Relationship between spatial and environmental factors of bathing and toilet. (Illustration: Author, based on the literature).

4.1.7. Study and Work Space

The occupant needs a disturbance-free space, with self-regulating individual noise, for working, learning and reading in a domestic setting. Without such a space for study, it may generate stress reactions and negative psychological emotions may be generated in the occupants. In a high-density domestic environment, living and other rooms can be utilised for this purpose with an interval according to the occupants' preferences [12]. According to several studies, privacy, relaxation, functionality and creativity are the main human psychological ambiances for this particular space [9,56]. Consequently, numerous studies related to the home office identified that feeling comfortable when working from home is the occupants' primary concern, as is a noiseless working zone with a comfortable furniture arrangement [44]. Ceiling height has an influential role in social engagement and impacts human focus ability [75]. Research findings noticed that when occupants reside in a room or space with low ceilings, they perform better on focused works, such as reading and studying. In contrast, high ceilings encourage imaginative thinking and influence social gatherings [8,56]. A recent study investigated ceiling height's impact on occupants' aesthetic perceptions and activity; spaces with high ceilings have higher aesthetic attractiveness scores than low height spaces [75].

Moreover, in the present pandemic (COVID-19) period, scenarios have changed. People now need to finish all office activities sitting at home. At present, the working environment inside the house and people's psychological relationship with it are critical when designing a domestic setting. Studies have shown that having an indoor green space enhances positive emotions for the occupants, particularly those who have socioeconomic constraints [76]. Insufficient indoor lighting levels, a variation in wall colors, noise and other physical environmental ambient properties stimulate mental stress by changing people's circadian rhythms and troublesome work cycles [27]. Besides these, indoor air pollution, excessive temperature and lack of ventilation also negatively impact human efficiency [1,2,77]. The overall relationship is illustrated in Figure 12.

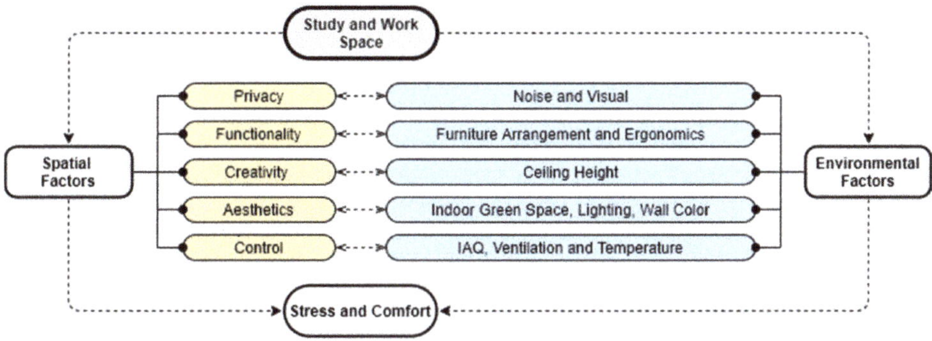

Figure 12. Relationship between spatial and environmental factors of study and working space. (Illustration: Author, based on literature).

4.1.8. Balcony

The domestic setting needs a connection to the outside through intermediate spaces (e.g., balconies) with views of a neighborhood or communal green areas [56]. A residential balcony creates a flexible space for the diverse activities of occupants. Significantly, a balcony permits a connection with the outdoor space without leaving the indoor environment. The balcony can be attached to a personal space (e.g., bedroom) and public space (e.g., living room) or semi-private spaces in a domestic setting and becomes a prominent transitional space between the outside and inside for refreshment, which affect the quality of life in a residential environment [67]. External vision plays an essential role in the indoor-outdoor relationships in a domestic setting, where balconies promote occupants' opportunity to connect with nature (e.g., during COVID-19). A balcony creates a spatial change in the dwelling plan and makes it more flexible for use as a third room [12].

According to Mridha, a balcony is a changeover space between the outdoors and indoors and a source of air and light [57]. Madsen mentioned Jacob's photograph and stated that a balcony with a proper overhang promotes a healthy balance between daylight and shadow and the natural ventilation system and indoor comfort in a house [44]. According to Kim and Kim, despite all the positive uses of sunlight, ultraviolet rays become harmful to human skin. By removing the balcony space, inhabitants' chances of being exposed to indirect sun in a residential apartment increase [78].

According to Kennedy et al., specific physical and spatial design characteristics related to occupants' everyday living functions, privacy and indoor environmental comfort are significant factors to consider in the context of adding a balcony to a domestic setting [67]. The author found that most residents described their functional utilisation of balconies for a varied range of household activities, such as preparing foods, gardening, exterior private space and other mixed attitudes, where privacy is the common phenomenon for all the inhabitants. Residents do not want to hear and engage with the other residents or the community noise from their private balconies. In that case, the balcony's location is a prominent design factor for the occupants [67]. Drying laundry and storage are also essential purposes of the balcony. Residents typically use their balconies for changes and everyday events, such as studying, relaxing, reading, care of pets, physical exercise or just inactive sitting. Contentiously, some inhabitants smoked on their balconies, annoying inhabitants of other residences [67]. Sometimes dust and other external pollution may enter through the balconies and dirty indoor spaces and furniture [67]. Lack of fencing and bar grills on balcony, windows and doors increased occupants' weakness to outside attacks and experience of distress, particularly during the night, due to insecurity [38].

The moisture and condensation consequences due to the balcony space's poor insulation have been stated in some studies. In a questionnaire survey in Seoul, most residents complained about mold and indoor dampness problems with a changed balcony. In that study, health problems among the children and residents were identified where balconies were removed [79]. Ozaki also discusses some ritual perceptions of a balcony that may impact the inhabitants, as the domestic environment is closely related to peoples' symbolic, private, secular life. The balcony is an individual's access to open spaces for personal utility purposes. Still, this space is frequently positioned at the front side of a building and some unsightly activities traditionally connected with a backyard, for example, laundry, drying, food preparation and washing, where the occupants prefer high privacy [80]. The overall relationship between the spatial and environmental design factors of a balcony is illustrated in Figure 13.

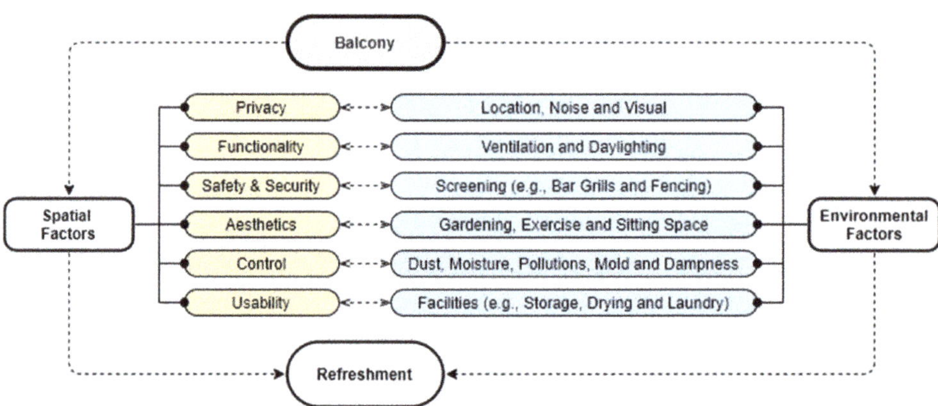

Figure 13. Relationship between spatial and environmental factors of balcony space. (Illustration: Author, based on the literature)

4.1.9. Lobby and Circulation Space

In 1986, Haber considered occupants' perceptions of a highrise housing lobby and identified that the need for communal interaction was a critical feature determining the occupants' assessment of this space [37]. It also seems that lobby design is an essential factor for occupants' communal experiences of highrise residential buildings and comprehensive research needs to be conducted to investigate influential lobby designs [37].

Consequently, circulation is vital for social interaction, where considering daylight and external views make circulation an enjoyable experience by offering spatial variation in the dwelling environment. The spatial enclosure enhances aesthetic and psychological responses to indoor environments. People generally feel secure in more open indoor spaces with better external visual connectivity [75,81]. Confirming suitable dimensions promotes circulation, enjoyable for less physically able people with an accessibility threshold criteria [56]. The lobby and circulation space are used for social interaction and occupants can also use this space for physical exercise, particularly children and older adults, who cannot go outside frequently in highrise residential apartment buildings. This area also promotes other household activities for women to enhance stress-free living [8]. The overall relationship is illustrated in Figure 14.

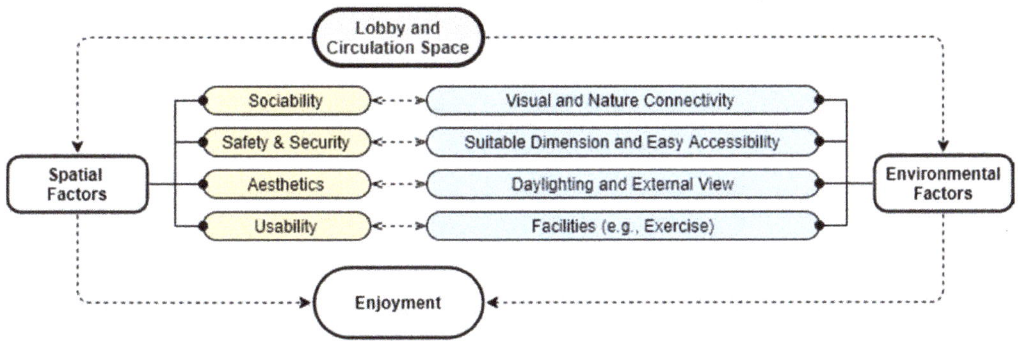

Figure 14. Relationship between spatial and environmental factors of lobby and circulation space. (Illustration: Author, based on the literature).

4.1.10. Storage and Utilities

Generally, a house's storage and utility facilities refer to a place where various household necessities are stored. This can be a separate room or an area in the domestic space. According to Alexander, storage has become critical in housing development in recent years because houses are getting smaller over time. It is challenging to allocate extra space for storage and utility facilities in a domestic area [34]. The type of storage facility employed depends on the inhabitants' needs and demands in their daily household activities. Alexander mentioned that designers identified that almost twenty-five percent of a domestic space should be dedicated to storage facilities [34]. Storage and utility provision in a domestic setting promotes stress-free functional living for the occupants. The functionality of the storage space depends on the shape, size and location of the room. In many cases, a small unoccupied area of a dwelling, such as the unused space under the stairs or the top of the kitchen or toilet, is usually used as a storage space [56].

According to various studies, humidity and ventilation inside the room are very important for such storage and utility spaces. Without the right amount of light and ventilation, storage spaces can accumulate various fungi and mold growth, which is subsequently dangerous for stored goods and technical accessories [28,44,60]. The overall relationship between the spatial and environmental design factors of storage and utilities is illustrated in Figure 15.

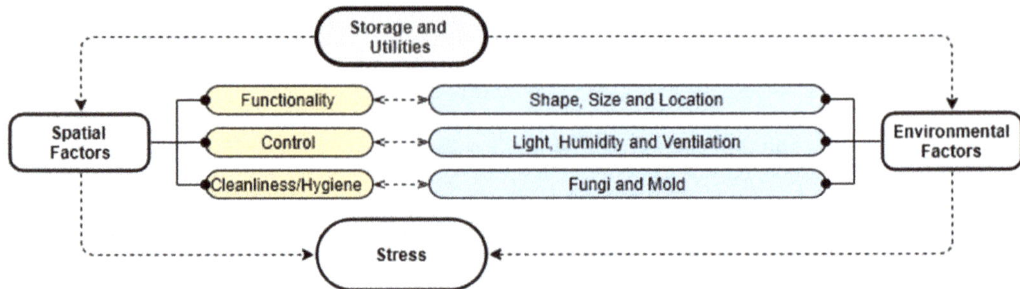

Figure 15. Relationship between spatial and environmental factors of storage and utilities. (Illustration: Author, based on the literature)

4.2. Domestic Environment and Occupants' Sociocultural Context

In this study, different domestic spaces have been conceptualised according to spatial and environmental factors, reflecting occupants' domestic experiences based on the academic literature. A domestic environment expresses a symbolic connection of occupants' contextual factors through the user's life journey [33]. Regardless of the spatial features and environmental design components, each domestic space provides its dwellers with senses that serve their individual needs and demands [35]. Therefore, a domestic environment comprises individual emotional expression, spatial requirements and contextual relationships [35,39].

Accordingly, several studies highlighted the status of macro-level features such as sociocultural factors (e.g., climate, profession, culture, education, religion, sex, age, household compositions) shaping individual perception in a domestic living environment. Hence, contextual factors affect occupant's behavior and perception due to diverse social and cultural aspects [35]. The sociocultural aspects of spatial characteristics are related to occupants' lifestyles, which indicate different environmental preferences under different domestic settings and circumstances [82,83]. Studies also showed that environmental factors are affected by the varying needs and choices of user groups. According to Lawrence, the space-use of a dwelling and morphological changes cannot be disconnected from differences in the sociocultural meaning, as well as household personalisation, which establishes changes in an occupant's relationship with the home environment [29]. Therefore, in a domestic setting, occupants' personalisation diverges concerning social and cultural factors and lifestyle behavior [29,33,35,82]. Thus, domestic living concepts synchronise users' diverse sociocultural relations, along with spatial and environmental preferences that may enhance occupants' wellbeing [6,29].

Nonetheless, social and cultural values play an essential role in defining space identity and rituals, which may affect occupants' perceptions in their domestic private spaces [26,29,31,37]. Therefore, identifying the core relationship between spatial, environmental and user contextual factors in domestic spaces is significant in architectural design to interpret occupants' experiences (Figure 16.). Several ethnographic studies illustrate that sociocultural phenomena' domestic spatial appearance is articulated in numerous ways, influenced by individual choice according to regional and social code variation. Hence, a different cluster of user activities may change domestic spaces' spatial and environmental preferences because of a complex sociocultural phenomenon [82,83]. However, this study mainly indicates the theoretical correlation between spatial and environmental factors within domestic spaces; combining users' sociocultural context with a pragmatic understanding of designing domestic space is suggested for future direction.

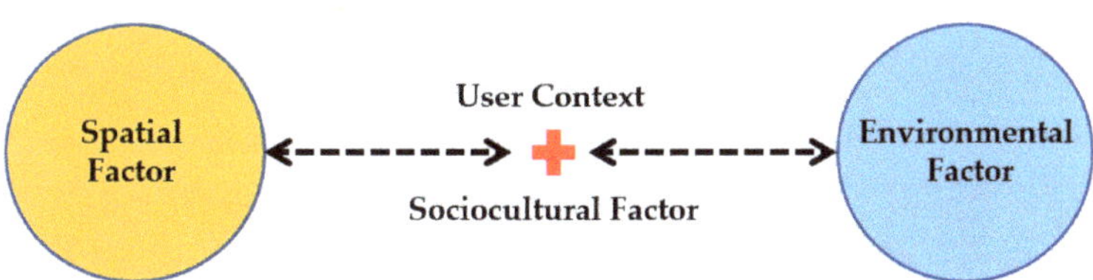

Figure 16. Relationship between spatial, environmental and sociocultural factors for a domestic setting. (Illustration: Author, based on the literature).

5. Discussion and Future Research Direction

From the above discussion, occupants living in each space in a household have different needs and preferences regarding spatial and environmental aspects related to their psychological and behavioral factors. However, sociocultural factors influence the occupants' spatial and environmental attributes, which are closely intertwined, and shape those relationships. Research into domestic environments' spatial and environmental factors usually relies on the inhabitants' behavioral and psychological attributes [42]. These factors need to be considered to understand domestic spaces' diversities to enhance occupants' feelings of satisfaction and comfort. According to several studies, adequate knowledge of spatial activities in a household environment tends to access the occupants' overall behavioral process within a domestic setting [6,9,12,42]. It is also imperative to determine the occupants' spatial and environmental experiences in different domestic settings and utilise them accordingly in the architectural design solution.

According to the literature, this review identified twelve spatial factors for the domestic environment according to occupants' needs and demands (Table 1). Consistent with different domestic spaces, attempts were made to explore the correlation of environmental factors with each spatial factor in view of occupants' experiences. From this review, it has been observed that privacy, functionality and aesthetics are the prominent spatial factors for different spaces in domestic environments that impact occupants' psychological experiences, considering their needs and demands. The overall discussion also suggests that any domestic area's privacy is related to its environmental design factors such as size, shape, location, visual connectivity to the external environment and sound, where the occupants' lifestyle has a significant impact. For functionality, ventilation, daylight and outdoor connectivity indicate essential design components, as well as the opening system, shape and location.

Additionally, space layout and ergonomics also have influential connectivity to occupants' psychological perceptions. However, in almost every domestic space, nature connectivity, natural and artificial light, interior color scheme, smell and the nature of the material used to accelerate the occupant's aesthetic perceptions. Subsequently, any place's aesthetic features profoundly affect the minds of the people living there. Most studies have identified that residents value the indoor environment's quality (e.g., daylight, noise, ventilation, artificial light spectrums) in domestic spaces, which are closely connected to the comfortable living concept. Other issues, such as indoor dust, mold, moisture, pollutants and dampness are directly related to occupants' physical and mental health. Occupants prefer to control these issues according to their choices regarding safety and hygiene. Looking at the safety–security measures in different domestic spaces, environmental design phenomena such as opening system, screening, quality of materials, accessibility, light and color generally indicate the most influential design components according to occupants' experiences.

Overall indoor environmental components such as light, color, temperature, materials, layout, shape, size, height, opening and greenery affect human interactions and emotions. Research has shown that furniture arrangement in a space significantly affects the behavior when living there. In that sense, all the elements of a room are closely related to each other.

In 1986, Pennartz mentioned that a room's spatial and environmental arrangement affects occupants' experience in a domestic environment [21].

Table 1. Occupants' spatial factors in the domestic environment.

Spatial Factors	En-trance	Living Room	Dining Room	Kitchen	Master Bed-room	Child Bed-room	Guest Bed-room	Study & Workspace	Bathing & Toilet	Bal-cony	Storage & Utilities	Lobby & Circu-lation
Control				■	■	■		■	■	■	■	
Privacy	■	■		■	■	■	■	■	■	■		
Functionality		■	■		■	■	■	■	■	■	■	
Usability			■	■						■		■
Flexibility		■				■	■					
Sociability		■	■	■		■						■
Accessibility	■		■	■			■		■			
Aesthetics		■	■	■	■	■		■	■	■		■
Creativity						■		■				
Variety of Choice					■							
Cleanliness and Hygiene				■					■		■	
Safety and Security	■	■		■	■				■	■		■

(Here, the color field indicates the factors' essentiality for different domestic spaces).

The above study also identified that, in a domestic environment, the public zone can be divided into subzones, separated by a movable light partition or door and allocated as the study area, guest room, prayer space or work station, etc. Besides, circulation spaces or lobbies can be used for children's play areas or for adults' physical exercise, where safety–security and psychological emotions are very important to consider. The essential aspects in the kitchen are flexibility, functionality and safety–security for users. A kitchen is a place where, usually, the family members of the house spend most of their time. As a result, occupant's emotional issues, as well as other practical and functional issues are strongly related to this place. The cooking space encourages interaction between family members. Research has also identified that cleanliness and hygiene are the most significant factors for residents, whereas adequate lighting, ventilation and layout provide peace of mind. Other design components, such as furniture, ergonomics and functionality are closely related to occupants' psychological satisfaction and comfort.

Furthermore, the relationship with the outside environment through windows, wall color, room shape and size, balcony and toilet attachments create a spiritual connection with the human mind. The occupants usually prioritise adequate ventilation systems, fixture layout, lighting and odor quality when considering bathrooms and toilets in a domestic environment. The windows of a house create a connection between the occupants' outside and inner world. Studies on highrise residential buildings have shown that large windows or window height negatively impact the occupants in many cases. This effect is especially evident in women and children who live on the top floor of highrise buildings. The balcony has recreational facilities and some ritual perceptions that may impact the occupants in their domestic environments. The overall conceptual parametric relationship between spatial and environmental design factors in different domestic spaces is illustrated, according to the literature, in Figure 17.

Moreover, during the COVID situation, working from home grew increasingly predominant. Every home needs an environment where work can be done peacefully. In that case, the bedroom, study area, dining space or living space can be utilised in a dual way considering their space usability, functionality and flexibility. Consideration of these design components depends on the occupants' contextual status. Users adjust their personal spaces according to their individual context and try to find pleasure or satisfaction in their living environments. However, consideration of different dwelling models and the pragmatic nature of spaces also influence numerous occupants' perceptions of workability because of their diverse shapes, sizes and multi-functionalities [84]. In general, residential satisfaction depends on occupants' needs and preferences, which are closely related to the spatial and environmental components in living environments. Consequently, occupants may achieve happiness by changing or modifying their living environments' physical characteristics to create more comfortable settings.

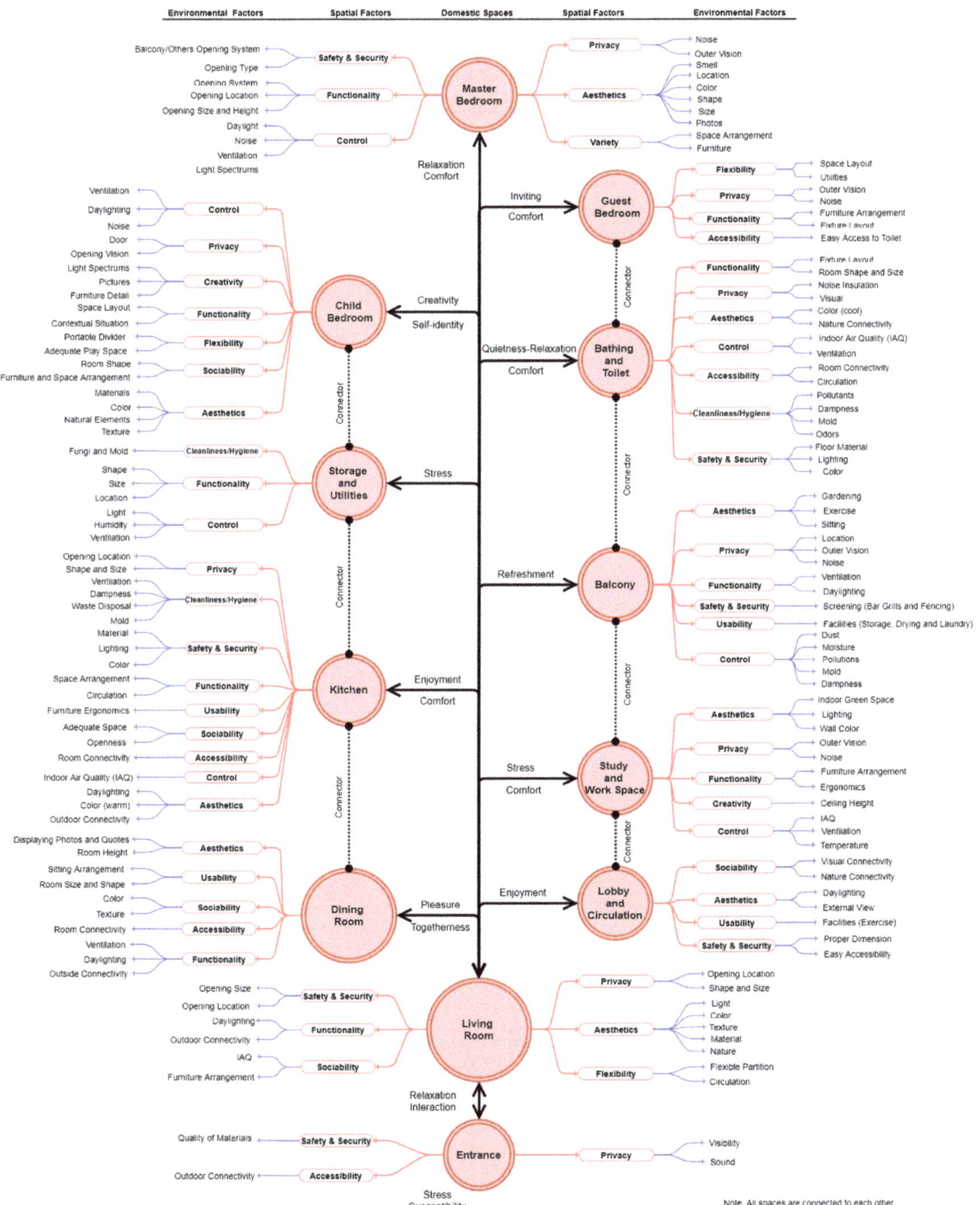

Figure 17. The overall conceptual parametric relationship of occupants' domestic environmental experiences.

After 1970, architectural psychological concepts emerged on human performance, wellbeing and control in built-environmental research [1,6]. However, these are mainly limited to evaluating institutional and health care spaces. According to studies, limited research is conducted on domestic environments and occupant experiences [6]. Nonetheless, studies of architecture and environmental psychology concerning residential settings mainly focus on indoor environmental quality related to occupants' behaviors and physical health risks [28].

From the reviews, it is clear that, in any living condition, occupants' behavior is affected not only by the spatial and environmental aspects but also the occupants' perceptions, feelings and needs, as well as the users' sociocultural context [85]. Occupants who use domestic spaces may have specific values and standards for a given area regarding meaningfulness, attachment and perceptions. Individuals' sociocultural contextual situations may impact different spaces or rooms in a domestic setting, where spatial behavior is related to various cultural and social factors and user preferences. Today's architectural design approaches do not adequately address the relationship between users' context, spatial and environmental design factors, along with occupants' psychological satisfaction and comfort [6,85].

From the literature, it has also been identified that there is a study gap between two current theories, "Environmental Deterministic"and "Social Constructivism" that drive users' experiences within the built environment [86,87]. Here, the "Environmental Deterministic" theory based on environmental psychology describes the physical environmental impacts on human behavior. The scope of explanation about user contexts is limited, to some extent, in this theory. Consequently, the "Social Constructivism" theory describes cultural and social perceptions as challenging to measure or correlate the effects of the built environment is limited [85]. However, the position of *"Environmental Experience Design (EXD)"* between the two spectra derives from users' physical and psychological experiences and addresses users' sociocultural, spatial and environmental design aspects (Figure 18).

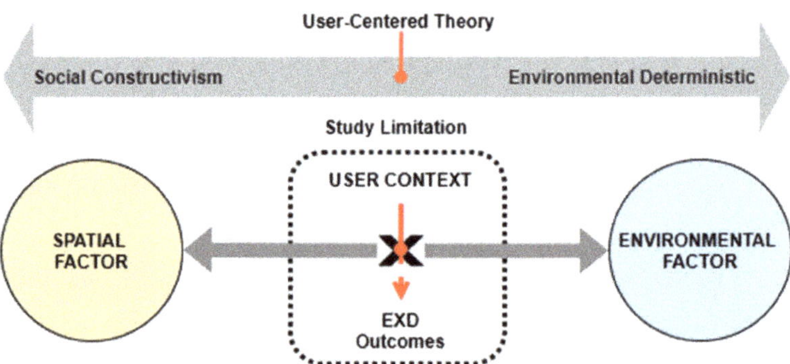

Figure 18. Research gap and future research direction.

This design approach may combine users' spatial preferences and environmental design factors, as well as users' sociocultural context, through their domestic experiences, which may improve occupants' mental wellbeing [6,7]. Therefore, this study establishes a theoretical relationship between spatial and environmental design factors based on the literature, focusing on occupants' experiences in different domestic spaces, where a users' contextual situation plays a critical role in enhancing their wellbeing in domestic settings. Therefore, combining occupants' preferences according to their sociocultural context with spatial and environmental design factors will be a future research direction to explore in the sustainable notion of *"Domestic Environmental Experience Design"*.

6. Conclusions

This study was conducted by a comprehensive literature review based on occupants' domestic experiences and identified that numerous spatial and environmental design factors affect occupants' psychological responses in a domestic setting. This study has illustrated that the theoretical associations between spatial and environmental design factors of different domestic spaces can stimulate occupants' satisfaction and comfort in domestic living. However, occupants' contextual situations impact their domestic living environment, where diverse sociocultural factors such as age, gender, religion, income, education, occupation and lifestyle shape their household needs and demands, which are beyond this study's scope. Each living space has multidimensional uses that are tailored to the sociocultural context of the occupants. Today's environmental design approaches, in the architectural design domain, fail to cohesively address the relationship between user context and spatial and environmental design factors that may enhance occupants' mental wellbeing in a domestic setting. The scope of explanation of users' contextual situation is also limited, to some extent, in environmental design theories. Without a clear perception of occupants' context, environmental design solutions may be harder to implement to enhance wellbeing. Thus, the concept of "*Environmental Experience Design (EXD)*" may combine users' spatial preferences and environmental design factors, along with user contextual factors through their experiences to improve occupants' mental wellbeing. Therefore, a combination of occupants' contextual factors, e.g., sociocultural factors, with spatial and environmental design factors will be the future research direction to explore the notion of "*Domestic Environmental Experience Design*" for the sustainable development of high-density housing sectors.

Author Contributions: Conceptualisation, S.C. and M.N.; Data curation, S.C.; Formal analysis, S.C.; Investigation, S.C.; Resources, S.C.; Methodology, S.C.; Project administration, M.N.; Supervision, M.N. and H.D.; Visualisation, S.C.; Writing—original draft, S.C.; Writing—review and editing, S.C. All authors have read and agreed to the published version of the manuscript.

Funding: This research received no external funding.

Institutional Review Board Statement: Not applicable.

Informed Consent Statement: Not applicable.

Acknowledgments: The authors would like to express their sincere gratitude to Melbourne School of Design, Faculty of Architecture, Building and Planning, The University of Melbourne for providing access to the facilities required for this research activity as well as a full PhD scholarship given to the first author of this paper.

Conflicts of Interest: The authors declare no conflict of interest.

References

1. Bluyssen, P.M. Towards an integrated analysis of the indoor environmental factors and its effects on occupants. *Intell. Build. Int.* **2020**, *12*, 199–207. [CrossRef]
2. Caan, S. *Rethinking Design and Interiors: Human beings in the Built Environment*; Laurence King: London, UK, 2011.
3. Mallgrave, H.F. *From Object to Experience: The New Culture of Architectural Design*; Bloomsbury Publishing: London, UK, 2018.
4. Miller, S.; Schlitt, J.K. *Interior Space: Design Concepts for Personal Needs*; Praeger Publishers: Westport, CT, USA, 1985.
5. Goldhagen, S.W. *Welcome to Your World: How the Built Environment Shapes Our Lives*; Harper Collins: New York, NY, USA, 2017.
6. Chowdhury, S.; Noguchi, M.; Doloi, H. Defining Domestic Environmental Experience for Occupants' Mental Health and Wellbeing. *Designs* **2020**, *4*, 26. [CrossRef]
7. Noguchi, M.; Ma, N.; Woo, C.M.M.; Chau, H.-W.; Zhou, J. The Usability Study of a Proposed Environmental Experience Design Framework for Active Ageing. *Buildings* **2018**, *8*, 167. [CrossRef]
8. Evans, G.W. The Built Environment and Mental Health. *J. Hered.* **2003**, *80*, 536–555. [CrossRef] [PubMed]
9. Graham, L.T.; Gosling, S.D.; Travis, C.K. The psychology of home environments: A call for research on residential space. Perspectives on Psychological Science. *Psychol. Couns.* **2015**, *10*, 346–356.
10. Ridley, D. *The Literature Review: A Step-by-Step Guide for Students*; Sage: New York, NY, USA, 2012.
11. Williams, C. Research methods. *J. Bus. Econ. Res.* **2007**, *5*, 65–70. [CrossRef]

12. Bao, M. Research on Space Diagram Under Behavior Psychology in High Density Indoor Residential Environment. In Proceedings of the 4th International Conference on Arts, Design and Contemporary Education (ICADCE 2018), Zhengzhou, China, 6–8 May 2018; Atlantis Press: Paris, France, 2018; pp. 526–529. [CrossRef]
13. McClure, W.R.; Bartuska, T.J.; Young, G.L. *The Built Environment: A Collaborative Inquiry into Design and Planning*; John Wiley & Sons Inc.: Hoboken, NJ, USA, 2011.
14. Blossom, N.H. Human Nature and the Near Environment. In *The Built Environment: A Collaborative Inquiry into Design and Planning*; John Wiley & Sons Inc.: Hoboken, NJ, USA, 2011.
15. Stedman, R.C. Toward a social psychology of place: Predicting behavior from place-based cognitions, attitude, and identity. *Environ. Behav.* **2002**, *34*, 561–581. [CrossRef]
16. Kopec, D.A. *Environmental Psychology for Design*, 3rd ed.; Bloomsbury Publishing Inc.: London, UK, 2018.
17. Stokols, D. A Social-Psychological Model of Human Crowding Phenomena. *J. Am. Inst. Plan.* **1972**, *38*, 72–83. [CrossRef]
18. Hayward, D.G. Psychological concepts of 'home'. *HUD Chall.* **1977**, *8*, 10–13.
19. Dovey, K.; Altman, I.; Werner, C. Home Environments. In *Human Behavior and Environment: Advances in Theory and Research*; Altman, I., Werner, C.M., Eds.; Plenum Press: New York, NY, USA, 1985.
20. Rapoport, A. Thinking about Home Environments. In *Home Environments*; Altman, I., Werner, C.M., Eds.; Springer International Publishing: New York, NY, USA, 1985; pp. 255–286.
21. Pennartz, P.J. Atmosphere at home: A qualitative approach. *J. Environ. Psychol.* **1986**, *6*, 135–153. [CrossRef]
22. Mallett, S. Understanding Home: A Critical Review of the Literature. *Sociol. Rev.* **2004**, *52*, 62–89. [CrossRef]
23. Douglas, M. The idea of a home: A kind of space. Social research. *Soc. Res.* **1991**, *58*, 287–307.
24. Kaplan, S. The restorative benefits of nature: Toward an integrative framework. *J. Environ. Psychol.* **1995**, *15*, 169–182. [CrossRef]
25. Ulrich, R.S.; Simons, R.F.; Losito, B.D.; Fiorito, E.; Miles, M.A.; Zelson, M. Stress recovery during exposure to natural and urban environments. *J. Environ. Psychol.* **1991**, *11*, 201–230. [CrossRef]
26. Lawlor, A. *A Home for the Soul: A Guide for Dwelling with Spirit and Imagination*; Clarkson Potter Publishers: New York, NY, USA, 1997.
27. Ergan, S.; Shi, Z.; Yu, X. Towards quantifying human experience in the built environment: A crowdsourcing based exper-iment to identify influential architectural design features. *J. Build. Eng.* **2018**, *20*, 51–59. [CrossRef]
28. Bluyssen, P.M. *The Indoor Environment Handbook: How to Make Buildings Healthy and Comfortable*; Taylor & Francis: London, UK, 2009.
29. Lawrence, R.J. Transition spaces and dwelling design. *J. Archit. Plan. Res.* **1984**, *1*, 261–271.
30. Amérigo, M.; Aragonés, J.I. A Theoretical and methodological approach to the study of residential satisfaction. *J. Environ. Psychol.* **1997**, *17*, 47–57. [CrossRef]
31. Moore, J. Placing home in context. *J. Environ. Psychol.* **2000**, *20*, 207–217. [CrossRef]
32. Shirazi, M. *Towards an Articulated Phenomenological Interpretation of Architecture: Phenomenal Phenomenology*; Routledge: New York, NY, USA, 2013.
33. Rybczynski, W. *Home: A Short History of an Idea*; Penguin Books: London, UK, 1987.
34. Alexander, H.H. *Analyzing the Interior Spaces in Your Home*; Agricultural Extension Service; The University of Minnesota: St. Paul, MI, USA, 1981.
35. Othman, Z.; Aird, R.; Buys, L. Privacy, modesty, hospitality, and the design of Muslim homes: A literature review. *Front. Arch. Res.* **2015**, *4*, 12–23. [CrossRef]
36. Altas, N.E.; Ozsoy, A. Spatial adaptability and flexibility as parameters of user satisfaction for quality housing. *Build. Environ.* **1998**, *33*, 315–323. [CrossRef]
37. Kalantari, S.; Shepley, M. Psychological and social impacts of high-rise buildings: A review of the post-occupancy evaluation literature. *Hous. Stud.* **2020**, 1–30. [CrossRef]
38. Ochodo, C.; Ndetei, D.M.; Moturi, W.N.; Otieno, J.O. External Built Residential Environment Characteristics that Affect Mental Health of Adults. *J. Hered.* **2014**, *91*, 908–927. [CrossRef]
39. Oswald, F.; Wahl, H.-W.; Schilling, O.; Nygren, C.; Fänge, A.; Sixsmith, A.; Sixsmith, J.; Széman, Z.; Tomsone, S.; Iwarsson, S. Relationships Between Housing and Healthy Aging in Very Old Age. *Gerontology* **2007**, *47*, 96–107. [CrossRef] [PubMed]
40. Amaturo, E.; Costagliola, S.; Ragone, G. Furnishing and status attributes: A sociological study of the living room. *Environ. Behav.* **1987**, *19*, 228–249. [CrossRef]
41. Walters, T. Facilitating well-being at the second home: The role of architectural design. *Leis. Stud.* **2016**, *36*, 493–504. [CrossRef]
42. Saruwono, M.; Zulkiflin, N.F.; Mohammad, N.M.N. Living in Living Rooms: Furniture Arrangement in Apartment-Type Family Housing. *Procedia Soc. Behav. Sci.* **2012**, *50*, 909–919. [CrossRef]
43. Mitton, M.; Nystuen, C. *Residential Interior Design: A Guide to Planning Spaces*; John Wiley & Sons: New York, NY, USA, 2016.
44. Madsen, L.V. The Comfortable Home and Energy Consumption. *Hous. Theory Soc.* **2017**, *35*, 329–352. [CrossRef]
45. Hendrassukma, D. The Influence of Room Colors in A House for Its Occupants. *Humaniora* **2016**, *7*, 37. [CrossRef]
46. Lee, S.; Alzoubi, H.H.; Kim, S. The Effect of Interior Design Elements and Lighting Layouts on Prospective Occupants' Perceptions of Amenity and Efficiency in Living Rooms. *Sustainability* **2017**, *9*, 1119. [CrossRef]
47. Banaei, M.; Yazdanfar, A.; Hatami, J.; Ahmadi, A. The Impacts of Sustainable Residential Interior Space on Inhabitant's Emotions. *Environ. Proc. J.* **2016**, *1*, 291–299. [CrossRef]

48. Petermans, A.; Pohlmeyer, A.E. Design for subjective wellbeing in interior architecture. In Proceedings of the Annual Architectural Research Symposium in Finland, Oulu, Finland, 29 November–2 December 2014; pp. 206–218. Available online: https://journal.fi/atut/article/view/45378 (accessed on 10 January 2021).
49. Kuo, F.E.; Bacaicoa, M.; Sullivan, W.C. Transforming inner-city landscapes: Trees, sense of safety, and preference. *Environ. Behav.* **1998**, *30*, 28–59. [CrossRef]
50. El-Zeiny, R.M.A. Biomimicry as a problem-solving methodology in interior architecture. *J. Soc. Behav. Sci.* **2012**, *50*, 502–512. [CrossRef]
51. Dreyer, B.C.; Coulombe, S.; Whitney, S.; Riemer, M.; Labbé, D. Beyond Exposure to Outdoor Nature: Exploration of the Benefits of a Green Building's Indoor Environment on Wellbeing. *Front. Psychol.* **2018**, *9*, 1583. [CrossRef] [PubMed]
52. Shibata, S.; Suzuki, N. Effects of an indoor plant on creative task performance and mood. *Scand. J. Psychol.* **2004**, *45*, 373–381. [CrossRef] [PubMed]
53. Pérez-Urrestarazu, L.; Kaltsidi, M.P.; Nektarios, P.A.; Markakis, G.; Loges, V.; Perini, K.; Fernández-Cañero, R. Particu-larities of having plants at home during the confinement due to the COVID-19 pandemic. *Urban For. Urban Green.* **2020**, in press. [CrossRef]
54. Li, X.; Zhang, C.; Li, W.; Kuzovkina, Y.A.; Weiner, D. Who lives in greener neighborhoods? The distribution of street greenery and its association with residents' socioeconomic conditions in Hartford, Connecticut, USA. *Urban For. Urban Green.* **2015**, *14*, 751–759. [CrossRef]
55. Zanjani, A.; Hilscher, M.C.; Cupchik, G.C. The Perception of Virtual Residential Spaces. *Empir. Stud. Arts* **2016**, *34*, 53–73. [CrossRef]
56. UK Green Building Council. Health and wellbeing in Homes. Available online: https://www.ukgbc.org/ukgbc-work/health-wellbeing-homes/ (accessed on 10 January 2021).
57. Mridha, M. Living in an apartment. *J. Environ. Psychol.* **2015**, *43*, 42–54. [CrossRef]
58. Ritterfeld, U.; Cupchik, G.C. Perceptions of interior spaces. *J. Environ. Psychol.* **1996**, *16*, 349–360. [CrossRef]
59. Couret, D.G.; Díaz, P.D.R.; De La Rosa, D.F.A. Influence of architectural design on indoor environment in apartment buildings in Havana. *Renew. Energy* **2013**, *50*, 800–811. [CrossRef]
60. Cho, S.H.; Lee, T.K.; Kim, J.T. Residents' Satisfaction of Indoor Environmental Quality in Their Old Apartment Homes. *Indoor Built Environ.* **2010**, *20*, 16–25.
61. Liu, S.; Cao, Q.; Zhao, X.; Lu, Z.; Deng, Z.; Dong, J.; Lin, X.; Qing, K.; Zhang, W.; Chen, Q. Improving indoor air quality and thermal comfort in residential kitchens with a new ventilation system. *Build. Environ.* **2020**, *180*, 107016. [CrossRef]
62. Adams, R.I.; Bateman, A.C.; Bik, H.M.; Meadow, J.F. Microbiota of the indoor environment: A meta-analysis. *Microbiome* **2015**, *3*, 1–18. [CrossRef] [PubMed]
63. Rautio, N.; Filatova, S.; Lehtiniemi, H.; Miettunen, J. Living environment and its relationship to depressive mood: A sys-tematic review. *Int.J. Soc. Psych.* **2018**, *64*, 92–103. [CrossRef]
64. Ori, K.; Bharti, A.; Kumar, S. Disposal of Kitchen Waste from High Rise Apartment. *J. Inst. Eng.* **2017**, *98*, 237–243. [CrossRef]
65. Marino, S.; Stasi, S. Urban-kitchen. Ergonomics and sustainability to the social complexity. Advances in Human Factors and Sustainable Infrastructure. In Proceedings of the International Conference on Applied Human Factors and Ergonomics and the Affiliated Conferences, Krakow, Poland, 19–23 July 2014.
66. Cromley, E.C. A History of American Beds and Bedrooms. *Perspect. Vernac. Arch.* **1991**, *4*, 177. [CrossRef]
67. Kennedy, R.; Buys, L.; Miller, E. Residents' experiences of privacy and comfort in multi-storey apartment dwellings in sub-tropical Brisbane. *Sustainability* **2015**, *7*, 7741–7761. [CrossRef]
68. Gosling, S.D.; Gifford, R.; Mccuan, L. Environmental perception and interior design. *Body Behav. Space* **2014**, *242*, 278–290.
69. Faizi, M.; Azari, A.K.; Maleki, S.N. Design Principles of Residential Spaces to Promote Children's Creativity. *J. Soc. Behav. Sci.* **2012**, *35*, 468–474. [CrossRef]
70. Edwards, C.P.; Springate, K.W. *Encouraging Creativity in Early Childhood Classrooms*; ERIC Digest: Urbana, IL, USA, 1995.
71. McCoy, J.M.; Evans, G.W. The Potential Role of the Physical Environment in Fostering Creativity. *Creat. Res. J.* **2002**, *14*, 409–426. [CrossRef]
72. Shafaei, M.; Madani, R. Design principles of educational facilities for children based on creativity. *J. Technol. Educ.* **2010**, *4*, 215–222.
73. Corradi, G.; Garcia-Garzon, E.; Barrada, J.R. The Development of a Public Bathroom Perception Scale. *Int. J. Environ. Res. Public Heal.* **2020**, *17*, 7817. [CrossRef]
74. Struckmeyer, L.; Morgan-Daniel, J.; Ahrentzen, S.; Ellison, C. Home Modification Assessments for Accessibility and Aesthetics: A Rapid Review. *Health Environ. Res. Des. J.* **2020**. [CrossRef] [PubMed]
75. Coburn, A.; Vartanian, O.; Kenett, Y.N.; Nadal, M.; Hartung, F.; Hayn-Leichsenring, G.; Navarrete, G.; González-Mora, J.L.; Chatterjee, A. Psychological and neural responses to architectural interiors. *Cortex* **2020**, *126*, 217–241. [CrossRef]
76. Larcombe, D.-L.; Van Etten, E.; Logan, A.; Prescott, S.L.; Horwitz, P. Etten High-Rise Apartments and Urban Mental Health—Historical and Contemporary Views. *Challenges* **2019**, *10*, 34. [CrossRef]
77. Krieger, J.; Higgins, D.L. Housing and Health: Time Again for Public Health Action. *Am. J. Public Heal.* **2002**, *92*, 758–768. [CrossRef]
78. Kim, G.; Kim, J.T. Healthy-daylighting design for the living environment in apartments in Korea. *Build. Environ.* **2010**, *45*, 287–294. [CrossRef]

79. Ribeiro, C.; Ramos, N.M.M.; Flores-Colen, I. A Review of Balcony Impacts on the Indoor Environmental Quality of Dwellings. *Sustainability* **2020**, *12*, 6453. [CrossRef]
80. Ozaki, R. House Design as a Representation of Values and Lifestyles: The Meaning of Use of Domestic Space. In *Housing, Space and Quality of Life*; García Mira, M., Uzzell, D., Real, J.E., Romey, J., Eds.; Routledge: London, UK, 2017; pp. 97–111.
81. Vartanian, O.; Navarrete, G.; Chatterjee, A.; Fich, L.B.; Gonzalez-Mora, J.L.; Leder, H.; Modroño, C.; Nadal, M.; Rostrup, N.; Skov, M. Architectural design and the brain: Effects of ceiling height and perceived enclosure on beauty judgments and approach-avoidance decisions. *J. Environ. Psychol.* **2015**, *41*, 10–18. [CrossRef]
82. Lawrence, R.J. The social classification of domestic space: A cross-cultural case study. *Anthropos* **1981**, *76*, 649–664.
83. Chiu, R.L.H. Socio-cultural sustainability of housing: A conceptual exploration. *Hous. Theory Soc.* **2004**, *21*, 65–76. [CrossRef]
84. Sand, J. *House and Home in Modern Japan: Architecture, Domestic Space, and Bourgeois Culture*; Harvard University Asia Center: Cambridge, MA, USA, 2005.
85. Vischer, J.C. Towards a user-centred theory of the built environment. *Build. Res. Inf.* **2008**, *36*, 231–240. [CrossRef]
86. Lawrence, D.L.; Low, S.M. The built environment and spatial form. *Annu. Rev.* **1990**, *19*, 453–505. [CrossRef]
87. Steg, L.; Van Den Berg, A.E.; De Groot, J.I.M. *Environmental Psychology: An Introduction*; Wiley-Blackwell: Chichester, UK, 2013.

Article

Sustainability at an Urban Level: A Case Study of a Neighborhood in Dubai, UAE

Sundus Shareef [1,*] and Haşim Altan [2,*]

[1] Faculty of Engineering & IT, British University in Dubai (BUID), Dubai 345015, United Arab Emirates
[2] Faculty of Design, Arkin University of Creative Art and Design (ARUCAD), Kyrenia 99300, Cyprus
* Correspondence: sundus.l.shareef@gmail.com (S.S.); hasimaltan@gmail.com (H.A.); Tel.: +971-50-638-9133 (S.S.); +90-533-835-9090 (H.A.)

Abstract: The United Arab Emirates is witnessing enormous growth and the sustainability attitude has become one of the most important priorities in this development. This paper aims to optimize the environmental sustainability of the Emirate of Dubai communities by adopting an existing community as a case study. The investigation of the case study is looking at sustainability levels that consists of two major factors in neighborhood sustainable design, such as livability and thermal performance. The strategy of enhancing and optimizing the communities' sustainability starts with an approach to the applicable modifications and solutions to the existed community master planning, where the modifications cover the two main urban design variables; (a) building design, and (b) open and landscape areas. The effect of the adopted scenarios is analyzed to find the improvement in environmental and thermal performance. The study has adopted two computer software packages, namely CityCAD and Integrated Environmental Solutions—Virtual Environment (IES-VE), to undertake the assessments. Furthermore, factors of urban sustainability are evaluated using the United States Green Building Council (USGBC)'s Leadership in Energy and Environmental Design (LEED) neighborhood assessment tool. The results have shown that the environmental sustainability levels can be increased after the adoption of certain suggested scenarios, in order to mitigate the likely weakness indicated in the livability aspects, covering land-use diversity, accessibility, transportation system, green and landscape areas, and energy efficiency, and the case study community can be turned toward "Sustainable Community" by implementing recommended actions and modifications.

Keywords: sustainability and livability of neighborhoods; sustainable urban environments; sustainable solar shading; building height diversity; United Arab Emirates

Citation: Shareef, S.; Altan, H. Sustainability at an Urban Level: A Case Study of a Neighborhood in Dubai, UAE. *Sustainability* **2021**, *13*, 4355. https://doi.org/10.3390/su13084355

Academic Editor: Abdollah Shafieezadeh

Received: 24 February 2021
Accepted: 8 April 2021
Published: 14 April 2021

1. Introduction

Cities are numbers of communities and neighborhoods where people can work live and have entertainment. Day by day cities offer tremendous opportunities for community, employment, education, excitement and interest. For all of these reasons, cities became attractive areas for living and more than half of the world's population are living in cities [1]. On the other hand, cities create problems of congestion, noise, and pollution, but most people do not have the choice, recognizing the trade-offs. How to live and getting the right balance are parts of the solution. Living in towns and low-density cities has some advantages, however people may like living in a compact and dense city as far as there is an equilibrium among the development elements; built area and open spaces, private and public transportation, using natural and artificial resources [2]. City, community or neighborhoods could be considered as a system of depending components [3]. The major variables or components that affect the design of any development are; urban form, transport, landscape, building design, waste management, energy and water supply. The most sustainable design is about equilibrium among these components [4]. In order to make cities or neighborhoods more suitable for people, all aspects of viable city and neighborhoods are required to involve and operate smoothly within design or system equation.

It is obvious that communities and city growth becomes a key issue in the global problems of climate change, global warming, greenhouse gas emissions, and depleting natural resources. Therefore, many studies and publications have explored the relationship between the sustainability levels required to be achieved and the urban planning of any development. Many of these studies concentrated on the main urban design factors, such as urban form, building design, liveability, land use, and transportation system to analyse, evaluate and develop the sustainability level of cities and developments [5,6]. However, some of these researchers studied sustainability on an urban scale from the aspect of resource conservation and pollution reduction. Cities and urban environment pollution are caused by different factors; density and transport within cities, human activities, construction and buildings' effects on nature and landscape areas, atmospheric pollution by CO2 emissions, and noise pollution [7]. Pollution influences human health and well-being and can make cities uneasy places to live. Greenhouse gas (GHG) averages constitute one of the most used air pollution indicators. Greenhouse gas emissions refer to all gases that trap heat in the atmosphere; the main greenhouse gases in the atmosphere are; carbon dioxide (CO2), methane (CH4), and nitrous oxide (N2O). These gases are the main reason for global warming, depleting the ozone layer, and climate change [8]. GHG are emitted through various fossil fuel burning processes. Fossil fuels (coal, natural gas, and oil) are burnt for heating, solid waste burning, trees, and wood products; the decay of organic waste in municipal solid waste landfills, chemical reactions, and manufacturing operations are some resources of GHG [9].

Furthermore, the building and construction industry, transportation, agriculture, and industry are the major recourses of these gases. Global warming, urban heat island (UHI), and the increase in global air temperature are a result of GHG emission. Buildings design, transportation systems, and open areas significantly affect the sustainability of the urban level. Buildings contribute to the CO2 emissions by 43%, while the transportation share is 32% [10]. Therefore, cities should be designed in a way that minimizes the GHG averages and pollution percentages. The new cities should be designed to keep their inhabitants healthy, secure, and happy. For this aim, a neighborhood must become greener and robust, with a stable ecosystem. Our built environment at the present time suffers enough and an integrated approach is urgently needed. Successful solutions depend on understanding the relationship among the involved sustainability elements; environmental, historical, social, and economic. The solutions should start from the individual building to the block, neighborhood, district, city, region, and up towards the globe. Furthermore, adopting active urban design strategies, such as using renewable energy (PV) solar panels at the urban level, will enhance the sustainability on an urban and city scale [11].

The terms "Neighborhood" or "Community" refer to a number of residential units and the related facilities that serve the resident's needs [12]. Livable, sustainable neighborhoods are one of the determining and essential factors for developing sustainable environments [13]. The sustainable neighborhood is a neighborhood that integrates the three sustainability pillars "Environment, Economy and Society". From the social aspect, providing the required open areas, landscaped areas, playgrounds, and community facilities will encourage sociality and people communications [14]. From the economic aspect, those sustainable neighborhoods that provide all the services and facilities will create livable, healthy independent communities that will have a positive effect on the individuals and the whole society [15]. However, "The Sustainable Urban Design" provides a high level of sustainability and efficiency in terms of major urban design dimensions, such as livability, land use, transportation, buildings design, landscaped areas, and environmental performance [16,17]. However, "The Sustainable Urban Design" provides a high level of sustainability and efficiency in terms of major urban design dimensions, such as liveability, land use, transportations, buildings design, landscaped areas, and environmental performance.

Urban sustainability is significant to the future of humans; it directly affects people's lifestyle, time, effort, health, wellbeing and welfare [18]. Transportation, resources conservation, indoor, and outdoor thermal comfort represent some of the sustainable urban design factors that have a direct effect on the livable community. The major challenge for the urban designer is to improve and optimize the relation among the three factors in urban geometry; density, movement and recourses. Sustainability at the urban level could be achieved through optimizing the three aspects of, and finding the best design for, the neighborhood, district and city [19]. The urban areas and communities include buildings, open and green spaces, water features and road networks. These are the urban design elements that should be organized in a way that provides vitality and improves the people's lifestyle [20].

Creating a liveable environment is one of the sustainable urban design principles, and the level of urban liveability could be considered important in achieving sustainability in the urban environment. Urban sustainability can be obtained by creating a liveable community, neighbourhood, and city [20]. Urban liveability covers a number of factors; it is a multi-dimensional construct that includes accessibility, number of public parks and open spaces, walkability, transportation planning, urban density, and land use diversity, all of which are design elements that could be improved to achieve high levels of liveability and sustainability [21]. However, it is difficult to define and measure the concept of urban liveability, and set some principles for liveability measurements, such as safety, equity, and continuity [12]. Moreover, when it comes to accessibility and inclusiveness of the previously mentioned indicators [22], accessibility, land use diversity, providing parks and green areas are of the strategies used when planning a sustainable neighborhood. Passive design also has an effective role in achieving a green and sustainable community by offering recourses efficiency [23,24]. The crucial roles of the green space on ecosystem have already been proven by some researchers [25]. The leakage in accessibility to these areas and other community services affects the community sustainability level; ensuring a good accessibility will improve the community sustainability, and this consequently improves the community social life [25]. The service within the green areas, including sport services, has an impressive impact on peoples' wellbeing, and is resulting with good social relations among the residences [25]. It has been proven that the design and the architecture of the buildings should collaborate with surrounding nature to create a harmony between the outdoor and indoor spaces. The concept of human community should be designed to positively influence the human behavior, health and culture [26]. Other than that, land use diversity is another factor that forms a sustainable community. Land use diversity and ensuring a good accessibility to the daily required services would improve peoples' lifestyle from one side, and have a positive effect on resource saving from the other side [27]. The reduction in the use of transportation and vehicle's journey will consequently have a positive impact environmentally, by reducing CO_2 emissions [28,29].

Passive design is one of the strategies that the urban planner can adopt for designing a sustainable community according to its direct effect on outdoor and indoor thermal performance [30]. The urban air temperature is rising in all cities around the world, as a result of global warming and the decrease in the natural and greenery area in cities. This rise in outdoor air temperature consequently affects the thermal performance of the inner space environment and increases the indoor air temperature averages [30]. The impact of buildings and urban geometry on the urban heat island phenomena and the outdoor thermal performance has been proven in many studies [31]. Increasing and enhancing the sustainability of our developments is an urgent matter when it comes to facing global warning, resources limitation, and pollution. Implementing the passive and active design elements on buildings and at the urban level represents a part of the solution [32].

Building design, orientation, and block density are of significant effects in development sustainability [33,34]. Creating a desired shading on urban level will have a positive thermal impact on both outdoor and indoor environments. In the hot climate conditions of the UAE, the reduction in air temperature and solar gain due to the orientation can reach 1.8 °C and 13% respectively. [34]. One of the rule of thumb in urban design is the belief that energy consumption decreases when the community or city density increases. This is a challenge to the urban designer to find the best balance between the two variables in urban planning; density and energy [35]. Furthermore, optimizing the indoor and thermal performance on an urban level will have a positive impact on livability, productivity, and indoor energy consumption [36].

The strong urban structure provides less use or need for transportation and reduces the path, the cycling transportation in the most preferred plan, and different types of transportation plays a significant role in changing the traditional urban structure. The vehicle flow, parking areas, street width, public transportation stations, and many others related to the transit system are the elements that should be well designed to obtain a strong structure [37,38]. Consequently, road planning affects the other urban factors such as gardens and open areas, and playgrounds, which should be counted on during early design stages. The sustainable land use planning is the significant factor in reducing the daily transporting cycle, and increasing walkability as one of the sustainable neighborhood requirements [39]. Furthermore, greenery and landscaped areas could be effective influences on increasing walkability from one side, reducing air temperatures and enhancing outdoor and indoor thermal performances from the other side [40].

The impact of communities and developments has been illustrated previously. The former studies proved the significant impact of sustainable urban design and sustainable developments on reducing the negative environmental effect caused by continuous urbanization. This study aims to contribute to this concept by investigating the potential of improving the performance of one of the Dubai community's performances towards sustainable performance. Hence, a community located in the city of Dubai in the United Arab Emirates (UAE) will be explored, evaluated, and optimized to achieve a sustainable community that follows the sustainable design standards. Dubai is located in the north of the UAE, and extended along the Arabian Gulf with a climate that is known with its humidity during summers, due to its location of the city on Dubai Creek. Generally, the weather in Dubai is sunny most days of the year; in winter, the average temperature is 25 °C, while in summer, the temperature may reach up to 38 °C, with a high percentage of humidity between 20–60%, and a low average of rainy days. The annual air temperature varied between 17 °C in winter and 35 °C by a Dubai weather file generated through the Integrated Environmental Solutions—Virtual Environment (IES-VE) software [41].

2. Methods

A case study method has been used to achieve the research aim and objectives. The study focused on exploring the urban sustainability in a selected residential community in Dubai through analysis and evaluation using two separate software packages; (i) IES-VE [41], and (ii) CityCAD [42]. The study will adopt the following steps:

- Analyzing livability of the community through presenting the quantity of land use, services and accessibility.
- Calculating the number of units on the long axis within 15 degrees of the east-west axis.
- Presenting virtual images, plans and reports for the existing case study and the modified scenarios that are suggested in order to optimize the community sustainability.
- Simulating sun path and solar shading analysis using SunCast application.
- Investigating the effect of the suggested modifications on solar gains of the community units in percentage and hours.
- Investigating the effect of the suggested modifications on air temperatures within the community units using ApacheSim application.

In addition to the CityCAD and IES-VE software packages, the community sustainability has been evaluated with the use of the United States Green Building Council (USGBC)'s Leadership in Energy and Environmental Design (LEED) rating system.

In this study, LEED for Neighborhood and Developments (ND), version 4 (2014), has been used [43]. The strategy for the modifications adopted passive urban design solutions to the community master plan, which was also applicable to the existing community. Five of the urban design parameters have been modified according to three scenarios in order to enhance the community livability and thermal performance. The modifications/scenarios covered; land use, accessibility, and walkability, building design, open and green areas. Moreover, the effect of the modifications has been analyzed to find out the improvements on environmental and thermal performance, as well as through enhancing solar gain performance.

The Existing Community as a Case Study

The case study of this research is represented by Al Waha community, which was developed by Dubai Properties Group (DPG). The existing residential community located in "Dubai land" adjacent to the Emirates Road with easy access to Sheikh Mohammed bin Zayed Road through Al Qudra Road. The community is close to the Arabian Ranches, Sport and Motor City communities as key developments (Figure 1). The case study "Al Waha" community consists of 206 semidetached villas where the villas are designed in three types according to bedroom numbers; two, three and four bedrooms (Figure 2). The facilities are very limited in the community, covering swimming pool, playground area, landscape and hardscape.

Figure 1. The case study location. Al Waha, Dubai [44].

The total area of the community is approximately 130,000 sqm, while the landscape covers 15,200 sqm from the community total area. The neighbor community is the Layan community, from the same developer (DPG), and contains seven G+2 residential buildings and 588 villas, with small facilities such as small shops and a supermarket.

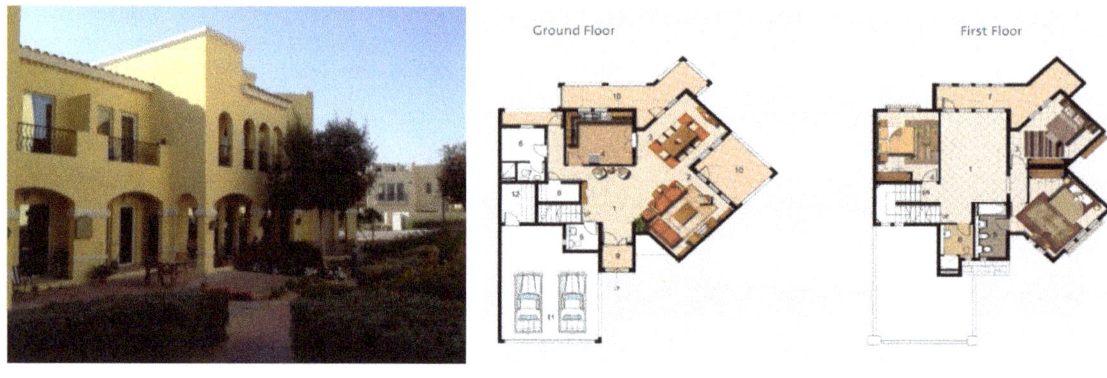

(a) (b)

Figure 2. Al Waha community villa view (**a**) and layout (**b**).

3. Background for the Analysis

3.1. Assessing Livability in the Existing Community

The observation and the assessment during the community site visit, and the use of the CityCAD software for livability analysis showed that there are weaknesses in many livability aspects. The major weakness is in land use diversity, as the existing land use variety is very limited. The community consists of three types of semidetached villas, playground area and a communal swimming pool, with hard and soft landscape. There is a clear absence of many services required, such as supermarket, laundry, pharmacy, school, healthcare center and amenity facilities. The livability analysis of the existing case study using CityCAD shows the average distance from community dwellings to some services and facilities (Table 1).

Table 1. Assessing this existing case study services and the adopted scenarios with additional services, namely new services.

	Average Distance (m)		
Services	**(Existing Case Study)**	**Scenario One**	**Scenario Two**
Green Spaces	30	30	30
Parking Spaces	10	10	10
Playground	50	50	50
Public Space and Swimming Pool	55	55	55
Shops	1000	200	300
Super market	1000	200	300
Hot Food and Takeaway	1000	200	300
Pharmacy	8000	200	300
Educational Services	8000	200	300
Metro Station (Emirates)	15,000	15,000	15,000
Shopping Mall (Emirates	15,000	15,000	15,000
Hospitals	12,000	12,000	12,000
	Average Distance (m)		
New Services	(Existing Case Study)	Scenario One	Scenario Two
Assembly and Leisure	8000	200	300
Laundry	8000	200	300
Restaurant and Cafe	8000	200	300
Financial Services	8000	200	300

The community, as a gated community, provides a good level of safety as one of the livability requirements [39]. On the other hand, the only one access through the Emirates Road indicated some weakness in accessibility, which could be enhanced and optimized by providing more than one access to improve transition and movement.

3.2. Assessing Thermal and Environmental Performance in the Existing Community

Analysis he community layout using integrated environmental solution—virtual environment (IES-VE) software, and adopting a sun path application and unit orientation showed that only 40% of the units are extended along the East-West axis. The benefit of the orientation along the East-West axis is to obtain a minimum amount of solar exposure, as the long facade is facing the North–South axis [43]. The IES-VE software was used to analyze the community shading performance and solar gains through the SunCast application. Figure 3 shows the community layout orientation and the sun path on a summer day, 1 June.

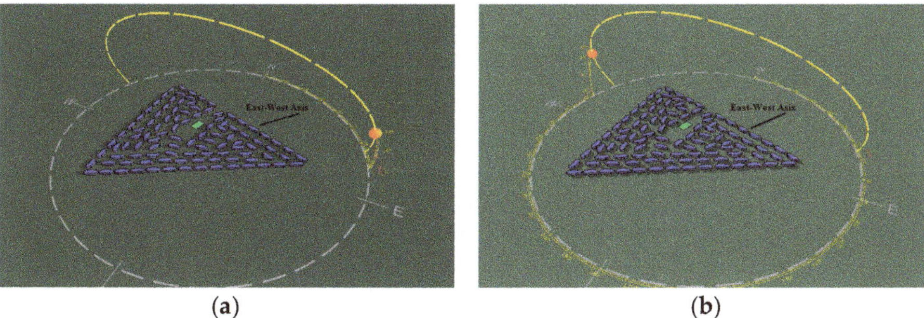

Figure 3. The Sun Path analysis during morning and evening along the East-West axis. (**a**) 1 June, 6:00 a.m.; (**b**) 1 June, 6:00 p.m.

The community plot is a triangle shape and one edge of the community plot is extended along the East-West axis, but only 40% of the units extend along the same direction. Thus, the community urban plan would be more sustainable if the units were arranged parallel to the side along the East-West axis in an early planning stage. Yet, it was observed that the compacted form provides more shading and less exposure to solar radiation for the inner units compared to the outer units (Figure 4).

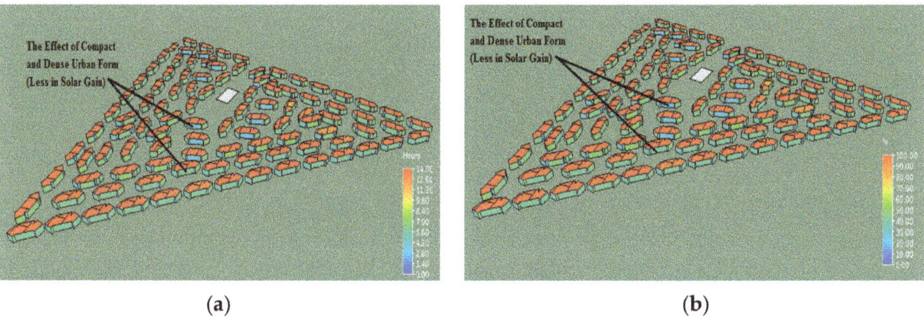

Figure 4. Solar attitude in the compact form community during 1 June. (**a**) Solar Gain in Hours; (**b**) Solar Gain in Percentage.

Furthermore, the thermal and environmental performance of the community could be improved through increasing green areas and planting empty/uncultivated areas, which are about 35% of the community landscape area (Figure 5).

Figure 5. Uncultivated areas in the community.

As part of the analysis, exploring the potential of enhancing the community performance towards sustainability, three scenarios were adopted to improve the Al Waha community sustainability. The thinking or the criteria behind these scenarios was to suggest an applicable practice to enhance the community sustainability. Community urban sustainability is improved from two aspects; livability and environmental or thermal performance. Enhancing livability covers a number of parameters; (1) land use, (2) accessibility, (3) walkability, and (4) open and green areas. While the thermal performance parameters represented by improving solar shading and reducing total solar gains through adopting (5) height diversity. The modifications consist of three adopted scenarios to enhance the community sustainability, which are simulated and analyzed by using CityCAD and IES-VE. In addition, using LEED (ND) v4 checklist as an overall and integrated urban sustainability evaluation and assessment tool was to find the sustainability level of the existing and modified case study.

4. Results

4.1. The Results of the Suggested Scenarios for Enhancing the Community Performance

4.1.1. Scenario One

The community livability could be improved by providing some daily required services such as shops, supermarket, pharmacy, restaurant and cafe, financial services, assembly and leisure, hot food and takeaway. This could be obtained by converting a number of the residential units in the community to provide the missing services. In addition, to add two stories for these units to increase building design diversity (one of LEED's requirement for sustainability) and height diversity as well.

Land use has been improved and a number of facilities were increased by converting some units into services for the daily important and missing facilities, such as adding supermarket, laundry and a pharmacy to be within 300 m–500 m for more than 50% of the community units to fulfill the LEED's land use diversity requirement.

4.1.2. Scenario Two

The second opportunity is enhancing the community services as well as the accessibility, by opening new access to the neighbor community, Layan community, as both of these communities are developed by the same developer (DPG). The new access will allow the residents to benefit from some services that are already existed in Layan community, such as supermarket, bookshop, and small cafe. Furthermore, opening new access to Al Quadra Road will enhance the accessibility and the movement entirely (Figure 6).

Opening new access to Layan community (Scenario Two) would improve the community livability, even though some services are still indicating a weak performance such as educational and medical services. This could be resolved by providing these services (primary school or medical center) in Layan community, as it is larger in area and has a number (7) of mid-rise buildings, which could be useful for this type of services (Table 1).

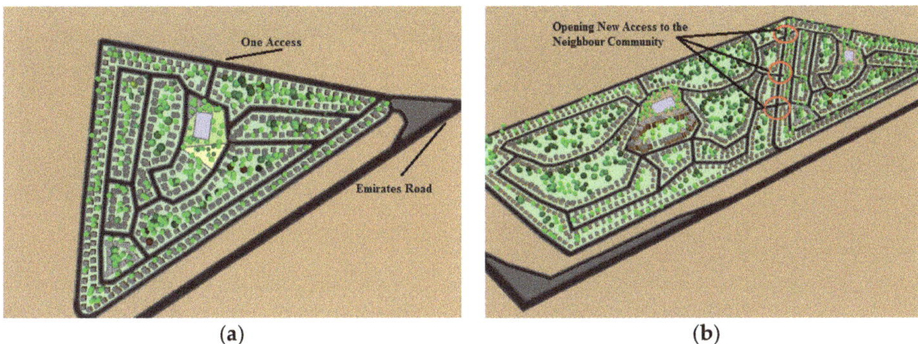

(**a**) (**b**)

Figure 6. (**a**) The existing gated community—one access; (**b**) New access to the neighbor community enhancing access.

Both scenarios 1 and 2 indicated improvement toward a more sustainable setting, as most of the mentioned services are within the LEED ND requirements (i.e., 200–300 m) (Table 1), even though the other services are still at a distance of 12–15 km from the two communities and therefore not able to fulfill the LEED ND sustainability credit requirements.

4.1.3. Scenario Three

Enhancing livability could also be achieved by increasing the open spaces and green areas; the open spaces are limited in the community but could be increased when adopting the new access to the Layan community. On the other hand, the green area could be increased by planting the uncultivated areas, which are calculated using site surveys and Google Earth (Pro) [44], while represented in CityCAD by 35% of the landscape area (Figure 7).

(**a**)

(**b**)

Figure 7. (**a**) The existing community walkway beside a road, (**b**) The pedestrian walkway that could be enhanced by adding rubber track, shading devices, and sport equipment.

In addition to increasing the number of trees and adding a green belt alongside the community boundary wall, adding some sports and kids playing equipment to provide the residents and the kids with a place for relaxation and amenity could allow for social communication, while also improving social sustainability (Figure 8).

Figure 8. Options of shading devices for walkways [45].

Furthermore, walkability could be increased by providing a rubber pathway, shaded walkway [45,46], and a number of benches to encourage people, especially elderly people, to walk and use the community green areas, as encouraging elderly people to walk is one of the social sustainability targets [47,48] (Figure 9).

Figure 9. Adding benches and enhancing empty areas by using hard and soft landscaping [45,46], and improving walkability for pedestrians by adding rubber walkways (1.1 m width) and outdoor equipment according to the LEED ND requirements

Moreover, adding a green belt around the community could provide more protection and shade areas further to increasing the total number of trees, which has an important role in enhancing the community environmental performance (Figure 10).

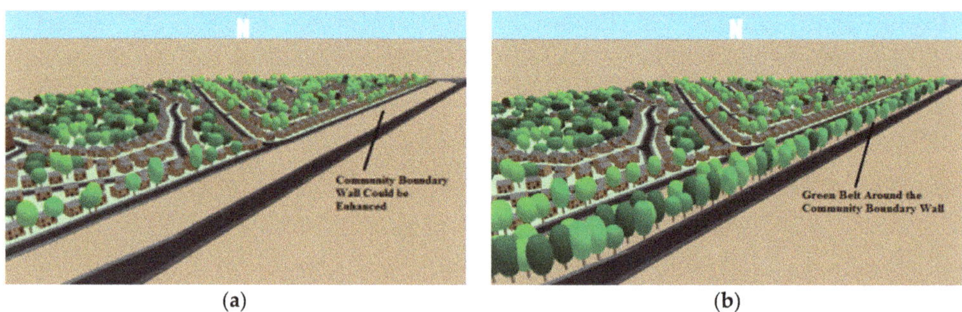

Figure 10. Planting the community boundary wall [42]. (**a**) Current Setting; (**b**) After Improvements.

4.2. The Effects of the Adopted Scenarios on Solar Shading and Solar Gains

Running IES-VE simulations with different heights of community units showed the importance of the height diversity in creating preferable shaded areas for walking people, in addition to the effect of reducing solar gains from the surrounding units. Figure 11 shows that the solar gains of the surrounding units decreased from 100% to 80% (80 h) when applying height diversity.

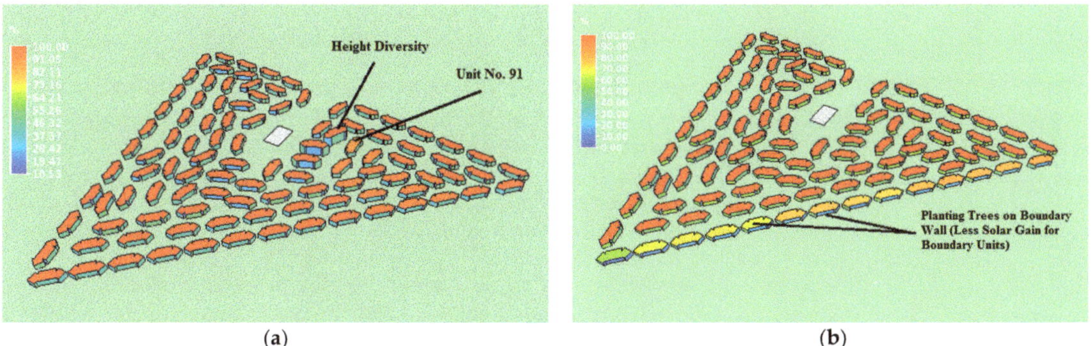

Figure 11. (**a**) Solar gains and the effect of diversity in building heights on shading parameters; (**b**) Solar gains and the effect of planting the boundary wall.

The effect of diversity in building heights for creating comfortable outdoor environments was proven by Edward (2010). The researcher explored the benefits of the diversity, dense, and compact form on the outdoor environment, presenting the "Environmental Diversity Map" to show the effect of diversity on the three microclimate parameters; temperature, shading and wind on the outdoor environment [23].

A model using IES-VE SunCast analysis to study the effect of planting height and dense trees along the boundary wall were simulated, as shown in Figure 11b. It is clear that the boundary units adjacent to the boundary wall are varied in solar gains, and the exposure percentage between 50–70% depending on the height and the dense of the trees, with 100% solar gains for the other units.

The effect of the first scenario and the modification in building height diversity analyzed and explored using one of the community units, Unit No. 91. Unit 91 was selected for this analysis due to being oriented toward west direction and has maximum solar gains with 100% in the existing case, and is close to the chosen building, to be converted to serve for the missing services, which is modified and increased in height by adding two more stories (to Unit No. 90), and the adjacent Unit 91 could be therefore less in solar exposure by 20% as it has only 80 h of exposure to the sun (Figure 12).

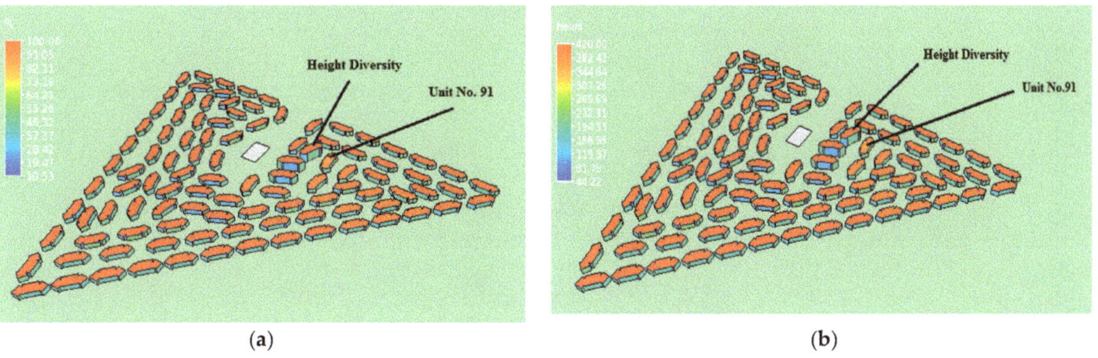

Figure 12. Building height diversity and shaded roofs adjacent to the modified building, Unit 91. Less in solar exposure by 20% and 80 h during the month of June. (**a**) SunCast solar shading analysis in percentage; (**b**) SunCast solar shading analysis in hours.

Moreover, using ApacheSim application within the IES-VE software showed that there is a reduction in solar gains by 20% for the modified case compared to the existing case

(Unit 91), and the solar exposure and solar gain hours are 80 h less in total during the month of June, with a reduction from 420 h to 340 h. Furthermore, the reduction in solar gains for Unit 91 showed a comparison between the existing case study (Unit 91) and the new shaded same unit with a reduction of 18.5% in total solar gains on 1 June.

4.3. The Community Assessment Using LEED (ND) Rating Tool

LEED Neighborhood as a sustainability assessment tool was used to evaluate the community sustainability or greenness, where the tool rating system consists of five categories, and each category covers a number of requirements. The requirements divided into mandatory requirements and optional requirements; for optional requirements, LEED allocates a number of points or credits for each category, as shown in Table 2.

Table 2. LEED (ND) allocated points.

Requirements	Points
Smart Location and Linkage	28 Points
Neighborhood Pattern and Design	41 Points
Green Infrastructure and Buildings	31 Points
Innovation and Design Process	6 Points
Regional Priority Credits	4 Points

The total number of points collected indicates the level of each community sustainability according to the following scale: Certified 40–49, Silver 50–59, Gold 60–69, Platinum 80+.

Using the LEED Neighborhood and Developments checklist to assess Al Waha community through each of the five categories resulted in the following:

- Smart Location and Linkage: The community fulfill all the five required items and obtain only 11 out of 25 points allocated to this category, as there is a clear weakness in community linkage and accessibility.
- Neighborhood Pattern and Design: With regards to the community design, the three required items related to the pattern are available in community design, and the community collected 16 out of 41 credit points. The demerits of the community design indicated in land use diversity, building design and affordability, and open and assembly areas.
- Green Infrastructure and Buildings: This category indicates a weakness in following the sustainable design requirements related to water and energy efficiency, solar orientation, the use of renewable energy, and the requirement of green building certification; only 4 points obtained out of 31 total points allocated.
- Innovation and Design Process: This category provides points to the new sustainable innovation not addressed in LEED, and none of the six innovation points were able to be collected.
- Regional Priority Credits: This category related to the regional practices and material, and only one regional point out of four was collected.

In total, the community collected only 32 points. This result indicates a low level of sustainability, and the community could not be certified as a green or sustainable community according to the LEED (ND) assessment tool. Generally, the weaknesses are indicated in land use, facilities, accessibility, transportation system, and water and energy efficiency.

The modified case study according to the three scenarios was assessed using the LEED assessment tool, and the results for each category are as follows (Figure 13):

- Smart Location and Linkage: In addition to the 11 points that were collected from the existing case study assessment, providing new access to the community to the main road and new access to the Layan community, adding four additional points, to be 15 points in total for this category.

- Neighborhood Pattern and Design: This category has been improved to collect 24 points as the modified community offer a required diversity in land use and building affordability, further to enhancing walkability, green and open areas.
- Green Infrastructure and Buildings: This category indicates a weakness in following the sustainable design requirement, only four points obtained out of 31 total points allocated.
- Innovation and Design Process: This category provides points to the new sustainable innovation not addressed in LEED, and none of the six innovation points were able to be collected.
- Regional Priority Credits: This category was related to the regional practices and material, and only one regional point out of four was collected.

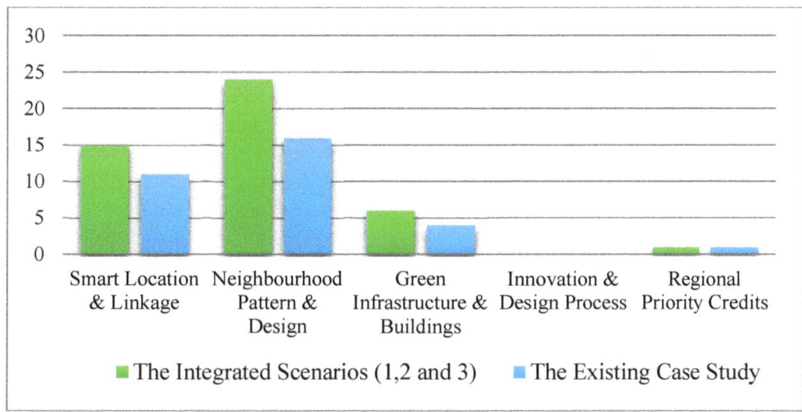

Figure 13. LEED assessment, comparison between existing and modified case study.

Moreover, it is worth mentioning that the aspect of building density has helped to improve and optimize relations among the factors in urban geometry [4,6], and therefore emphasizing on this aspect as a key solution in this analyzed context, which should also be taken into account in different climatic zones.

In total, the community collected 46 points and could be certified as a green community (Appendix A). This result shows the effect of the applicable practices and modifications to enhance the community sustainability.

5. Discussion

This study aimed to assess and optimize sustainability on the urban scale by selecting and evaluating one of Dubai's communities. The study investigated the potential of enhancing the case study neighborhood towards sustainability according to the sustainability rating system standards. The selected case study community was evaluated, and the likely weakness was indicated in the livability aspects covering land-use diversity, accessibility, transportation system, green and landscape area, and energy efficiency. Moreover, assessing the community showed a capability of enhancing the thermal performance, in addition to the community livability.

Different scenarios were applied in order to improve the community's sustainability on two levels; livability and thermal performance (Figure 13). The adopted scenarios targeted improving the livability by enhancing (1) walkability, (2) accessibility, (3) facilities, and (4) land use diversity [5,13,17]. The thermal performance has been enhanced by increasing the (1) shading effect (2) and improving greenery and landscape areas [34,39]. However, applying the suggested scenarios resulted in improving the community sustainability and resulted in upgrading the community rating level by increasing the number of points that can be achieved by applying LEED (ND) standards [43]. The community gained an additional 14 points, and resulted to be certified as a sustainable community. The study proved the capability of enhancing the sustainability level of the communities by implementing sustainable design strategies at the urban level. Improving the community performance will result in a positive impact on the total environmental performance [34,39,43].

6. Conclusions

Recently, and as part of the future vision, there is a strong trend toward passive design as an effective part in sustainable design. This is a result of real consciousness in limited resources, global warning and pollution problems, where sustainability is the only solution for sustaining our very future in the world. In line with this context, this study aimed to explore and improve a neighborhood in Dubai, UAE toward a more livable, sustainable community.

The Al Waha community in Dubai was selected for the analysis, and three scenarios were adopted for developing and obtaining more sustainable community. Analysis of the community covered two of the sustainable urban design dimensions (1) livability and (2) thermal and environmental performance. Livability analysis of the existing case study showed some clear weaknesses in land use diversity, accessibility, walkability, landscaped area and building design diversity. These weaknesses were covered in the adopted scenarios, through analyses using the two software packages, CityCAD and IES-VE, and the LEED (ND) checklist, showing clear improvements in all mentioned parameters. Community improvement strategy and adopted scenarios covered a number of urban design parameters including; (1) land use diversity, (2) accessibility, (3) walkability, (4) open public area and green spaces, and (5) building height and design variety, which directly affected environmental or thermal performance parameters covering solar gains and air temperatures. The effects of the adopted scenarios (with modified computer models) on solar gains and thermal performances have been studied and analyzed using the IES-VE applications SunCast and ApacheSim—Vista Pro. The results showed that the livability level of the community was increased by enhancing the land use diversity, accessibility, walkability, building height diversity and the green areas. The aspect of building density has played a key role in the analysed context to help the community livability. Furthermore, the effect of the adopted scenarios to enhance the community livability showed a clear and positive effect on environmental and thermal performance by increasing the shading effect and reducing indoor solar gains and air temperatures. Finally, the modified community that integrated in the three adopted scenarios have been evaluated using the LEED Neighborhood and Developments (ND) assessment tool v4, and the community was able to be certified as a "Sustainable Green Community" through implementing all of the applicable practices.

Author Contributions: S.S. and H.A. conceived and designed the concept and the paper outline; S.S. conducted the analyses and wrote the paper; and H.A. supervised, provided direction, sources, comments, and major edits to the paper. All authors have read and agreed to the published version of the manuscript.

Funding: This research received no external funding.

Institutional Review Board Statement: Not applicable.

Informed Consent Statement: Not applicable.

Data Availability Statement: Not applicable.

Conflicts of Interest: The authors declare no conflict of interest.

Appendix A. LEED Neighborhood & Development, (ND) V4 Checklist

LEED v4 for Neighbourhood Development Plan

Project Name: "Al Waha Community", Dubai, UAE
The Existing Case Study

Yes	?	No			
11	**0**	**0**	**Smart Location & Linkage**		**28**
Y			Prereq	Smart Location	Required
Y			Prereq	Imperiled Species and Ecological Communities	Required
Y			Prereq	Wetland and Water Body Conservation	Required
Y			Prereq	Agricultural Land Conservation	Required
Y			Prereq	Floodplain Avoidance	Required
5			Credit	Preferred Locations	10
			Credit	Brownfield Remediation	2
3			Credit	Access to Quality Transit	7
1			Credit	Bicycle Facilities	2
			Credit	Housing and Jobs Proximity	3
1			Credit	Steep Slope Protection	1
			Credit	Site Design for Habitat or Wetland and Water Body Conservation	1
			Credit	Restoration of Habitat or Wetlands and Water Bodies	1
1			Credit	Long-Term Conservation Management of Habitat or Wetlands and Water Bodies	1
16	**0**	**0**	**Neighbourhood Pattern & Design**		**41**
Y			Prereq	Walkable Streets	Required
Y			Prereq	Compact Development	Required
Y			Prereq	Connected and Open Community	Required
4			Credit	Walkable Streets	9
4			Credit	Compact Development	6
			Credit	Mixed-Use Neighbourhoods	4
3			Credit	Housing Types and Affordability	7
1			Credit	Reduced Parking Footprint	1
1			Credit	Connected and Open Community	2
			Credit	Transit Facilities	1
			Credit	Transportation Demand Management	2
1			Credit	Access to Civic & Public Space	1
			Credit	Access to Recreation Facilities	1
1			Credit	Visitability and Universal Design	1
			Credit	Community Outreach and Involvement	2
			Credit	Local Food Production	1
1			Credit	Tree-Lined and Shaded Streetscapes	2
			Credit	Neighbourhood Schools	1
4	**0**	**0**	**Green Infrastructure & Buildings**		**31**
Y			Prereq	Certified Green Building	Required
Y			Prereq	Minimum Building Energy Performance	Required
Y			Prereq	Indoor Water Use Reduction	Required
Y			Prereq	Construction Activity Pollution Prevention	Required
			Credit	Certified Green Buildings	5
			Credit	Optimize Building Energy Performance	2
1			Credit	Indoor Water Use Reduction	1
1			Credit	Outdoor Water Use Reduction	2
			Credit	Building Reuse	1
			Credit	Historic Resource Preservation and Adaptive Reuse	2
1			Credit	Minimized Site Disturbance	1
			Credit	Rainwater Management	4
			Credit	Heat Island Reduction	1
			Credit	Solar Orientation	1
			Credit	Renewable Energy Production	3
			Credit	District Heating and Cooling	2
			Credit	Infrastructure Energy Efficiency	1
1			Credit	Wastewater Management	2
			Credit	Recycled and Reused Infrastructure	1
			Credit	Solid Waste Management	1
			Credit	Light Pollution Reduction	1
0	**0**	**0**	**Innovation & Design Process**		**6**
			Credit	Innovation	5
			Credit	LEED® Accredited Professional	1
1	**0**	**0**	**Regional Priority Credits**		**4**
1			Credit	Regional Priority Credit: Region Defined	1
			Credit	Regional Priority Credit: Region Defined	1
			Credit	Regional Priority Credit: Region Defined	1
			Credit	Regional Priority Credit: Region Defined	1
32	**0**	**0**	**PROJECT TOTALS (Certification estimates)**		110

Certified: 40-49 points, **Silver:** 50-59 points, **Gold:** 60-79 points, **Platinum:** 80+ points

LEED v4 for Neighbourhood Development Plan

Project Name: "Al Waha Community", Dubai, UAE
The Integrated Scenarios and Modified Case

Yes	?	No			
15	**0**	**0**	**Smart Location & Linkage**		**28**
Y			Prereq	Smart Location	Required
Y			Prereq	Imperiled Species and Ecological Communities	Required
Y			Prereq	Wetland and Water Body Conservation	Required
Y			Prereq	Agricultural Land Conservation	Required
Y			Prereq	Floodplain Avoidance	Required
6			Credit	Preferred Locations	10
			Credit	Brownfield Remediation	2
5			Credit	Access to Quality Transit	7
2			Credit	Bicycle Facilities	2
			Credit	Housing and Jobs Proximity	3
1			Credit	Steep Slope Protection	1
			Credit	Site Design for Habitat or Wetland and Water Body Conservation	1
			Credit	Restoration of Habitat or Wetlands and Water Bodies	1
1			Credit	Long-Term Conservation Management of Habitat or Wetlands and Water Bodies	1
24	**0**	**0**	**Neighbourhood Pattern & Design**		**41**
Y			Prereq	Walkable Streets	Required
Y			Prereq	Compact Development	Required
Y			Prereq	Connected and Open Community	Required
4			Credit	Walkable Streets	9
4			Credit	Compact Development	6
2			Credit	Mixed-Use Neighbourhoods	4
5			Credit	Housing Types and Affordability	7
1			Credit	Reduced Parking Footprint	1
2			Credit	Connected and Open Community	2
			Credit	Transit Facilities	1
			Credit	Transportation Demand Management	2
1			Credit	Access to Civic & Public Space	1
1			Credit	Access to Recreation Facilities	1
1			Credit	Visitability and Universal Design	1
1			Credit	Community Outreach and Involvement	2
			Credit	Local Food Production	1
1			Credit	Tree-Lined and Shaded Streetscapes	2
1			Credit	Neighbourhood Schools	1
6	**0**	**0**	**Green Infrastructure & Buildings**		**31**
Y			Prereq	Certified Green Building	Required
Y			Prereq	Minimum Building Energy Performance	Required
Y			Prereq	Indoor Water Use Reduction	Required
Y			Prereq	Construction Activity Pollution Prevention	Required
			Credit	Certified Green Buildings	5
2			Credit	Optimize Building Energy Performance	2
1			Credit	Indoor Water Use Reduction	1
1			Credit	Outdoor Water Use Reduction	2
			Credit	Building Reuse	1
			Credit	Historic Resource Preservation and Adaptive Reuse	2
1			Credit	Minimized Site Disturbance	1
			Credit	Rainwater Management	4
			Credit	Heat Island Reduction	1
			Credit	Solar Orientation	1
			Credit	Renewable Energy Production	3
			Credit	District Heating and Cooling	2
			Credit	Infrastructure Energy Efficiency	1
1			Credit	Wastewater Management	2
			Credit	Recycled and Reused Infrastructure	1
			Credit	Solid Waste Management	1
			Credit	Light Pollution Reduction	1
0	**0**	**0**	**Innovation & Design Process**		**6**
			Credit	Innovation	5
			Credit	LEED® Accredited Professional	1
1	**0**	**0**	**Regional Priority Credits**		**4**
1			Credit	Regional Priority Credit: Region Defined	1
			Credit	Regional Priority Credit: Region Defined	1
			Credit	Regional Priority Credit: Region Defined	1
			Credit	Regional Priority Credit: Region Defined	1
46	**0**	**0**	**PROJECT TOTALS (Certification estimates)**		**110**

Certified: 40-49 points, **Silver:** 50-59 points, **Gold:** 60-79 points, **Platinum:** 80+ points

References

1. La Rosa, D.; Privitera, R. Characterization of non-urbanized areas for land-use planning of agricultural and green infrastructure in urban contexts. *Landsc. Urban Plan.* **2013**, *109*, 94–106. [CrossRef]
2. Litman, T. Sustainability and Livability: Summary of Definitions, Goals, Objectives and Performance Indicators. 2010. Available online: https://www.vtpi.org/sus_liv.pdf (accessed on 15 February 2020).
3. Pandolfini, E.; Bemposta, A.; Sbardella, M.; Simonetta, G.; Toschi, L. Well-being, Landscape and Sustainability of Communication. *Agric. Agric. Sci. Procedia* **2016**, *8*, 602–608. [CrossRef]
4. Ng, E. *Designing High-Density Cities for Social and Environmental Sustainability*; Earthscan: London, UK, 2010.
5. Badland, H.; Whitzman, C.; Lowe, M.; Davern, M.; Aye, L.; Butterworth, I.; Hes, D.; Giles-Corti, B. Urban Livability: Emerging lessons from Australia for exploring the potential for indicators to measure the social determinants of health. *Soc. Sci. Med.* **2014**, *111*, 64–73. [CrossRef] [PubMed]
6. Dempsey, N.; Brown, C.; Bramley, G. The Key to Sustainable Urban Development in UK Cities? The Influence of Density on Social Sustainability. *Prog. Plan.* **2015**, *77*, 89–141. Available online: https://www.google.com/earth/ (accessed on 20 February 2019).
7. Kent, J.; Thompson, S. The Three Domains of Urban Planning for Health and Well-being. *J. Plan. Lit.* **2014**, *29*, 239–256. [CrossRef]
8. Kruger, E.; Minella, F.; Rasia, F. Impact of urban geometry on outdoor thermal comfort and air quality from field measurements in Curitiba, Brazil. *Build. Environ.* **2011**, *46*, 621–634. [CrossRef]
9. Karl, T. Modern Global Climate Change. *Science* **2003**, *302*, 1719–1723. [CrossRef] [PubMed]
10. National Aeronautics and Space Administration (NASA). Science Briefs: Greenhouse Gases—Refining the Role of Carbon Dioxide 1998. 1998. Available online: https://data.giss.nasa.gov/ (accessed on 20 December 2016).
11. Maleki, S.; Bell, S.; Hosseini, S.; Faizi, M. Developing and testing a framework for the assessment of neighborhood liveability in two contrasting countries: Iran and Estonia. *Ecol. Indic.* **2015**, *48*, 263–271. [CrossRef]
12. U.S. Pew Center. As U.S. Energy Production Grows Public Policy Views Show Little Change. 2013. Available online: http://www.people-press.org/2014/12/18/as-u-s-energy-production-grows-public-policy-views-show-little-change/ (accessed on 15 January 2020).
13. Shareef, S.; Altan, H. Assessing the Implementation of Renewable Energy Policy within the UAE by Adopting the Australian 'Solar Town' Program. *Future Cities Environ.* **2019**, *5*, 1–11. [CrossRef]
14. Kearns, A.; Parkinson, M. The Significance of Neighborhood. *Urban Study* **2001**, *38*, 2103–2110. [CrossRef]
15. Wheeler, S. Livable Communities: Creating Safe and Livable Neighborhoods, Towns, and Regions in California. 2001. Available online: https://escholarship.org/uc/item/8xf2d6jg (accessed on 15 February 2020).
16. Yamagata, Y.; Seya, H. Simulating a Future Smart City: An Integrated Land Use Energy Model. *Appl. Energy* **2013**, *112*, 1466–1474. [CrossRef]
17. Arnberger, A.; Allex, B.; Eder, R.; Ebenberger, M.; Wanka, A.; Kolland, F.; Hutter, H. Elderly resident's uses of and preferences for urban green spaces during heat periods. *Urban For. Urban Green.* **2017**, *21*, 102–115. [CrossRef]
18. Zhang, W.; Huang, B.; Luo, D. Effects of land use and transportation on carbon sources and carbon sinks: A case study in Shenzhen, China. *Landsc. Urban Plan.* **2014**, *122*, 175–185. [CrossRef]
19. Jusuf, S.K.; Wong, N.H.; Hagen, E.; Anggoro, R.; Hong, Y. The influence of land use on the urban heat island in Singapore. *Habitat Int.* **2007**, *31*, 232–242. [CrossRef]
20. Lang, W.; Chen, T.; Chan, E.H.; Yung, E.H.; Lee, T.C. Understanding livable dense urban form for shaping the landscape of community facilities in Hong Kong using fine-scale measurements. *Cities* **2019**, *84*, 34–45. [CrossRef]
21. Amado, M.; Poggi, F.; Amado, A.R. Energy efficient city: A model for urban planning. *Sustain. Cities Soc.* **2016**, *26*, 476–485. [CrossRef]
22. Lehmann, S. *The Principles of Green Urbanism*; Earthscan: London, UK, 2010.
23. Ritchie, A.; Thomas, R. *Sustainable Urban Design*; Taylor & Francis: London, UK; New York, NY, USA, 2009.
24. Oberlink, M. Livable communities for adults with disabilities. *Policy Brief (Cent. Home Care Policy Res.)* **2006**, *29*, 1–6.
25. Mushtaha, E.; Shareef, S.; Alsyouf, I.; Mori, T.; Kayed, A.; Abdelrahim, M.; Albannay, S. A Study of the Impact of Major Urban Heat Island Factors in a Hot Climate Courtyard: The Case of the University of Sharjah, UAE. *Sustain. Cities Soc.* **2021**, *69*, 102844. [CrossRef]
26. Li, Z.; Quan, S.; Yang, P. Energy performance simulation for planning a low carbon neighborhood urban district: A Case Study in the City of Macau. *Habitat Int.* **2016**, *53*, 206–214. [CrossRef]
27. Virtudes, A. Benefits of Greenery in Contemporary City. *IOP Conf. Ser. Earth Environ. Sci.* **2016**, *44*, 032020. [CrossRef]
28. Croome, D.J. The determinants of architectural form in modern buildings within the Arab world. *Build. Environ.* **1991**, *26*, 62. [CrossRef]
29. Belmeziti, A.; Cherqui, F.; Kaufmann, B. Improving the multi-functionality of urban green spaces: Relations between components of green spaces and urban services. *Sustain. Cities Soc.* **2018**, *43*, 1–10. [CrossRef]
30. La Rosa, D. Accessibility to green spaces: GIS based indicators for sustainable planning in a dense urban context. *Ecol. Indic.* **2014**, *42*, 122–134. [CrossRef]
31. Jones, P.J.; Lannon, S.; Williams, J. Modeling building energy use at urban scale. In Proceedings of the Seventh International IBPSA Conference, Rio de Janeiro, Brazil, 13–15 August 2001.

32. Shareef, S. The impact of urban morphology and building's height diversity on energy consumption at urban scale. The case study of Dubai. *Build. Environ.* **2021**, *194*, 107675. [CrossRef]
33. Oke, T. *Boundary Layer Climate*, 2nd ed.; Routledge: London, UK, 1987.
34. Shen, L.Y.; Ochoa, J.J.; Shah, M.N.; Zhang, X. The application of urban sustainability indicators—A comparison between various practices. *Habitat Int.* **2011**, *35*, 17–29. [CrossRef]
35. Armson, D.; Stringer, P.; Ennos, A. The effect of tree shade and grass on surface and globe temperatures in an urban area. *Urban For. Urban Green.* **2012**, *11*, 245–255. [CrossRef]
36. Shareef, S.; Abu-Hijleh, B. The effect of building height diversity on outdoor microclimate conditions in hot climate. A case study of Dubai-UAE. *Urban Clim.* **2020**, *32*, 100611. [CrossRef]
37. Vine, D.; Buys, L.; Aird, R. The use of amenities in high density neighborhoods by older urban Australian residents. *Landsc. Urban Plan.* **2012**, *107*, 159–171. [CrossRef]
38. Adunola, A. Evaluation of urban residential thermal comfort in relation to indoor and outdoor air temperatures in Ibadan, Nigeria. *Build. Environ.* **2014**, *75*, 190–205. [CrossRef]
39. Lo, A. Small is green? Urban form and Sustainable Consumption in Selected OECD Metropolitan Areas. *Land Use Policy* **2016**, *54*, 212–220. [CrossRef]
40. Ode Sang, A.; Knez, I.; Gunnarsson, B.; Hedblom, M. The Effects of Naturalness, Gender, and Age on How Urban Green Space Is Perceived and Used. *Urban For. Urban Green.* **2016**, *18*, 268–276. Available online: https://www.sciencedirect.com/science/article/abs/pii/S1618866715300765 (accessed on 20 February 2019).
41. Integrated Environmental Solutions—Virtual Environment (IES-VE). Available online: https://www.iesve.com (accessed on 20 February 2019).
42. CityCAD. Available online: http://www.holisticcity.co.uk/ (accessed on 20 February 2019).
43. LEED Neighborhood & Developments, V4. (2009). USGBC. Updated October (2014). Available online: http://www.usgbc.org/resources/leed-neighborhood-development-v2009-current-version (accessed on 15 February 2019).
44. Google Earth Pro (2020). Available online: https://www.google.com/intl/en-GB/earth/ (accessed on 15 January 2020).
45. Sierra Surfaces Products. Available online: http://www.sierrasurfaces.com/?page_id=1026 (accessed on 15 February 2019).
46. Maglin, Outdoor Furniture. Available online: http://www.maglin.com/products/bench/hbsfseries.html (accessed on 20 February 2019).
47. Guzman, S.; Harrell, R. Increasing Community Livability for People of All Ages. *Public Policy Aging Rep.* **2014**, *25*, 28. [CrossRef]
48. Takano, T.; Nakamura, K.; Watanabe, M. Urban residential environments and senior citizens' longevity in megacity areas: The importance of walkable green spaces. *J. Epidemiol. Community Health* **2002**, *56*, 913–918. [CrossRef] [PubMed]

MDPI
St. Alban-Anlage 66
4052 Basel
Switzerland
Tel. +41 61 683 77 34
Fax +41 61 302 89 18
www.mdpi.com

Sustainability Editorial Office
E-mail: sustainability@mdpi.com
www.mdpi.com/journal/sustainability

www.ingramcontent.com/pod-product-compliance
Lightning Source LLC
LaVergne TN
LVHW070107170726
843515LV00007B/1633